Integral Transformation, Operational Calculus and Their Applications

Integral Transformation, Operational Calculus and Their Applications

Editor

Hari Mohan Srivastava

MDPI • Basel • Beijing • Wuhan • Barcelona • Belgrade • Manchester • Tokyo • Cluj • Tianjin

Editor
Hari Mohan Srivastava
University of Victoria
Canada

Editorial Office
MDPI
St. Alban-Anlage 66
4052 Basel, Switzerland

This is a reprint of articles from the Special Issue published online in the open access journal *Symmetry* (ISSN 2073-8994) (available at: https://www.mdpi.com/journal/symmetry/special_issues/Integral_Transformation_Operational_Calculus_Their_Applications).

For citation purposes, cite each article independently as indicated on the article page online and as indicated below:

LastName, A.A.; LastName, B.B.; LastName, C.C. Article Title. *Journal Name* **Year**, *Volume Number*, Page Range.

ISBN 978-3-0365-5481-5 (Hbk)
ISBN 978-3-0365-5482-2 (PDF)

Contents

About the Editor

Hari Mohan Srivastava

Hari Mohan Srivastava is Professor Emeritus in the Department of Mathematics and Statistics at the University of Victoria in Canada. He currently holds several advisory, honorary, visiting and chair professorships at universities and research institutes around the world. Prof. Hari Mohan Srivastava's current research interests include (for example) real and complex analysis, fractional calculus and its applications, integral equations and integral transformations, higher transcendental functions and their applications, q-series and q-polynomials, analytic and geometric inequalities, probability and statistics, and inventory modeling and optimization. Further biographical and professional details about Professor Hari Mohan Srivastava are available at the following link: https://www.math.uvic.ca/~harimsri/

Preface to "Integral Transformation, Operational Calculus and Their Applications"

This volume consists of a collection of a total of 12 peer-reviewed and accepted submissions (including several invited feature articles) from all over the world to the Special Issue of the MDPI journal Symmetry on the general subject area of integral transformation, operational calculus and their applications.

For this Special Issue, we cordially invited and welcomed reviews and expository and original research articles dealing with the recent state-of-the-art advances on the topics of integral transformations and operational calculus as well as their multidisciplinary applications with some relevance to the aspect of symmetry. The theory and applications of integral transformations and associated operational calculus are remarkably widespread in many diverse areas of the mathematical, physical, chemical, engineering and statistical sciences. The topics of interest covered in this Special Issue include the following: integral transformations and integral equations as well as other related operators, including their symmetry properties and characteristics; applications involving mathematical (or higher transcendental) functions, including their symmetry properties and characteristics; applications involving fractional-order differential and integral equations and their associated symmetry; applications involving symmetrical aspects of the geometric function theory of complex analysis; applications involving q-series and q-polynomials and their associated symmetry; applications involving special functions of mathematical physics and applied mathematics and their symmetrical aspect; applications involving analytic number theory and symmetry.

Finally, it gives me great pleasure to thank all of the participating authors, referees and reviewers for their invaluable contributions toward the remarkable success of this Special Issue. I do also express my appreciation for the editorial and managerial help and assistance provided efficiently and generously by Mr. Philip Li and other colleagues and associates in the Editorial Office of Symmetry. The dedicated and wholehearted support and help of one and all are indeed greatly appreciated.

Hari Mohan Srivastava
Editor

symmetry

MDPI

Article

Unified Integrals of Generalized Mittag–Leffler Functions and Their Graphical Numerical Investigation

Nabiullah Khan [1], Mohammad Iqbal Khan [1], Talha Usman [2], Kamsing Nonlaopon [3,*] and Shrideh Al-Omari [4]

[1] Department of Applied Mathematics, Aligarh Muslim University, Aligarh 202002, India; nukhanmath@gmail.com (N.K.); miqbalkhan1971@gmail.com (M.I.K.)
[2] Department of General Requirements, University of Technology and Applied Sciences-Sur, Muscat 133, Oman; talhausman.maths@gmail.com
[3] Department of Mathematics, Faculty of Science, Khon Kaen University, Khon Kaen 40002, Thailand
[4] Department of Scientific Basic Sciences, Faculty of Engineering Technology, Al-Balqa Applied University, Amman 11134, Jordan; shridehalomari@bau.edu.jo
* Correspondence: nkamsi@kku.ac.th

Abstract: In this article, we obtain certain finite integrals concerning generalized Mittag–Leffler functions, which are evaluated in terms of the generalized Fox–Wright function. The integrals of concern are unified in nature and thereby yield some new integral formulas as special cases. Moreover, we numerically compute some integrals using the Gaussian quadrature formula and draw a comparison with the main integrals by using graphical numerical investigation.

Keywords: Fox–Wright function; generalized hypergeometric function; Mittag–Leffler function

Citation: Khan, N.; Khan, M.I.; Usman, T.; Nonlaopon, K.; Al-Omari, S. Unified Integrals of Generalized Mittag–Leffler Functions and Their Graphical Numerical Investigation. *Symmetry* **2022**, *14*, 869. https://doi.org/10.3390/sym14050869

Academic Editor: Boris Malomed

Received: 15 March 2022
Accepted: 14 April 2022
Published: 23 April 2022

1. Introduction

In mathematics, functions and symmetric functions are very common in theory and applications. They have been applied to various fields including group theory, Lie algebras, and algebraic geometry, to mention but a few. In applied mathematics, many functions are defined via integrals or series (or infinite products), which are usually referred to as special functions [1–6]. One of them is the Mittag–Leffler function, which was introduced in connection with a method of summation of some divergent series. The Mittag–Leffler function has recently received the interest of scientists due to its wide applications in pure as well as applied mathematics. It is noted that the importance of the Mittag–Leffler function has been envisaged during the last two decades due to its entanglement in physics, chemistry, biology, engineering and applied sciences. The Mittag–Leffler function naturally occurs as a solution of fractional order differential equations or fractional order integral equations. Problems of physics and applied mathematics involve a notable numerical implementation of the Mittag-Leffler function in general and modified forms; therefore, it remains an engaging object of applied research. The implementation of Mittag-Leffler functions is required in a wide variety of problems of physics and mathematics. Because of their crucial requirement, many research works have been dedicated to them, and various representations and generalizations of Mittag-Leffler functions can be found in the literature. Among the most popular special function of fractional calculus is the simplest ${}_p\Psi_q$ function and p = 0, q = 1, called the Wright function or the Bessel–Maitland function or the Wright–Bessel function. From this point of view, the Mittag–Leffler function, expressible in terms of the Fox–Right function, is a special function of fractional calculus. Therefore the Mittag–Leffler function is called the queen function of fractional calculus. The results obtained in the manuscript, connected with a generalized Mittag=-Leffler function that will be used to solve a variety of problems of fractional calculus, for example, Riemann-–Liouville fractional integrals and derivatives, Laplace and Sumudu fractional and integral derivatives and Marichev–Saigo–Maeda fractional integrals and derivatives, etc. Recently, fractional calculus associated with some special functions has proved itself to be a useful tool for applications in many fields of research such as physical systems, biomedicine, nonlinear electronic circuits, chaos-based cryptography, and image encryption. Examples of systems that can be precisely described by fractional-order differential equations (FODEs) involve viscoelastic material models, electrical components, electronic circuits, diffusion waves, the propagation of waves in non-local elastic continua, hydro-logic systems, earthquakes' nonlinear oscillations, models of world economies, fractional viscoelastic models and continuous random walk and equations of muscular blood vessels (see [7–12]). In the past few years, several integral formulas having a variety of special functions have been achieved by many authors (see [13–30]). The present paper provides the study of finite integrals of the generalized Mittag-Leffler function and investigates some useful formulas. We have computed many new results involving integral transforms of the Mittag–Leffler function and plotted three graphs as the major novelty of our work. The results derived in this paper are of general character and likely to find certain applications in the theory of special functions. Additionally, the results provide unification and extension of known results given earlier by various researchers. We compare the results of analytically evaluated integrals with integrals evaluated numerically using the Gaussian quadrature formula. We conclude that the results obtained will provide a significant step in the theory of integral formulas and can yield some potential applications in the field of classical and applied mathematics. Motivated by the aforementioned research and success of the application of integral formulas, we evaluate a new type of integral formulas involving the generalized Mittag–Leffler function (GMLF) expressed in terms of the Fox–Wright function. The Mittag–Leffler function [31,32] is defined as

$$E_{\sigma}(w) = \sum_{n=0}^{\infty} \frac{w^n}{\Gamma(\sigma n + 1)}, \quad \sigma \in \mathbb{C},\ Re(\sigma) > 0 \tag{1}$$

where ω ia a complex variable and $\Gamma(.)$ is the gamma function [25].

In 1905, A. Wiman [33] established a generalization of $E_{\sigma}(w)$, as follows:

$$E_{\sigma,\mu}(w) = \sum_{n=0}^{\infty} \frac{w^n}{\Gamma(\sigma n + \mu)}, \quad (\sigma, \mu \in \mathbb{C},\ \ Re(\mu) > 0,\ Re(\sigma) > 0). \tag{2}$$

In 1971, Prabhakar [23] came up with a further generalization of $E_{\sigma,\mu}(w)$ in the form

$$E^{\gamma}_{\sigma,\mu}(w) = \sum_{n=0}^{\infty} \frac{(\gamma)_n}{\Gamma(\sigma n + \mu)} \frac{w^n}{n!} \quad (\gamma, \mu, \sigma \in \mathbb{C},\ Re(\sigma) > 0, Re(\gamma) > 0, Re(\mu) > 0), \tag{3}$$

where $(\gamma)_n$ is known as the Pochhammer symbol [25]. The underlying generalization of the Mittag–Leffler function is given by Shukla and Prajapati (2007) [29] as

$$E^{\gamma,b}_{\sigma,\mu}(w) = \sum_{n=0}^{\infty} \frac{(\gamma)_{bn}}{\Gamma(\sigma n + \mu)} \frac{w^n}{n!} \tag{4}$$

and expressed by Salim (2009) [26] in the form

$$E^{\gamma,\delta}_{\sigma,\mu}(w) = \sum_{n=0}^{\infty} \frac{(\gamma)_n\, w^n}{\Gamma(\sigma n + \mu)(\delta)_n}. \tag{5}$$

A certain further generalization of the Mittag–Leffler function was given by Salim and Faraj (2012) [27] as

$$E^{\gamma,\delta,b}_{\sigma,\mu,a}(w) = \sum_{n=0}^{\infty} \frac{(\gamma)_{bn}\, w^n}{\Gamma(\sigma n + \mu)(\delta)_{an}}. \tag{6}$$

On the other hand, Khan and Ahmad introduced a new generalization of the Mittag–Leffler function (2013) [34] as

$$E^{\gamma,b}_{\sigma,\mu,\delta}(w) = \sum_{n=0}^{\infty} \frac{(\gamma)_{bn}\, w^n}{\Gamma(\sigma n + \mu)\ (\delta)_n}, \tag{7}$$

where $\sigma,\ \mu,\ \gamma,\ \delta \in \mathbb{C};\ Re(\sigma) > 0, Re(\mu) > 0, Re(\gamma) > 0, Re(\delta) > 0;\ b \in (0,1) \cup \mathbb{N}$.

Consequently, they have introduced a generalization of (7) in the following form [34]

$$E^{\xi,\lambda,\gamma,b}_{\sigma,\mu,\nu,\phi,\delta,a}(w) = \sum_{n=0}^{\infty} \frac{(\xi)_{\lambda n}\,(\gamma)_{bn}\,w^n}{\Gamma(\sigma n+\mu)(\nu)_{\phi n}(\delta)_{an}}, \tag{8}$$

where $\sigma, \mu, \nu, \phi, \delta, \xi, \lambda, \gamma \in \mathbb{C}$; $\min\{Re(\sigma),\ Re(\mu),\ Re(\nu), Re(\phi), Re(\delta), Re(\xi),\ Re(\lambda), Re(\gamma)\} > 0; a, b > 0, b \le Re(\sigma) + a$.

Above all, (8) is the most generalized definition of all the above formalizations introduced in (1)–(7). Upon substituting $\xi = \nu, \lambda = \phi$ and $a = 1$ in (8), it becomes (7), which has been established by Khan and Ahmad (2013) [34]. Upon substituting $\xi = \nu$ and $\lambda = \phi$, in (8), it becomes a special case (6), which has been established by Salim and Faraj (2012) [27]. Upon substituting $\xi = \nu, \lambda = \phi$ and $b = a = 1$ in (8), it becomes (5), which has been discussed by Salim (2009) [26]. Upon substituting $\xi = \nu, \lambda = \phi$ and $\delta = a = 1$ in equation (8), it is a special case (4); see Shukla and Prajapati (2007) [29]. If $b = 1$, it becomes a special case (3) of Prabhakar (1971) [23]. On substituting $\xi = \nu, \lambda = \phi$ and $\gamma = \delta = a = b = 1$ in (8), it becomes a special case (2) established by A. Wiman (1905) [33]. Furthermore, if $\mu = 1$, we get the Mittag–Leffler function $E_\sigma(w)$ defined in (1). Finally, on setting $\delta = a = b = 1$ in (8), we establish a new generalization of the Mittag–Leffler function in the form

$$E^{\xi,\lambda,\gamma}_{\sigma,\mu,\nu,\phi}(w) = \sum_{n=0}^{\infty} \frac{(\xi)_{\lambda n}\,(\gamma)_{n}}{\Gamma(\sigma n+\mu)(\nu)_{\phi n}} \frac{w^n}{n!}, \tag{9}$$

where $\sigma, \mu, \nu, \phi, \xi, \lambda, \gamma \in \mathbb{C}$; $Re(\mu) > 0, Re(\sigma) > 0, Re(\nu) > 0, Re(\phi) > 0, Re(\xi) > 0$, $Re(\lambda) > 0$ *and* $Re(\gamma) > 0$.

The Fox–Wright function ${}_r\Psi_s[w]$ (see [35–42]), is defined by

$${}_r\Psi_s[w] = {}_r\Psi_s\left[\begin{matrix}(\lambda_1,\acute{\lambda}_1),\ldots,(\lambda_r,\acute{\lambda}_r); \\ \\ (l_1,\acute{l}_1),\ldots,(l_s,\acute{l}_s);\end{matrix}\quad w\right] \tag{10}$$

$$= \sum_{k=0}^{\infty} \frac{\Gamma(\lambda_1+\acute{\lambda}_1 k),\ldots,\Gamma(\lambda_r+\acute{\lambda}_r k)}{\Gamma(l_1+\acute{l}_1 k),\ldots,\Gamma(l_s+\acute{l}_s k)} \frac{w^k}{k!} \tag{11}$$

$$= H^{1,r}_{r,s+1}\left[-w\left|\begin{matrix}(1-\lambda_1,\acute{\lambda}_1),\ldots,(1-\lambda_r,\acute{\lambda}_r) \\ \\ (0,1),(1-l_1,\acute{l}_1),\ldots,(1-l_s,\acute{l}_s)\end{matrix}\right.\right], \tag{12}$$

where $H^{1,r}_{r,s+1}[w]$ represents the Fox-H function [38]. When $\acute{\lambda}_1,\ldots,\acute{\lambda}_r = 1, \acute{l}_1,\ldots,\acute{l}_s = 1$ in (10), the Fox–Wright function reduces to the generalized hypergeometric function ${}_rF_s$[w] (see [41])

$${}_r\Psi_s\left[\begin{matrix}(\lambda_1,1),\ldots,(\lambda_r,1); \\ (l_1,1),\ldots,(l_s,1);\end{matrix}\quad w\right] = \frac{\Gamma(\lambda)_1,\ldots,\Gamma(\lambda)_r}{\Gamma(l)_1,\ldots,\Gamma(l)_s}\,{}_rF_s(\lambda_1,\ldots,\lambda_r;l_1,\ldots,l_s;w). \tag{13}$$

Here, we recall the result due to Prudnikov et al. [24] (see also [39], p. 250 (2.8)), by means of which we have established our main result in the present article

$$\int_p^q \frac{(x-p)^{\alpha-1}(q-x)^{\beta-1}}{[(q-p)+r_1(x-p)+r_2(q-x)]^{\alpha+\beta}}\,dx$$

$$= \frac{\Gamma(\alpha)\Gamma(\beta)}{\Gamma(\alpha+\beta)}\quad (q-p)^{-1}(1+r_1)^{-\alpha}(1+r_2)^{-\beta}, \tag{14}$$

provided that $Re(\alpha) > 0$, $Re(\beta) > 0$, $q \neq p$ and the constants r_1 and r_2 are such that none of the expression $1+r_1, 1+r_2, [(q-p)+r_1(x-p)+r_2(q-x)]$, where $p \le x \le q$ is zero.

2. Main Results

Theorem 1. *Let α and β exist such that $Re(\alpha) > 0, Re(\beta) > 0, q \neq p$ and the constants r_1 and r_2 are such that none of the expressions $1 + r_1, 1 + r_2, [(q-p) + r_1(x-p) + r_2(q-x)]$, where $p \leq x \leq q$ is zero. Let $\sigma, \mu, \nu, \phi, \delta, \xi, \lambda, \gamma \in \mathbb{C}$; if $\min\{Re(\sigma), Re(\mu), Re(\nu), Re(\phi), Re(\delta), Re(\xi), Re(\lambda), Re(\gamma)\} > 0$; $a, b > 0$, $b \leq Re(\sigma) + a$, then the following identity holds:*

$$\int_p^q \frac{(x-p)^{\alpha-1}(q-x)^{\beta-1}}{[(q-p)+r_1(x-p)+r_2(q-x)]^{\alpha+\beta}} E^{\xi,\lambda,\gamma,b}_{\sigma,\mu,\nu,\phi,\delta,a}\left[w\left\{\frac{(x-p)(q-x)}{[(q-p)+r_1(x-p)+r_2(q-x)]^2}\right\}^m\right]dx$$
$$= \frac{1}{(q-p)\,(1+r_1)^{\alpha}\,(1+r_2)^{\beta}} \frac{\Gamma(\nu)\Gamma(\delta)}{\Gamma(\xi)\Gamma(\gamma)}$$
$$\times\ {}_5\Psi_4\left[\begin{matrix}(\xi,\lambda),\ (\gamma,b),\ (\alpha,m),\ (\beta,m), & (1,1); \\ (\mu,\sigma),\ (\nu,\phi),\ (\delta,a), & (\alpha+\beta,2m);\end{matrix}\ \frac{w}{(1+r_1)^m(1+r_2)^m}\right], \tag{15}$$

where $E^{\xi,\lambda,\gamma,b}_{\sigma,\mu,\nu,\phi,\delta,a}(w)$ is a GMLF given by (8).

Proof. Denoting the left hand side of (15) by I, writing $E^{\xi,\lambda,\gamma,b}_{\sigma,\mu,\nu,\phi,\delta,a}(w)$ in its summation formula in the integrand with the help of (8), we obtain

$$I = \int_p^q \frac{(x-p)^{\alpha-1}(q-x)^{\beta-1}}{[(q-p)+r_1(x-p)+r_2(q-x)]^{\alpha+\beta}}$$
$$\times \sum_{n=0}^{\infty} \frac{(\xi)_{\lambda n}(\gamma)_{bn}\,(1)_n\,w^n\,(x-p)^{mn}(q-x)^{mn}}{\Gamma(\sigma n+\mu)\,(\nu)_{\phi n}\,(\delta)_{an}\,[(q-p)+r_1(x-p)+r_2(q-x)]^{2mn}\,n!}\,dx, \tag{16}$$

which, by further simplification, yields

$$I = \sum_{n=0}^{\infty} \frac{(\xi)_{\lambda n}\,(\gamma)_{bn}}{\Gamma(\sigma n+\mu)\,(\nu)_{\phi n}\,(\delta)_{an}} \frac{w^n}{n!} \int_p^q \frac{(x-p)^{mn+\alpha-1}\,(q-x)^{mn+\beta-1}}{[(q-p)+r_1(x-p)+r_2(q-x)]^{2mn+\alpha+\beta}}\,dx. \tag{17}$$

We apply the result of (14), and, through simplifying, this yields

$$I = \frac{1}{(q-p)\,(1+r_1)^{\alpha}\,(1+r_2)^{\beta}} \frac{\Gamma(\nu)\,\Gamma(\delta)}{\Gamma(\xi)\,\Gamma(\gamma)}$$
$$\times \sum_{n=0}^{\infty} \frac{\Gamma(\xi+\lambda n)\,\Gamma(\gamma+bn)\,\Gamma(\alpha+mn)\,\Gamma(\beta+mn)\,\Gamma(1+n)\left(\frac{w}{(1+r_1)^m\,(1+r_2)^m}\right)^n}{\Gamma(\mu+\sigma n)\,\Gamma(\nu+\lambda n)\,\Gamma(\delta+an)\,\Gamma(\alpha+\beta+2mn)\,n!}. \tag{18}$$

Finally, after summing up, with the help of (11), we arrive at (15). This completes the proof of Theorem 1. □

Corollary 1. *For $b = \delta = a = 1$ and all the conditions already stated in (15), the following identity holds:*

$$\int_p^q \frac{(x-p)^{\alpha-1}(q-x)^{\beta-1}}{[(q-p)+r_1(x-p)+r_2(q-x)]^{\alpha+\beta}} E^{\xi,\lambda,\gamma}_{\sigma,\mu,\nu,\phi}\left[w\left\{\frac{(x-p)(q-x)}{[(q-p)+r_1(x-p)+r_2(q-x)]^2}\right\}^m\right]dx$$
$$= \frac{1}{(q-p)\,(1+r_1)^{\alpha}\,(1+r_2)^{\beta}} \frac{\Gamma(\nu)}{\Gamma(\xi)\Gamma(\gamma)}$$
$$\times\ {}_4\Psi_3\left[\begin{matrix}(\xi,\lambda),\ (\gamma,1),\ (\alpha,m), & (\beta,m); \\ (\mu,\sigma),\ (\nu,\phi), & (\alpha+\beta,2m);\end{matrix}\ \frac{w}{(1+r_1)^m(1+r_2)^m}\right]. \tag{19}$$

Theorem 2. *Let α and β be such that $Re(\alpha) > 0$, $Re(\beta) > 0$, $q \neq p$ and the constants r_1 and r_2 are such that none of the expressions $1 + r_1, 1 + r_2, [(q-p) + r_1(x-p) + r_2(q-x)]$, where $p \leq x \leq q$ is zero. Let $\sigma, \mu, \nu, \phi, \delta, \xi, \lambda, \gamma \in \mathbb{C}$; if $min\{Re(\sigma), Re(\mu), Re(\nu), Re(\phi), Re(\delta), Re(\xi), Re(\lambda), Re(\gamma)\} > 0$; $a, b > 0$, $b \leq Re(\sigma) + a$, the following identity holds:*

$$\int_p^q \frac{(x-p)^{\alpha-1}(q-x)^{\beta-1}}{[(q-p)+r_1(x-p)+r_2(q-x)]^{\alpha+\beta}} E^{\xi,\lambda,\gamma,b}_{\sigma,\mu,\nu,\phi,\delta,a}\left[\left\{\frac{(x-p)(q-x)}{[(q-p)+r_1(x-p)+r_2(q-x)]^2}\right\}^m w\right] dx$$
$$= \frac{1}{(q-p)\,(1+r_1)^{\alpha}\,(1+r_1)^{\beta}} \frac{\Gamma(\alpha)\Gamma(\beta)}{\Gamma(\alpha+\beta)\Gamma(\mu)} {}_{2m+\lambda+b+1}F_{2m+\sigma+\phi+a}\left[\begin{array}{llllll} \Delta(m;\alpha), & \Delta(m;\beta), & \Delta(\lambda;\xi), & \Delta(b;\gamma), & 1; & \\ \Delta(\sigma;\mu), & \Delta(\phi;\nu), & \Delta(a;\delta), & \Delta(2m;\alpha+\beta); & & \end{array} \frac{w\,\lambda^{\lambda}b^{b}}{\sigma^{\sigma}\phi^{\phi}a^{a}4^{m}(1+r_1)^m(1+r_2)^m}\right], \tag{20}$$

where $\Delta(m;\lambda)$ abbreviates the arrangement of m parameters $\frac{\lambda}{m}\ \frac{\lambda+1}{m} \dots \frac{\lambda+m-1}{m}$ and $m \geq 1$.

Proof. By using the formulas

$$\Gamma(\lambda+n) = \Gamma(\lambda)(\lambda)_n \tag{21}$$

and

$$(\lambda)_{mn} = m^{mn}\left(\frac{\lambda}{m}\right)_n \left(\frac{\lambda+1}{m}\right)_n \cdots \left(\frac{\lambda+m-1}{m}\right)_n, \tag{22}$$

and after a little simplification, the required result (20) can be obtained. Therefore, we omit the proof. □

Corollary 2. *On putting $a = b = \delta = 1$ under the condition already set out in (20), the following identity holds:*

$$\int_p^q \frac{(x-p)^{\alpha-1}(q-x)^{\beta-1}}{[(q-p)+r_1(x-p)+r_2(q-x)]^{\alpha+\beta}} E^{\xi,\lambda,\gamma}_{\sigma,\mu,\nu,\phi}\left[w\left\{\frac{(x-p)(q-x)}{[(q-p)+r_1(x-p)+r_2(q-x)]^2}\right\}^m\right] dx$$
$$= \frac{1}{(q-p)\,(1+r_1)^{\alpha}\,(1+r_2)^{\beta}} \frac{\Gamma(\alpha)\Gamma(\beta)}{\Gamma(\alpha+\beta)\Gamma(\mu)}$$
$$\times {}_{2m+\lambda+1}F_{2m+\sigma+\phi}\left[\begin{array}{lllll} \Delta(m;\alpha), & \Delta(m;\beta), & \Delta(\lambda;\xi), & \gamma; & \\ \Delta(\sigma;\mu), & \Delta(\phi;\nu), & & \Delta(2m;\alpha+\beta); & \end{array} \frac{w\,\lambda^{\lambda}}{\sigma^{\sigma}\phi^{\phi}4^{m}(1+r_1)^m(1+r_2)^m}\right]. \tag{23}$$

3. Special Cases

Here, we compute certain integral formulas as special cases of our key results.

(i) On setting $\xi = \nu$, $\lambda = \phi$ in (15), the following identity holds:

$$\int_p^q \frac{(x-p)^{\alpha-1}(q-x)^{\beta-1}}{[(q-p)+r_1(x-p)+r_2(q-x)]^{\alpha+\beta}} E^{\gamma,\delta,b}_{\sigma,\mu,a}\left[w\left\{\frac{(x-p)(q-x)}{[(q-p)+r_1(x-p)+r_2(q-x)]^2}\right\}^m\right] dx$$
$$= \frac{1}{(q-p)\,(1+r_1)^{\alpha}\,(1+r_2)^{\beta}} \frac{\Gamma(\delta)}{\Gamma(\gamma)}$$
$$\times {}_4\Psi_3\left[\begin{array}{llll} (\gamma,b), & (\alpha,\, m), & (\beta,\, m) & (1,1); \\ (\mu,\sigma), & (\delta,a), & & (\alpha+\beta,\, 2m); \end{array} \frac{w}{(1+r_1)^m(1+r_2)^m}\right]; \tag{24}$$

(ii) Setting $\xi = \nu$, $\lambda = \phi$ and $a = 1$ in (15), the following identity holds:

$$\int_p^q \frac{(x-p)^{\alpha-1}(q-x)^{\beta-1}}{[(q-p)+r_1(x-p)+r_2(q-x)]^{\alpha+\beta}} E^{\gamma,b}_{\sigma,\mu,\delta}\left[w\left\{\frac{(x-p)(q-x)}{[(q-p)+r_1(x-p)+r_2(q-x)+]^2}\right\}^m\right] dx$$

$$= \frac{1}{(q-p)\,(1+r_1)^{\alpha}\,(1+r_2)^{\beta}}\,\frac{\Gamma(\delta)}{\Gamma(\gamma)}$$

$$\times\ {}_4\Psi_3\left[\begin{matrix}(\gamma,b),\ (\alpha,\ m),\ (\beta,\ m) & (1,1); \\ (\mu,\sigma),\ (\delta,1), & (\alpha+\beta,\ 2m);\end{matrix}\ \frac{w}{(1+r_1)^m(1+r_2)^m}\right]; \tag{25}$$

(iii) Setting $\xi = \nu$, $\lambda = \phi$ and $a = b = 1$ in (15), the following identity holds:

$$\int_p^q \frac{(x-p)^{\alpha-1}(q-x)^{\beta-1}}{[(q-p)+r_1(x-p)+r_2(q-x)]^{\alpha+\beta}}\, E_{\sigma,\mu}^{\gamma,\delta}\left[w\left\{\frac{(x-p)(q-x)}{[(q-p)+r_1(x-p)+r_2(q-x)]^2}\right\}^m\right]dx$$

$$= \frac{1}{(q-p)\,(1+r_1)^{\alpha}\,(1+r_2)^{\beta}}\,\frac{\Gamma(\delta)}{\Gamma(\gamma)}$$

$$\times\ {}_4\Psi_3\left[\begin{matrix}(\gamma,1),\ (\alpha,\ m),\ (\beta,\ m) & (1,1); \\ (\mu,\sigma),\ (\delta,1), & (\alpha+\beta,\ 2m);\end{matrix}\ \frac{w}{(1+r_1)^m(1+r_2)^m}\right]; \tag{26}$$

(iv) Setting $\xi = \nu$, $\lambda = \phi$ and $a = \delta = 1$ in (15), the following identity holds:

$$\int_p^q \frac{(x-p)^{\alpha-1}(q-x)^{\beta-1}}{[(q-p)+r_1(x-p)+r_2(q-x)]^{\alpha+\beta}}\, E_{\sigma,\mu}^{\gamma,b}\left[w\left\{\frac{(x-p)(q-x)}{[(q-p)+r_1(x-p)+r_2(q-x)]^2}\right\}^m\right]dx$$

$$= \frac{1}{(q-p)\,(1+r_1)^{\alpha}\,(1+r_2)^{\beta}}\,\frac{1}{\Gamma(\gamma)}$$

$$\times\ {}_3\Psi_2\left[\begin{matrix}(\gamma,b),\ (\alpha,\ m), & (\beta,\ m); \\ (\mu,\sigma), & (\alpha+\beta,\ 2m);\end{matrix}\ \frac{w}{(1+r_1)^m(1+r_2)^m}\right]; \tag{27}$$

(v) Setting $\xi = \nu$, $\lambda = \phi$ and $a = b = \delta = 1$ in (15), the following identity holds:

$$\int_p^q \frac{(x-p)^{\alpha-1}(q-x)^{\beta-1}}{[(q-p)+r_1(x-p)+r_2(q-x)]^{\alpha+\beta}}\, E_{\sigma,\mu}^{\gamma}\left[w\left\{\frac{(x-p)(q-x)}{[(q-p)+r_1(x-p)+r_2(q-x)]^2}\right\}^m\right]dx$$

$$= \frac{1}{(q-p)\,(1+r_1)^{\alpha}\,(1+r_2)^{\beta}}\,\frac{1}{\Gamma(\gamma)}$$

$$\times\ {}_3\Psi_2\left[\begin{matrix}(\gamma,1),\ (\alpha,\ m), & (\beta,\ m); \\ (\mu,\sigma), & (\alpha+\beta,\ 2m);\end{matrix}\ \frac{w}{(1+r_1)^m(1+r_2)^m}\right]; \tag{28}$$

(vi) Setting $\xi = \nu$, $\lambda = \phi$ and $a = b = \gamma = \delta = 1$ in (15), the following identity holds:

$$\int_p^q \frac{(x-p)^{\alpha-1}(q-x)^{\beta-1}}{[(q-p)+r_1(x-p)+r_2(q-x)]^{\alpha+\beta}}\, E_{\sigma,\mu}\left[w\left\{\frac{(x-p)(q-x)}{[(q-p)+r_1(x-p)+r_2(q-x)]^2}\right\}^m\right]dx$$

$$= \frac{1}{(q-p)\,(1+r_1)^{\alpha}\,(1+r_2)^{\beta}}$$

$$\times\ {}_3\Psi_2\left[\begin{matrix}(\alpha,\ m),\ (\beta,\ m),\ (1,1); & \\ (\mu,\sigma), & (\alpha+\beta,\ 2m);\end{matrix}\ \frac{w}{(1+r_1)^m(1+r_2)^m}\right]; \tag{29}$$

(vii) Setting $\xi = \nu$, $\lambda = \phi$ and $a = b = \gamma = \delta = \mu = 1$ in (15), the following identity holds:

$$\int_p^q \frac{(x-p)^{\alpha-1}(q-x)^{\beta-1}}{[(q-p)+r_1(x-p)+r_2(q-x)]^{\alpha+\beta}}\, E_{\sigma}\left[w\left\{\frac{(x-p)(q-x)}{[(q-p)+r_1(x-p)+r_2(q-x)]^2}\right\}^m\right]dx$$

$$= \frac{1}{(q-p)\,(1+r_1)^{\alpha}\,(1+r_2)^{\beta}}$$

$$\times\ {}_3\Psi_2\left[\begin{matrix}(\alpha,m), & (\beta,m), & (1,1); & \\ & (1,\sigma), & (\alpha+\beta,2m); & \dfrac{w}{(1+r_1)^m(1+r_2)^m}\end{matrix}\right]; \tag{30}$$

(viii) Setting $\xi = \nu$, $\lambda = \phi$ and $a = b = \gamma = \delta = \mu = \sigma = 1$ in (15), the following identity holds:

$$\int_p^q \frac{(x-p)^{\alpha-1}(q-x)^{\beta-1}}{[(q-p)+r_1(x-p)+r_2(q-x)]^{\alpha+\beta}}\, e^{\left[w\left\{\frac{(x-p)(q-x)}{[(q-p)+r_1(x-p)+r_2(q-x)]^2}\right\}^m\right]}\, dx$$
$$= \frac{1}{(q-p)\,(1+r_1)^{\alpha}\,(1+r_2)^{\beta}}$$
$$\times\ {}_2\Psi_1\left[\begin{matrix}(\alpha,m), & (\beta,m), & \\ & (\alpha+\beta,2m); & \dfrac{w}{(1+r_1)^m(1+r_2)^m}\end{matrix}\right]; \tag{31}$$

(ix) Setting $\xi = \nu$, $\lambda = \phi$ and $a = b = \gamma = \delta = \mu = 1$, $\sigma = 0$ in (15), the following identity holds:

$$\int_p^q \frac{(x-p)^{\alpha-1}(q-x)^{\beta-1}}{[(q-p)+r_1(x-p)+r_2(q-x)]^{\alpha+\beta}}\, \frac{1}{\left[1-w\left\{\frac{(x-p)(q-x)}{[(q-p)+r_1(x-p)+r_2(q-x)]^2}\right\}^m\right]}\, dx$$
$$= \frac{1}{(q-p)\,(1+r_1)^{\alpha}\,(1+r_2)^{\beta}}$$
$$\times\ {}_3\Psi_2\left[\begin{matrix}(\alpha,m), & (\beta,m), & (1,1); & \\ & (1,0), & (\alpha+\beta,2m); & \dfrac{w}{(1+r_1)^m(1+r_2)^m}\end{matrix}\right]; \tag{32}$$

(x) Setting $\xi = \nu$, $\lambda = \phi$ in (20), the following identity holds:

$$\int_p^q \frac{(x-p)^{\alpha-1}(q-x)^{\beta-1}}{[(q-p)+r_1(x-p)+r_2(q-x)]^{\alpha+\beta}}\, E^{\gamma,\delta,b}_{\sigma,\mu,a}\left[w\left\{\frac{(x-p)(q-x)}{[(q-p)+r_1(x-p)+r_2(q-x)]^2}\right\}^m\right] dx$$
$$= \frac{1}{(q-p)\,(1+r_1)^{\alpha}\,(1+r_2)^{\beta}}\,\frac{\Gamma(\alpha)\Gamma(\beta)}{\Gamma(\alpha+\beta)\Gamma(\mu)}$$
$$\times\ {}_{2m+b+1}F_{2m+\sigma+a}\left[\begin{matrix}\Delta(m;\alpha), & \Delta(m;\beta),\ \Delta(b;\gamma), & 1; & \\ \Delta(\sigma;\mu), & \Delta(a;\delta), & \Delta(2m;\alpha+\beta); & \dfrac{w\,b^b}{4^m\sigma^\sigma a^a(1+r_1)^m(1+r_2)^m}\end{matrix}\right]; \tag{33}$$

(xi) Setting $\xi = \nu$, $\lambda = \phi$ and $a = 1$ in (20), the following identity holds:

$$\int_p^q \frac{(x-p)^{\alpha-1}(q-x)^{\beta-1}}{[(q-p)+r_1(x-p)+r_2(q-x)]^{\alpha+\beta}}\, E^{\gamma,b}_{\sigma,\mu,\delta}\left[w\left\{\frac{(x-p)(q-x)}{[(q-p)+r_1(x-p)+r_2(q-x)]^2}\right\}^m\right] dx$$
$$= \frac{1}{(q-p)\,(1+r_1)^{\alpha}\,(1+r_2)^{\beta}}\,\frac{\Gamma(\alpha)\Gamma(\beta)}{\Gamma(\alpha+\beta)\Gamma(\mu)}$$
$$\times\ {}_{2m+b+1}F_{2m+\sigma+1}\left[\begin{matrix}\Delta(m;\alpha), & \Delta(m;\beta),\ \Delta(b;\gamma), & 1; & \\ \Delta(\sigma;\mu), & \Delta(2m;\alpha+\beta), & \delta; & \dfrac{w\,b^b}{4^m\sigma^\sigma(1+r_1)^m(1+r_2)^m}\end{matrix}\right]; \tag{34}$$

(xii) Setting $\xi = \nu$, $\lambda = \phi$ and $a = b = 1$ in (20), the following identity holds:

$$\int_p^q \frac{(x-p)^{\alpha-1}(q-x)^{\beta-1}}{[(q-p)+r_1(x-p)+r_2(q-x)]^{\alpha+\beta}}\, E^{\gamma,\delta}_{\sigma,\mu}\left[w\left\{\frac{(x-p)(q-x)}{[(q-p)+r_1(x-p)+r_2(q-x)]^2}\right\}^m\right] dx$$
$$= \frac{1}{(q-p)\,(1+r_1)^{\alpha}\,(1+r_2)^{\beta}}\,\frac{\Gamma(\alpha)\Gamma(\beta)}{\Gamma(\alpha+\beta)\Gamma(\mu)}$$

$$\times\ {}_{2m+2}F_{2m+\sigma+1}\left[\begin{array}{llll}\Delta(m;\alpha), & \Delta(m;\beta), \ \gamma, & 1; & \dfrac{w}{4^m\sigma^\sigma(1+r_1)^m(1+r_2)^m}\\ \Delta(\sigma;\mu), & \Delta(2m;\alpha+\beta), & \delta; & \end{array}\right]; \tag{35}$$

(xiii) Setting $\xi = \nu$, $\lambda = \phi$ and $a = \delta = 1$ in (20), the following identity holds:

$$\int_p^q \frac{(x-p)^{\alpha-1}(q-x)^{\beta-1}}{[(q-p)+r_1(x-p)+r_2(q-x)]^{\alpha+\beta}}\, E_{\sigma,\mu}^{\gamma,b}\left[w\left\{\frac{(x-p)(q-x)}{[+(q-p)r_1(x-p)+r_2(q-x)]^2}\right\}^m\right]dx$$

$$= \frac{1}{(q-p)\,(1+r_1)^\alpha\,(1+r_2)^\beta}\,\frac{\Gamma(\alpha)\Gamma(\beta)}{\Gamma(\alpha+\beta)\Gamma(\mu)}$$

$$\times\ {}_{2m+b}F_{2m+\sigma}\left[\begin{array}{lll}\Delta(m;\alpha), & \Delta(m;\beta), \ \Delta(b;\gamma); & \dfrac{w\, b^b}{4^m\sigma^\sigma(1+r_1)^m(1+r_2)^m}\\ \Delta(\sigma;\mu), & \Delta(2m;\alpha+\beta); & \end{array}\right]; \tag{36}$$

(xiv) Setting $\xi = \nu$, $\lambda = \phi$ and $a = b = \delta = 1$ in (20), the following identity holds:

$$\int_p^q \frac{(x-p)^{\alpha-1}(q-x)^{\beta-1}}{[(q-p)+r_1(x-p)+r_2(q-x)]^{\alpha+\beta}}\, E_{\sigma,\mu}^{\gamma}\left[w\left\{\frac{(x-p)(q-x)}{[(q-p)+r_1(x-p)+r_2(q-x)]^2}\right\}^m\right]dx$$

$$= \frac{1}{(q-p)\,(1+r_1)^\alpha\,(1+r_2)^\beta}\,\frac{\Gamma(\alpha)\Gamma(\beta)}{\Gamma(\alpha+\beta)\Gamma(\mu)}$$

$$\times\ {}_{2m+1}F_{2m+\sigma}\left[\begin{array}{lll}\Delta(m;\alpha), & \Delta(m;\beta), \ \gamma; & \dfrac{w}{4^m\sigma^\sigma(1+r_1)^m(1+r_2)^m}\\ \Delta(\sigma;\mu), & \Delta(2m;\alpha+\beta); & \end{array}\right]; \tag{37}$$

(xv) Setting $\xi = \nu$, $\lambda = \phi$ and $a = b = \gamma = \delta = 1$ in (20), the following identity holds:

$$\int_p^q \frac{(x-p)^{\alpha-1}(q-x)^{\beta-1}}{[(q-p)+r_1(x-p)+r_2(q-x)]^{\alpha+\beta}}\, E_{\sigma,\mu}\left[w\left\{\frac{(x-p)(q-x)}{[(q-p)+r_1(x-p)+r_2(q-x)]^2}\right\}^m\right]dx$$

$$= \frac{1}{(q-p)\,(1+r_1)^\alpha\,(1+r_2)^\beta}\,\frac{\Gamma(\alpha)\Gamma(\beta)}{\Gamma(\alpha+\beta)\Gamma(\mu)}$$

$$\times\ {}_{2m+1}F_{2m+\sigma}\left[\begin{array}{lll}\Delta(m;\alpha), & \Delta(m;\beta), \ 1; & \dfrac{w}{4^m\sigma^\sigma(1+r_1)^m(1+r_2)^m}\\ \Delta(\sigma;\mu), & \Delta(2m;\alpha+\beta); & \end{array}\right]; \tag{38}$$

(xvi) Setting $\xi = \nu$, $\lambda = \phi$ and $a = b = \gamma = \delta = \mu = 1$ in (20), the following identity holds:

$$\int_p^q \frac{(x-p)^{\alpha-1}(q-x)^{\beta-1}}{[(q-p)+r_1(x-p)+r_2(q-x)]^{\alpha+\beta}}\, E_{\sigma}\left[w\left\{\frac{(x-p)(q-x)}{[(q-p)+r_1(x-p)+r_2(q-x)]^2}\right\}^m\right]dx$$

$$= \frac{\Gamma(\alpha)\Gamma(\beta)}{\Gamma(\alpha+\beta)}\,\frac{1}{(q-p)\,(1+r_1)^\alpha\,(1+r_2)^\beta}$$

$$\times\ {}_{2m+1}F_{2m+\sigma}\left[\begin{array}{lll}\Delta(m;\alpha), & \Delta(m;\beta), \ 1; & \dfrac{w}{4^m\sigma^\sigma(1+r_1)^m(1+r_2)^m}\\ \Delta(\sigma;1), & \Delta(2m;\alpha+\beta); & \end{array}\right]; \tag{39}$$

(xvii) Setting $\xi = \nu$, $\lambda = \phi$ and $a = b = \gamma = \delta = \mu = \sigma = 1$ in (20), the following identity holds:

$$\int_p^q \frac{(x-p)^{\alpha-1}(q-x)^{\beta-1}}{[(q-p)+r_1(x-p)+r_2(q-x)]^{\alpha+\beta}}\, e^{\left[w\left\{\frac{(x-p)(q-x)}{[(q-p)+r_1(x-p)+r_2(q-x)]^2}\right\}^m\right]}dx$$

$$= \frac{\Gamma(\alpha)\Gamma(\beta)}{\Gamma(\alpha+\beta)}\,\frac{1}{(q-p)\,(1+r_1)^\alpha\,(1+r_2)^\beta}$$

$$\times\ {}_{2m}F_{2m}\left[\begin{array}{ll}\Delta(m;\alpha), & \Delta(m;\beta);\\ \Delta(2m;\alpha+\beta); & \end{array}\ \frac{w}{4^m\sigma^\sigma(1+r_1)^m(1+r_2)^m}\right]. \tag{40}$$

4. Graphical Representation

Here, in terms of the parameter β, we illustrate Equations (14) and (15) using graphical simulations. For this, we evaluate the integrals numerically using the Gaussian quadrature Method (see [37]) and compare this with the main results. We choose $k = 5$ and $n = 8$ to get more precise results.

5. Conclusions

It is worth stressing that the generalized Mittag–Leffler function obtained and the integral formulas computed are amenable to further generalizations and future investigation. We have attempted to exploit the close connection of the generalized Mittag–Leffler functions with several important special functions and compute the integrals of the functions mentioned above in the form of the generalized Mittag–Leffler, linking different families of special functions. Our main results (15) and (20) and some special cases (24)–(32) can yield several new integrals in terms of Fox-H functions obtained from Equations (11) and (13). For instance, we write

$$\int_p^q \frac{(x-p)^{\alpha-1}(q-x)^{\beta-1}}{[(q-p)+r_1(x-p)+r_2(q-x)]^{\alpha+\beta}}\, E^{\xi,\lambda,\gamma,b}_{\sigma,\mu,\nu,\phi,\delta,a}\left[w\left\{\frac{(x-p)(q-x)}{[(q-p)+r_1(x-p)+r_2(q-x)]^2}\right\}^m\right]dx$$

$$= \frac{1}{(q-p)\,(1+r_1)^\alpha\,(1+r_2)^\beta}\,\frac{\Gamma(\nu)\,\Gamma(\delta)}{\Gamma(\xi)\,\Gamma(\gamma)}$$

$$\times\ {}_5\Psi_4\left[\begin{array}{ll}(\xi,\lambda),\ (\gamma,b),\ (\alpha,m),\ (\beta,m), & (1,1);\\ (\mu,\sigma),\ (\nu,\phi),\ (\delta,a), & (\alpha+\beta,2m);\end{array}\ \frac{w}{(1+r_1)^m(1+r_2)^m}\right] \tag{41}$$

$$= \frac{1}{(q-p)\,(1+r_1)^\alpha\,(1+r_2)^\beta}\,\frac{\Gamma(\nu)\,\Gamma(\delta)}{\Gamma(\xi)\,\Gamma(\gamma)}$$

$$\times\ H^{1,5}_{5,5}\left[\frac{-w}{(1+r_1)^m(1+r_2)^m}\,\middle|\,\begin{array}{ll}(1-\xi,\lambda),\ (1-\gamma,b),\ (1-\alpha,m),\ (1-\beta,m), & (0,1)\\ (0,1),\ (1-\mu,\sigma),\ (1-\nu,\phi),\ (1-\delta,a), & \{1-(\alpha+\beta,2m)\}\}\end{array}\right], \tag{42}$$

with all the conditions prescribed in Theorem 1. We have also proved that Figures 1–3 show a good compatibility of the numerical solution obtained by the Gaussian quadrature method and the analytic expression. We conclude that the results obtained will provide a significant step in the theory of integral transforms and can yield some potential applications in the field of the classical and applied mathematics.

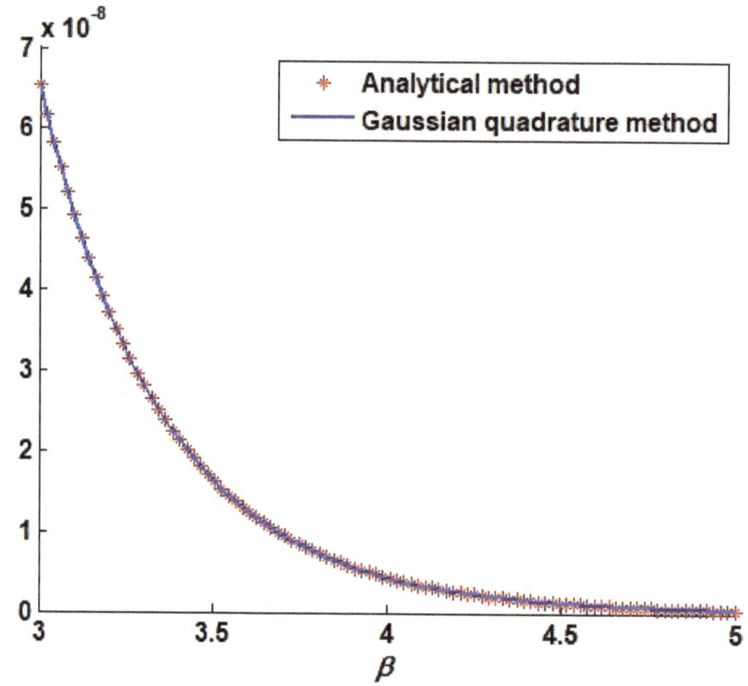

Figure 1. Solution of (14) for $\alpha = 6, r_1 = 2, r_2 = 2, p = 0$ and $q = 1$.

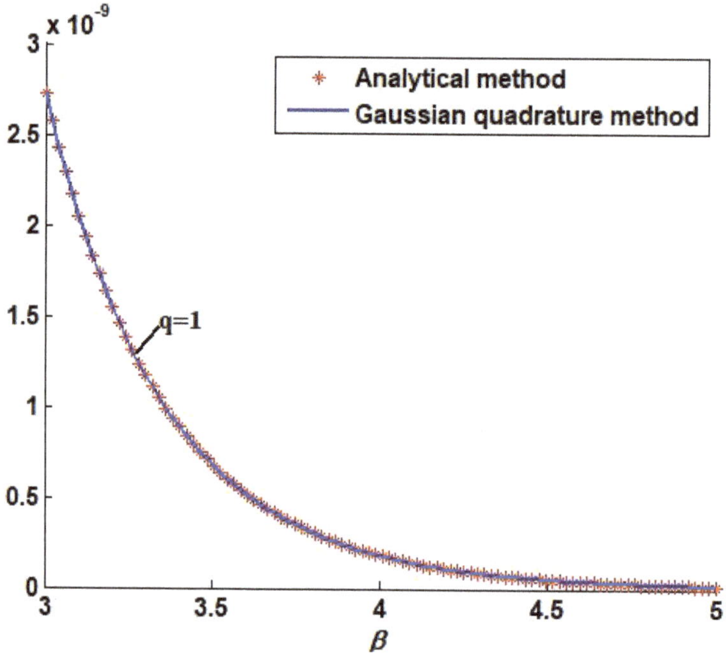

Figure 2. Solution of (15) (for $q = 1$) for $\alpha = 6, r_1 = 2, r_2 = 4, p = 0, \xi = 1, \lambda = 2, \gamma = 2\ b = 3, \sigma = 2, \mu = 5, v = 2, \phi = 3, \delta = 2, a = 4, w = 3$ and $m = 2$.

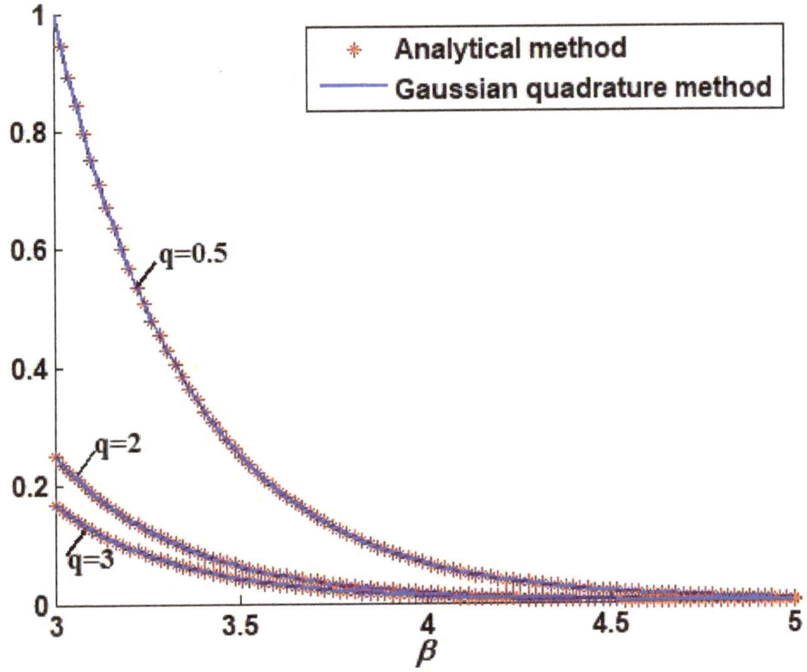

Figure 3. Solution of (15) (for all q) for $\alpha = 6, r_1 = 2, r_2 = 4, p = 0, \xi = 1, \lambda = 2, \gamma = 2\, b = 3, \sigma = 2,$ $\mu = 5, v = 2, \phi = 3, \delta = 2, a = 4, w = 3$ and $m = 2$.

Author Contributions: Conceptualization, N.K. and S.A.-O.; methodology, T.U.; software, M.I.K.; validation, K.N. and M.I.K.; formal analysis, T.U.; investigation, N.K.; resources, S.A.-O.; data curation, S.A.-O.; writing—original draft preparation, T.U.; writing—review and editing, M.I.K.; visualization, N.K.; supervision, M.I.K.; project administration, K.N.; funding acquisition, K.N.; All authors have read and agreed to the published version of the manuscript.

Funding: This research has received funding support from the National Science, 43 Research and Innovation Fund (NSRF), Thailand.

Institutional Review Board Statement: Not applicable.

Informed Consent Statement: Not applicable.

Data Availability Statement: Not applicable.

Acknowledgments: The authors are thankful to the editors and reviewers for their valuable suggestions and comments.

Conflicts of Interest: The authors declare that they have no competing interest.

References

1. Jain, S.; Agarwal, P.; Ahmad, B.; Al-Omari, S. Certain recent fractional integral inequalities associated with the hypergeometric operators. *J. King Saud Univ.-Sci.* **2016**, *28*, 82–86. [CrossRef]
2. Al-Omari, S.; Baleanu, D. On the Generalized Stieltjes transform of Fox's kernel function and its properties in the space of generalized functions. *J. Comput. Anal. Appl.* **2017**, *23*, 108–118.
3. Agarwal, P.; Jain, S.; Kıymaz, I.O.; Chand, M.; Al-Omari, S. Certain sequence of functions involving generalized hypergeometric functions. *Math. Sci. Appl. E-Notes* **2015**, *3*, 45–53. [CrossRef]
4. Khan, N.; Usman,T.; Aman, M.; Al-Omari, S.; Choi, J. Integral transforms and probality distributions involving generalized hypergeometric function. *Georgian J. Math.* **2021**, *28*, 2021–2105. [CrossRef]
5. Chandak, S.; Al-Omari, S.K.Q.; Suthar, D.L. Unified integral associated with the generalized V-function. *Adv. Differ. Equ.* **2020**, *2020*, 560. [CrossRef]

6. Choi,J. and Agarwal, P. A note on generalized integral operator associated with multiindex Mittag-Leffler function, Filomat 30, 1931–1939. *Adv. Differ. Equ.* **2020**, *448*, 1–11. [CrossRef]
7. Gorenflo, R.; Kilbas, A.A.; Mainardi, F.; Rogosin, S.V. *Mittag-Leffler Functions, Related Topics and Applications*; Springer: New York, NY, USA, 2020; p. 540.
8. Haubold, H.J.; Mathai, A.M.; Saxena, R.K. Mittag-Leffler Functions and Their Applications. *J. Appl. Math.* **2011**, *2011*, 298628. [CrossRef]
9. Kiryakova, V. *Generalized Fractional Calculus and Applications*; CRC Press: Boca Raton, FL, USA, 1993.
10. Kiryakova, V. The multi-index Mittag-Leffler functions as an important class of special functions of fractional calculus. *Comput. Math. Appl.* **2010**, *59*, 1885–1895. [CrossRef]
11. Kochubei, A.; Luchko, Y. Fractional Differential Equations. In *Handbook of Fractional Calculus with Applications*; De Gruyter: Berlin, Germany, 2019; Volume 2.
12. Mainardi, F. Why the Mittag-Leffler Function Can Be Considered the Queen Function of the Fractional Calculus? *Entropy* **2020**, *22*, 1359. [CrossRef] [PubMed]
13. Agarwal, P.; Choi, J.; Jain, S.; Rashidi, M.M. Certain integrals associated with generalized mittag-leffler function. *Commun. Korean Math. Soc.* **2017**, *32*, 29–38. [CrossRef]
14. Almalahi, M.A.; Ghanim, F.; Botmart, T.; Bazighifan, O.; Askar, S. Qualitative Analysis of Langevin Integro-Fractional Differential Equation under Mittag–Leffler Functions Power Law. *Fractal Fract.* **2021**, *5*, 266. [CrossRef]
15. Kamarujjama, M.; Khan, N.; Khan, O. Estimation of certain integrals with extended multi-index Bessel function. *Malaya J. Mat.* **2019**, *7*, 206–212. [CrossRef]
16. Khan, N.; Usman, T.; Aman, M.; Al-Omari, S.; Araci, S. Computation of certain integral formulas involving generalized Wright function. *Adv. Differ. Equ.* **2020**, *2020*, 491. [CrossRef]
17. Khan, N.; Usman, T.; Aman, M. Some properties concerning the analysis of generalized Wright function. *J. Comput. Appl. Math.* **2020**, *376*, 112840. [CrossRef]
18. Khan, N.; Khan, S. Integral transform of generalized K-Mittag-Lefller function. *J. Fract. Calc. Appl.* **2018**, *9*, 13–21.
19. Khan, N.; Ghayasuddin, M.; Shadab, M. Some Generating Relations of Extended Mittag-Leffler Functions. *Kyungpook Math. J.* **2019**, *59*, 325–333.
20. Khan, N.; Husain, S. A note on extended beta function involving generalized Mittag-Leffler function and its applications. *TWMS J. App. Eng. Math.* **2022**, *12*, 71–81.
21. Khan, O.; Khan, N.; Sooppy, K.A. Unified approach to the certain integrals of k-Mittag-Leffler type function of two variables. *Trans. Natl. Acad. Sci. Azerb. Ser. Phys.-Tech. Math. Sci. Math.* **2019**, *39*, 98–108.
22. Mihai, M.V.; Awan, M.U.; Noor, M.A.; Du, T.; Kashuri, A.; Noor, K.I. On Extended General Mittag–Leffler Functions and Certain Inequalities. *Fractal Fract.* **2019**, *3*, 32. [CrossRef]
23. Prabhakar, T.R. A Singular Integral Equation with a Generalized Mittag-Leffler Function in the Kernel. *Yokohama Math. J.* **1971**, *19*, 7–15.
24. Prudnikov, A.P.; Brychkov, Y.A.; Marichev, O.I. *Integral and Series V.1. More Special Functio*; Gordon and Breach: New York, NY, USA; London, UK, 1992.
25. Rainville, E.D. *Special Functions*; The Macmillan Company: New York, NY, USA, 1960.
26. Salim, T.O. Some properties relating to the generalized Mittag-Leffler function. *Adv. Appl. Math. Anal.* **2009**, *4*, 21–30.
27. Salim, T.O.; Faraj, A.W. A generalization of Mittag-Leffler function and integral operator associated with fractional calculus. *J. Fract. Calc. Appl.* **2012**, *3*, 1–13.
28. Shukla, A.; Prajapati, J. On a generalization of Mittag-Leffler function and its properties. *J. Math. Anal. Appl.* **2007**, *336*, 797–811. [CrossRef]
29. Singh, P.; Jain, S.; Cattani, C. Some Unified Integrals for Generalized Mittag-Leffler Functions. *Axioms* **2021**, *10*, 261. [CrossRef]
30. Suthar, D.L.; Amsalu, H.; Godifey, K. Certain integrals involving multivariate Mittag-Leffler function. *J. Inequalities Appl.* **2019**, *2019*, 208–224. [CrossRef]
31. Mittag-Leffler, G.M. Sur la nouvelle fonction $E_{\alpha(x)}$. *CR Acad. Sci. Paris* **1903**, *137*, 554–558.
32. Rahman, G.; Suwan, I.; Nisar, K.S.; Abdeljawad, T.; Samraiz, M.; Ali, A. A basic study of a fractional integral operator with extended Mittag-Leffler kernel. *AIMS Math.* **2021**, *6*, 12757–12770. [CrossRef]
33. Wiman, A. Uber den fundamental Satz in der Theories der Funktionen $E_{\alpha(z)}$. *Acta Math.* **1905**, *29*, 191–201. [CrossRef]
34. Khan, M.A.; Ahmed, S. On some properties of the generalized Mittag-Leffler function. *SpringerPlus* **2013**, *2*, 337. [CrossRef] [PubMed]
35. Wright, E.M. The asymptotic expansion of integral functions defined by Taylor series. *Philos. Trans. R. Soc. London. Ser. A Math. Phys. Sci.* **1940**, *238*, 423–451. [CrossRef]
36. Al-Omari, S. Estimation of a modified integral associated with a special function kernel of Fox's H-function type. *Commun. Korean Math. Soc.***2020**, *35*, 125–136.
37. Abramowitz, M.; Stegun, I.A. (Eds.) *Handbook of Mathematical Functions with Formulas, Graphs, and Mathematical Tables*; US Government Printing Office: Washington, DC, USA, 1948; Volume 55.
38. Fox, C. The Asymptotic Expansion of Generalized Hypergeometric Functions. *Proc. Lond. Math. Soc.* **1928**, *2*, 389–400. [CrossRef]
39. Al-Omari, S. A revised version of the generalized Krätzel-Fox integral operators. *Mathematics* **2018**, *6*, 222. [CrossRef]

40. Al-Omari, S. On a Class of Generalized Meijer-Laplace Transforms of Fox Function Type Kernels and Their Extension to a Class of Boehmians. *Georgian Math. J.* **2017**, *25*, 1–8. [CrossRef]
41. Wright, E.M. The asymptotic expansion of the generalized hypergeometric function. *Proc. Lond. Math. Soc.* **1940**, *2*, 389–408. [CrossRef]
42. Wright, E.M. The asymptotic expansion of the generalized hypergeometric function. *J. Lond. Math. Soc.* **1935**, *1*, 286–293. [CrossRef]

symmetry MDPI

Article

Non-Separable Linear Canonical Wavelet Transform

Hari M. Srivastava [1,2,3,4,*], Firdous A. Shah [5], Tarun K. Garg [6], Waseem Z. Lone [5] and Huzaifa L. Qadri [5]

1. Department of Mathematics and Statistics, University of Victoria, Victoria, BC V8W 3R4, Canada
2. Department of Medical Research, China Medical University Hospital, China Medical University, Taichung 40402, Taiwan
3. Department of Mathematics and Informatics, Azerbaijan University, 71 Jeyhun Hajibeyli Street, Baku AZ1007, Azerbaijan
4. Section of Mathematics, International Telematic University Uninettuno, I-00186 Rome, Italy
5. Department of Mathematics, University of Kashmir, Anantnag 192101, India; fashah@uok.edu.in (F.A.S.); lwaseem.scholar@kashmiruniversity.net (W.Z.L.); huzaifaqadri37@gmail.com (H.L.Q.)
6. Department of Mathematics, Satyawati College, University of Delhi, Delhi 110052, India; tkgarg@satyawati.du.ac.in

* Correspondence: harimsri@math.uvic.ca

Abstract: This study aims to achieve an efficient time-frequency representation of higher-dimensional signals by introducing the notion of a non-separable linear canonical wavelet transform in $L^2(\mathbb{R}^n)$. The preliminary analysis encompasses the derivation of fundamental properties of the novel integral transform including the orthogonality relation, inversion formula, and the range theorem. To extend the scope of the study, we formulate several uncertainty inequalities, including the Heisenberg's, logarithmic, and Nazorav's inequalities for the proposed transform in the linear canonical domain. The obtained results are reinforced with illustrative examples.

Keywords: non-separable linear canonical wavelet; symplectic matrix; non-separable linear canonical transform; uncertainty principle

MSC: 42C40; 42B10; 53D22; 65R10

Citation: Srivastava, H.M.; Shah, F.A.; Garg, T.K.; Lone, W.Z.; Qadri, H.L. Non-Separable Linear Canonical Wavelet Transform. *Symmetry* **2021**, *13*, 2182. https://doi.org/10.3390/sym13112182

Academic Editor: Marin Marin

Received: 23 September 2021
Accepted: 29 October 2021
Published: 9 Novemeber 2021

1. Introduction

The origin of the multi-dimensional linear canonical transform (LCT) dates back to the early 1970s with the foundational work of Moshinsky and Quesne [1] in quantum mechanics to study the linear maps of phase space. Soon after its inception in quantum mechanics, the linear canonical transform has been exclusively studied both in theory and applications [2,3]. The theory of multi-dimensional non-separable LCT involving a general $2n \times 2n$ real, symplectic matrix $\mathbf{M} = (A, B : C, D)$ with $n(2n+1)$ independent parameters offers a canonical formalism for the representation of several physical systems in a lucid and insightful way. For any $f \in L^2(\mathbb{R}^n)$, the non-separable LCT with respect to a real, symplectic matrix $\mathbf{M}$ is given by [4,5]

$$\mathcal{F}^{\mathbf{M}}[f](\mathbf{w}) = \frac{1}{|\det B|^{1/2}} \int_{\mathbb{R}^n} f(\mathbf{t})\, e^{i\pi\left(\mathbf{w}^T D B^{-1}\mathbf{w} - 2\mathbf{w}^T B^{T^{-1}}\mathbf{t} + \mathbf{t}^T B^{-1} A \mathbf{t}\right)}\, d\mathbf{t}, \quad |\det B| \neq 0. \quad (1)$$

The importance of the arbitrary real symplectic matrices involved in Equation (1) lies in the fact that an appropriate choice of the matrix can be taken to inculcate a sense of rotation and shift into both the time and frequency axes, resulting in an efficient representation of the chirp-like signals, which are ubiquitous both in nature and man-made systems. Due to the extra degrees of freedom, the non-separable LCT has been successfully employed in diverse problems arising in various branches of science and engineering, such as harmonic analysis, reproducing kernel Hilbert spaces, optical systems, quantum mechanics, sampling, image processing, and so on [6,7].

Undoubtedly, wavelet transforms have fascinated the scientific, engineering, and research communities both with their versatile applicability and lucid mathematical framework [8,9]. In recent years, the classical wavelet transform has been extended and employed in different domains. The most prompt ones are the fractional wavelet transform [10], linear canonical wavelet transform [11,12], special affine wavelet transform [13,14], quaternion linear canonical wavelet transform [15], and quadratic-phase wavelet transform [16]. Unfortunately, all these transforms only perform well at representing point singularities and are incompetent at handling the distributed singularities, such as curves or edges in higher-dimensional signals [17–20]. The intuitive reason for this inadequacy is that wavelets are isotropic entities generated by isotropically dilating the mother wavelet, and as such, they ignore the geometric properties of the structures to be analyzed. Therefore, the conventional wavelet approach is inadequate while dealing with multi-dimensional signals, wherein the primary interest is to efficiently capture the geometric features, such as edges and corners, appearing due to the spatial occlusion between different objects. As such, the key problem in multi-dimensional signal analysis is to extract and characterize the relevant geometric information regarding the occurrence of curves and boundaries in signals. Subsequently, a higher-dimensional variant of the standard wavelet transform has been proposed, which serves as a potent tool for representing non-transient multi-dimensional signals in the time-frequency domain. Mathematically, the multi-dimensional wavelet transform of any $f \in L^2(\mathbb{R}^n)$ is defined by [21]

$$\mathcal{W}_\psi[f](a, \mathbf{b}) = \frac{1}{\sqrt{a}} \int_{\mathbb{R}^n} f(\mathbf{t})\, \overline{\psi\left(\frac{\mathbf{t}-\mathbf{b}}{a}\right)}\, e^{-i\mathbf{w}\cdot\mathbf{t}}\, d\mathbf{t}, \quad a \in \mathbb{R}^+, \mathbf{b} \in \mathbb{R}^n, \tag{2}$$

where a is called the scaling parameter, which controls the degree of compression or scale, and $\mathbf{b}$ is the translation parameter that determines the time location of the wavelet. The multi-dimensional wavelet transform in Equation (2) has found numerous applications across diverse fields of science and engineering, particularly in video image processing, medical imaging, singular detection problems, fluid dynamics, shape recognition, and so on [21,22]. In the context of higher-dimensional wavelet theory, the symmetry property of wavelets is often desirable in practical applications, and as such, wavelets can reveal different patterns and singularities of highly nonstationary signals, such as Brownian motions, patterns on the water surfaces, fractal properties of the velocity field, computations of Renyi dimensions, Hurst and H¨older exponents. Some prominent examples of the symmetric wavelets include biorthogonal wavelets, quincunx wavelets, and carinal B-splines.

Keeping in view the profound characteristics of the multi-dimensional wavelet transform and more degrees of freedom of non-separable linear canonical transforms, we are deeply motivated to intertwine these integral transforms into a novel integral transform coined as a non-separable linear canonical wavelet transform. The novel integral transform can efficiently localize any non-transient signal in the time-frequency plane with more degrees of freedom. With major modifications to the existing multi-dimensional wavelet transform in Equation (2), we propose the non-separable linear canonical wavelet transform of any $f \in L^2(\mathbb{R}^n)$ concerning the free symplectic matrix $\mathbf{M} = (A, B : C, D)$ as

$$\mathcal{W}_\psi^{\mathbf{M}}[f](a, \mathbf{b}) = \frac{1}{|\det B|^{1/2}} \int_{\mathbb{R}^n} f(\mathbf{t})\, \overline{\psi\left(\frac{\mathbf{t}-\mathbf{b}}{a}\right)}\, e^{-\pi i\left(\Lambda_a^T D B^{-1} \Lambda_a - 2\Lambda_a^T B^{T^{-1}} \mathbf{t} + \mathbf{t}^T B^{-1} A \mathbf{t}\right)}\, d\mathbf{t}, \tag{3}$$

where $\Lambda_a = (a\ a\ \ldots\ a)^T$. Besides studying all the fundamental properties of the novel wavelet transform, we derive some well-known theorems, including the Rayleigh's theorem, inversion formula, and range theorem. In the sequel, we also formulate several uncertainty inequalities such as the Heisenberg's, logarithmic, and Nazorav-type inequalities for the non-separable linear canonical wavelet transform in Equation (3).

The rest of the article is structured as follows: Section 2 is concerned with the preliminary aspects of the study and the formulation of the non-separable linear canonical wavelet transform. Section 3 is devoted to formulating several variants of the uncertainty principles, such as Heisenberg's, logarithmic, and Nazorav-type inequalities, for the proposed transform. Finally, a conclusion is extracted in Section 4.

2. Non-Separable Linear Canonical Wavelet Transform in $L^2(\mathbb{R}^n)$

In this section, we first provide a healthy overview of the non-separable linear canonical transform. Then, we introduce the notion of the non-separable linear canonical wavelet transform in $L^2(\mathbb{R}^n)$, followed by some fundamental properties of the proposed transform, including the orthogonality relation, energy preserving relation, range theorem, and the inversion formula.

2.1. Non-Separable Linear Canonical Transform

For typographical convenience, we shall denote a real $2n \times 2n$ matrix

$$\mathbf{M} = \begin{pmatrix} A & B \\ C & D \end{pmatrix} = \begin{pmatrix} a_{11} & a_{12} & \dots & a_{1n} & b_{11} & b_{12} & \dots & b_{1n} \\ a_{21} & a_{22} & \dots & a_{2n} & b_{21} & b_{22} & \dots & b_{2n} \\ \vdots & \vdots & \ddots & \vdots & \vdots & \vdots & \ddots & \vdots \\ a_{n1} & a_{n2} & \dots & a_{nn} & b_{n1} & b_{n2} & \dots & b_{nn} \\ c_{11} & c_{12} & \dots & c_{1n} & d_{11} & d_{12} & \dots & d_{1n} \\ c_{21} & c_{22} & \dots & c_{2n} & d_{21} & d_{22} & \dots & d_{2n} \\ \vdots & \vdots & \ddots & \vdots & \vdots & \vdots & \ddots & \vdots \\ c_{n1} & c_{n2} & \dots & c_{nn} & d_{n1} & d_{n2} & \dots & d_{nn} \end{pmatrix} \tag{4}$$

as $\mathbf{M} = (A, B : C, D)$, where $A, B, C,$ and D are $n \times n$ sub-matrices with real entries. Moreover, the matrix $\mathbf{M} = (A, B : C, D)$ is said to be free symplectic if $\mathbf{M}^T\mathbf{J}\mathbf{M} = \mathbf{J}$ and $|\det B| \neq 0$, where $\mathbf{J} = (\mathbf{0}, I_n : -I_n, \mathbf{0})$, and I_n denotes the n-dimensional identity matrix. Furthermore, the sub-matrices corresponding to the free symplectic matrix $\mathbf{M} = (A, B : C, D)$ satisfy

$$AB^T = BA^T, CD^T = DC^T, AD^T - BC^T = I_n, \tag{5}$$

or equivalently

$$A^TC = C^TA, B^TD = D^TB, A^TD - C^TB = I_n. \tag{6}$$

The transpose and inverse corresponding to the free symplectic matrix $M = (A, B : C, D)$ are given by $\mathbf{M}^T = (A^T, C^T : B^T, D^T)$ and $\mathbf{M}^{-1} = (D^T, -B^T : -C^T, A^T)$, respectively. Moreover, we have

$$\begin{aligned} \mathbf{M}\mathbf{M}^{-1} &= \begin{pmatrix} A & B \\ C & D \end{pmatrix} \begin{pmatrix} D^T & -B^T \\ -C^T & A^T \end{pmatrix} \\ &= \begin{pmatrix} AD^T - BC^T & -AB^T + BA^T \\ CD^T - DC^T & -CB^T + DA^T \end{pmatrix} \\ &= \begin{pmatrix} I_n & 0 \\ 0 & I_n \end{pmatrix}. \end{aligned}$$

A typical example of a 4×4 free symplectic matrix is given below

$$\mathbf{M} = \begin{pmatrix} 1 & 2 & -1/2 & -1/2 \\ -2 & 1 & -1/2 & 1/2 \\ -1 & -3 & 1 & 1 \\ -1 & 0 & -1/2 & 1/2 \end{pmatrix}.$$

Definition 1. *Given a free symplectic matrix* $\mathbf{M} = (A, B : C, D)$, *the non-separable linear canonical transform of any* $f \in L^2(\mathbb{R}^n)$ *is denoted by* $\mathcal{F}^{\mathbf{M}}[f]$ *and is defined as*

$$\mathcal{F}^{\mathbf{M}}[f](\mathbf{w}) = \int_{\mathbb{R}^n} f(\mathbf{t})\, \mathcal{K}^{\mathbf{M}}(\mathbf{t}, \mathbf{w})\, d\mathbf{t}, \tag{7}$$

where the kernel $\mathcal{K}^{\mathbf{M}}(\mathbf{t}, \mathbf{w})$ *is given by*

$$\mathcal{K}^{\mathbf{M}}(\mathbf{x}, \mathbf{w}) = \frac{1}{|\det B|^{1/2}} \exp\Big\{\pi i\big(\mathbf{w}^T D B^{-1}\mathbf{w} - 2\mathbf{w}^T B^{T^{-1}}\mathbf{t} + \mathbf{t}^T B^{-1} A\mathbf{t}\big)\Big\}, \quad |\det B| \neq 0. \tag{8}$$

The additive property of the non-separable LCT (Equation (7)) is very crucial for its understanding and application and is given by

$$\mathcal{F}^{\mathbf{M}_1}\Big[\mathcal{F}^{\mathbf{M}_2}[f(\mathbf{t})]\Big](\mathbf{w}) = \mathcal{F}^{\mathbf{M}_1\mathbf{M}_2}[f(\mathbf{t})](\mathbf{w}).$$

The Plancheral and inversion formulae corresponding to Equation (7) are given by

$$\langle f, g\rangle_2 = \Big\langle \mathcal{F}^{\mathbf{M}}[f], \mathcal{F}^{\mathbf{M}}[g]\Big\rangle_2, \quad \forall f, g \in L^2(\mathbb{R}^n) \quad \text{and} \tag{9}$$

$$f(\mathbf{t}) = \mathcal{F}^{\mathbf{M}^{-1}}\Big[\mathcal{F}^{\mathbf{M}}[f](\mathbf{w})\Big](\mathbf{x}) = \int_{\mathbb{R}^n} \mathcal{F}^{\mathbf{M}}[f](\mathbf{w})\, \mathcal{K}^{\mathbf{M}^{-1}}(\mathbf{w}, \mathbf{t})\, d\mathbf{w}, \tag{10}$$

respectively, where $\mathbf{M}^{-1} = \big(D^T, -B^T : -C^T, A^T\big)$. Furthermore, the kernel in Equation (8) satisfies the following properties:

(i) $\mathcal{K}^{\mathbf{M}^{-1}}(\mathbf{w}, \mathbf{t}) = \overline{\mathcal{K}^{\mathbf{M}}(\mathbf{t}, \mathbf{w})}$,

(ii) $\int_{\mathbb{R}^n} \mathcal{K}^{\mathbf{M}}(\mathbf{t}, \mathbf{w})\, \mathcal{K}^{\mathbf{M}^{-1}}(\mathbf{t}, \mathbf{z})\, d\mathbf{t} = \delta(\mathbf{z} - \mathbf{w})$,

(iii) $\int_{\mathbb{R}^n} \mathcal{K}^{\mathbf{M}}(\mathbf{t}, \mathbf{w})\, \mathcal{K}^{\mathbf{M}^{-1}}(\mathbf{z}, \mathbf{w})\, d\mathbf{w} = \delta(\mathbf{z} - \mathbf{t})$,

(iv) $\int_{\mathbb{R}^n} \mathcal{K}^{\mathbf{M}}(\mathbf{t}, \mathbf{w})\, \mathcal{K}^{\mathbf{N}}(\mathbf{t}, \mathbf{z})\, d\mathbf{t} = \mathcal{K}^{\mathbf{MN}}(\mathbf{w}, \mathbf{z})$.

The non-separable linear canonical transform (Equation (7)) encompasses several well-known integral transforms, including the Fourier transform (FT), fractional Fourier transform (FrFT), linear canonical transform (LCT), and the Fresnel transforms (FrT) [4]. Table 1 shows some special cases of the non-separable linear canonical transform.

Table 1. Some special cases of the non-separable linear canonical transform.

Free Symplectic MATRIX $\mathbf{M} = (A, B : C, D)$	Free Metaplectic Transformation
• $A = D = \mathbf{0}, B = -C = I_n$	n-dimensional FT
• $A = \text{diag}(a_{11}, \cdots, a_{nn})$, $B = \text{diag}(b_{11}, \cdots, b_{nn})$, $C = \text{diag}(c_{11}, \cdots, c_{nn})$, $D = \text{diag}(d_{11}, \cdots, d_{nn})$	n-dimensional separable LCT
• $A = D = \text{diag}(\cos\theta_1, \cdots, \cos\theta_n)$, $B = -C = \text{diag}(\sin\theta_1, \cdots, \sin\theta_n)$	n-dimensional separable FrFT
• $A = D = I_n \cos\theta, B = -C = I_n \sin\theta$	n-dimensional non-separable FrFT
• $A = D = I_n, B = \text{diag}(b_{11}, \cdots, b_{nn}), C = \mathbf{0}$	n-dimensional separable FrT
• $A = D = I_n, C = \mathbf{0}$	n-dimensional non-separable FrT

2.2. *Non-Separable Linear Canonical Wavelet Transform*

Wavelets act as window functions whose radius increases in time (reduces in frequency) while resolving the low-frequency contents and decreases in time (increases in frequency) while resolving high-frequency contents of a non-transient signal. Mathematically, a doubly indexed family of wavelets $\psi_{a,\mathbf{b}}$ is generated by restricting the scaling parameter a belonging to $\mathbb{R}^+$ and the translation parameter $\mathbf{b}$ belonging to $\mathbb{R}^n$ as [8]:

$$\psi_{a,\mathbf{b}}(\mathbf{t}) = \frac{1}{\sqrt{a}}\psi\left(\frac{\mathbf{t}-\mathbf{b}}{a}\right), \quad a \in \mathbb{R}^+, b \in \mathbb{R}^n. \tag{11}$$

The scaling parameter a measures the degree of compression or scale, whereas the translation parameter $\mathbf{b}$ determines the location of the wavelet. With major modifications of the family (Equation (4)), we define a new family of functions $\psi^{\mathbf{M}}_{a,\mathbf{b}}(\mathbf{t})$ with respect to a free symplectic matrix $\mathbf{M} = (A, B : C, D)$ as:

$$\psi^{\mathbf{M}}_{a,\mathbf{b}}(\mathbf{t}) = \frac{1}{\sqrt{a}}\psi\left(\frac{\mathbf{t}-\mathbf{b}}{a}\right)\mathcal{K}^{\mathbf{M}}(\mathbf{t}, a), \tag{12}$$

where

$$\mathcal{K}^{\mathbf{M}}(\mathbf{t}, a) = \frac{1}{|\det B|^{1/2}}\exp\left\{\pi i\left(\Lambda_a^T D B^{-1}\Lambda_a - 2\Lambda_a^T B^{T^{-1}}\mathbf{t} + \mathbf{t}^T B^{-1} A\mathbf{t}\right)\right\}, \tag{13}$$

where $\Lambda_a = (a \ a \ \dots \ a)^T$. Having formulated a family of analyzing functions, we are now ready to introduce the definition of the non-separable linear canonical wavelet transform in $L^2(\mathbb{R}^n)$.

Definition 2. *For any $f \in L^2(\mathbb{R}^n)$, the non-separable linear canonical wavelet transform of f with respect to an analyzing wavelet ψ and the free symplectic matrix $\mathbf{M} = (A, B : C, D)$ is defined by*

$$\mathcal{W}^{\mathbf{M}}_{\psi}[f](a, \mathbf{b}) = \frac{1}{\sqrt{a|\det B|}}\int_{\mathbb{R}^n} f(\mathbf{t})\,\overline{\psi\left(\frac{\mathbf{t}-\mathbf{b}}{a}\right)}\, e^{-\pi i\left(\Lambda_a^T D B^{-1}\Lambda_a - 2\Lambda_a^T B^{T^{-1}}\mathbf{t} + \mathbf{t}^T B^{-1} A\mathbf{t}\right)} d\mathbf{t}. \tag{14}$$

Definition 2 allows us to make the following comments:

(i) The non-separable linear canonical wavelet transform can be written in the inner-product form as

$$\mathcal{W}^{\mathbf{M}}_{\psi}[f](a, \mathbf{b}) = \left\langle f, \psi^{\mathbf{M}}_{a,\mathbf{b}}\right\rangle,$$

where $\psi^{\mathbf{M}}_{a,\mathbf{b}}(\mathbf{t})$ is given by Equation (12).

(ii) It is worth noticing that the proposed transform in Equation (7) encompasses several existing integral transforms, such as the classical wavelet transform, fractional wavelet transform, linear canonical wavelet transform, and so on [8,9]. The corresponding wavelet transforms can be obtained by choosing an appropriate symplectic matrix $\mathbf{M} = (A, B : C, D)$.

We now present an example for the lucid illustration of the proposed non-separable linear canonical wavelet transform in Equation (14).

Example 1. *(a) Consider the function $f(\mathbf{t}) = e^{\left(\frac{12}{11}t_1 - \frac{30}{11}t_2\right)^2}$ and the 2D-Morlet function $\psi(\mathbf{t}) = e^{i\Lambda\cdot\mathbf{t} - |\mathbf{t}|^2/2}$, $\Lambda = (\lambda_1, \lambda_2) > 0$. Then, the translated and scaled versions of $\psi(\mathbf{t})$ are given by*

$$\psi\left(\frac{\mathbf{t}-\mathbf{b}}{a}\right) = \exp\left\{-\frac{i(\lambda_1 b_1 + \lambda_2 b_2)}{a} - \frac{(b_1^2 + b_2^2)}{2a^2}\right\}\exp\left\{-\frac{t_1^2}{2a^2} + \left(\frac{i\lambda_1}{a} + \frac{b_1}{a^2}\right)t_1\right\}$$
$$\times \exp\left\{-\frac{t_2^2}{2a^2} + \left(\frac{i\lambda_2}{a} + \frac{b_2}{a^2}\right)t_2\right\}. \tag{15}$$

Consequently, the family of non-separable linear canonical wavelets $\psi^{\mathbf{M}}_{a,\mathbf{b}}(\mathbf{t})$ *is obtained as:*

$$\psi^{\mathbf{M}}_{a,\mathbf{b}}(\mathbf{t}) = \frac{1}{\sqrt{a\,|\det B|}}\,\psi\left(\frac{\mathbf{t}-\mathbf{b}}{a}\right)\exp\left\{i\pi\left(\Lambda_a^T D B^{-1}\Lambda_a - 2\Lambda_a^T B^{T^{-1}}\mathbf{t} + \mathbf{t}^T B^{-1} A\mathbf{t}\right)\right\}. \tag{16}$$

To compute the non-separable linear canonical wavelet transform of $f(\mathbf{t})$ *with respect to the window function* $\psi(\mathbf{t})$, $\Lambda = (1,1)$, *and a real symplectic matrix*

$$\mathbf{M} = \begin{pmatrix} A & B \\ C & D \end{pmatrix} = \begin{pmatrix} 1/6 & 1 & -2 & 1/6 \\ -5/6 & -1/6 & 1/6 & 5/3 \\ 1 & 0 & 12/29 & -31/29 \\ -6/29 & -36/29 & 36/29 & 0 \end{pmatrix},$$

we proceed as:

$$\begin{aligned} \mathcal{W}^{\mathbf{M}}_{\psi}[f](a,\mathbf{b}) &= \frac{1}{\sqrt{a\,|\det B|}}\exp\left\{\frac{i(b_1+b_2)}{a} - \frac{(b_1^2+b_2^2)}{2a^2}\right\} \\ &\quad\times \int_{\mathbb{R}^n} e^{\left(\frac{12}{11}t_1 - \frac{30}{11}t_2\right)^2}\exp\left\{-\frac{t_1^2}{2a^2} + \left(\frac{b_1}{a^2} - \frac{i}{a}\right)t_1\right\}\exp\left\{-\frac{t_2^2}{2a^2} + \left(\frac{b_2}{a^2} - \frac{i}{a}\right)t_2\right\} \\ &\quad\times \exp\left\{-\pi i\left(\Lambda_a^T D B^{-1}\Lambda_a - 2\Lambda_a^T B^{T^{-1}}\mathbf{t} + \mathbf{t}^T B^{-1} A\mathbf{t}\right)\right\} dt_1\, dt_2. \end{aligned} \tag{17}$$

Moreover, we have

$$\begin{aligned} \Lambda_a^T D B^{-1}\Lambda_a &= -\frac{36}{121}\begin{pmatrix} a & a \end{pmatrix}\begin{pmatrix} 12/29 & -31/29 \\ 36/29 & 0 \end{pmatrix}\begin{pmatrix} 5/3 & -1/6 \\ -1/6 & -2 \end{pmatrix}\begin{pmatrix} a \\ a \end{pmatrix} \\ &= -\frac{36}{121}\begin{pmatrix} a & a \end{pmatrix}\begin{pmatrix} 5579/6786 & 1720/1131 \\ 60/29 & -6/29 \end{pmatrix}\begin{pmatrix} a \\ a \end{pmatrix} \\ &= -\frac{756\,a^2}{605}, \end{aligned} \tag{18}$$

$$\begin{aligned} \Lambda_a^T B^{T^{-1}}\mathbf{t} &= -\frac{36}{121}\begin{pmatrix} a & a \end{pmatrix}\begin{pmatrix} 5/3 & -1/6 \\ -1/6 & -2 \end{pmatrix}\begin{pmatrix} t_1 \\ t_2 \end{pmatrix} \\ &= -\frac{6a}{121}\begin{pmatrix} 9t_1 & 13t_2 \end{pmatrix}, \end{aligned} \tag{19}$$

$$\begin{aligned} \mathbf{t}^T B^{-1} A\mathbf{t} &= -\frac{36}{121}\begin{pmatrix} t_1 & t_2 \end{pmatrix}\begin{pmatrix} 5/3 & -1/6 \\ -1/6 & -2 \end{pmatrix}\begin{pmatrix} 1/6 & 1 \\ -5/6 & -1/6 \end{pmatrix}\begin{pmatrix} t_1 \\ t_2 \end{pmatrix} \\ &= -\frac{36}{121}\begin{pmatrix} t_1 & t_2 \end{pmatrix}\begin{pmatrix} 5/2 & 61/6 \\ 59/6 & 1 \end{pmatrix}\begin{pmatrix} t_1 \\ t_2 \end{pmatrix} \\ &= -\frac{36}{121}\left(\frac{5t_1^2}{2} + 20t_1t_2 + t_2^2\right). \end{aligned} \tag{20}$$

Implementing Equations (18)–(20) in Equation (17), we obtain

$$\begin{aligned}\mathcal{W}_{\psi}^{\mathbf{M}}[f](a,\mathbf{b}) &= \frac{11}{6\sqrt{a}}\exp\left\{\frac{i(b_1+b_2)}{a}-\frac{(b_1^2+b_2^2)}{2a^2}+\frac{756\pi i a^2}{605}\right\}\\ &\quad\times\int_{\mathbb{R}}\exp\left\{-\left(\frac{1}{2a^2}-\frac{90\pi i+144}{121}\right)t_1^2+\left(\frac{b_1}{a^2}-\frac{i}{a}-\frac{12a\pi i}{121}\right)t_1\right\}dt_1\\ &\quad\times\int_{\mathbb{R}}\exp\left\{-\left(\frac{1}{2a^2}-\frac{36\pi i+900}{121}\right)t_2^2+\left(\frac{b_2}{a^2}-\frac{i}{a}-\frac{156a\pi i}{121}\right)t_2\right\}dt_2\\ &= \frac{11\pi}{6\sqrt{a\left(\frac{1}{2a^2}-\frac{90\pi i+144}{121}\right)\left(\frac{1}{2a^2}-\frac{36\pi i+900}{121}\right)}}\\ &\quad\times\exp\left\{\frac{i(b_1+b_2)}{a}-\frac{(b_1^2+b_2^2)}{2a^2}+\frac{756\pi i a^2}{605}\right\}\\ &\quad\times\exp\left\{\frac{\left(\frac{b_1}{a^2}-\frac{i}{a}-\frac{12a\pi i}{121}\right)^2}{4\left(\frac{1}{2a^2}-\frac{90\pi i+144}{121}\right)}\right\}\exp\left\{\frac{\left(\frac{b_2}{a^2}-\frac{i}{a}-\frac{156a\pi i}{121}\right)^2}{4\left(\frac{1}{2a^2}-\frac{36\pi i+900}{121}\right)}\right\}. \end{aligned} \tag{21}$$

For different values of a and $\mathbf{b}$, the corresponding non-separable linear canonical wavelet transforms are plotted in Figures 1–3.

(b) Consider the constant function $f(\mathbf{t})=K$ and the two-dimensional Morlet wavelet $\psi(\mathbf{t})=e^{i\Lambda\cdot\mathbf{t}-|\mathbf{t}|^2/2}$, $\Lambda=(\lambda_1,\lambda_2)>0$. Then, the non-separable linear canonical wavelet transform of $f(\mathbf{t})$ with respect to the real symplectic matrix

$$\mathbf{M}=\begin{pmatrix}A & B\\ C & D\end{pmatrix}=\begin{pmatrix}2 & -1/8 & 1/4 & -1\\ 1/8 & -2 & 1 & 1/4\\ -2/15 & -31/30 & 1 & 0\\ 1 & 2/3 & -4/5 & -14/15\end{pmatrix}$$

is given by

$$\begin{aligned}\mathcal{W}_{\psi}^{\mathbf{M}}[f](a,\mathbf{b}) &= \frac{1}{\sqrt{a\,|\det B|}}\exp\left\{\frac{i(\lambda_1 b_1+\lambda_2 b_2)}{a}-\frac{(b_1^2+b_2^2)}{2a^2}\right\}\\ &\quad\times\int_{\mathbb{R}^n}K\exp\left\{-\frac{t_1^2}{2a^2}+\left(\frac{b_1}{a^2}-\frac{i\lambda_1}{a}\right)t_1\right\}\exp\left\{-\frac{t_2^2}{2a^2}+\left(\frac{b_2}{a^2}-\frac{i\lambda_2}{a}\right)t_2\right\}\\ &\quad\times\exp\left\{-i\pi\left(\Lambda_a^T DB^{-1}\Lambda_a-2\Lambda_a^T B^{T^{-1}}\mathbf{t}+\mathbf{t}^T B^{-1}A\mathbf{t}\right)\right\}d\mathbf{t}.\end{aligned} \tag{22}$$

Moreover, we have

$$\Lambda_a^T DB^{-1}\Lambda_a=\frac{16}{15}\begin{pmatrix}a & a\end{pmatrix}\begin{pmatrix}1 & 0\\ -4/15 & -14/15\end{pmatrix}\begin{pmatrix}-1/4 & 1\\ -1 & 1/4\end{pmatrix}\begin{pmatrix}a\\ a\end{pmatrix}=\frac{12a^2}{25}, \tag{23}$$

$$\Lambda_a^T B^{T^{-1}}\mathbf{t}=\frac{16}{15}\begin{pmatrix}a & a\end{pmatrix}\begin{pmatrix}-1/4 & 1\\ -1 & 1/4\end{pmatrix}\begin{pmatrix}t_1\\ t_2\end{pmatrix}=-\frac{4a}{3}(t_1-t_2), \tag{24}$$

$$\mathbf{t}^T B^{-1}A\mathbf{t}=\frac{16}{15}\begin{pmatrix}t_1 & t_2\end{pmatrix}\begin{pmatrix}-1/4 & 1\\ -1 & 1/4\end{pmatrix}\begin{pmatrix}2 & -1/8\\ 1/8 & -2\end{pmatrix}\begin{pmatrix}t_1\\ t_2\end{pmatrix}=-\frac{2}{5}\left(t_1^2+t_2^2\right). \tag{25}$$

Implementing Equations (23)–(25) in Equation (22), we obtain

$$\begin{aligned}
\mathcal{W}_{\psi}^{\mathbf{M}}[f](a,\mathbf{b}) &= K\sqrt{\frac{15}{16a}}\exp\left\{\frac{i(\lambda_1 b_1+\lambda_2 b_2)}{a}-\frac{(b_1^2+b_2^2)}{2a^2}-\frac{12\pi i a^2}{25}\right\}\\
&\quad\times\int_{\mathbb{R}}\exp\left\{-\left(\frac{1}{2a^2}-\frac{2\pi i}{5}\right)t_1^2+\left(\frac{b_1}{a^2}-\frac{i\lambda_1}{a}+\frac{8a\pi i}{3}\right)t_1\right\}dt_1\\
&\quad\times\int_{\mathbb{R}}\exp\left\{-\left(\frac{1}{2a^2}-\frac{2\pi i}{5}\right)t_2^2+\left(\frac{b_2}{a^2}-\frac{i\lambda_2}{a}-\frac{8a\pi i}{3}\right)t_2\right\}dt_2\\
&= \frac{K\pi}{\left(\frac{1}{2a^2}-\frac{2}{5}\right)}\sqrt{\frac{15}{16a}}\exp\left\{\frac{i(\lambda_1 b_1+\lambda_2 b_2)}{a}-\frac{(b_1^2+b_2^2)}{2a^2}-\frac{12\pi i a^2}{25}\right\}\\
&\quad\times\exp\left\{\frac{\left(\frac{b_1}{a^2}-\frac{i\lambda_1}{a}+\frac{8a\pi i}{3}\right)^2}{4\left(\frac{1}{2a^2}-\frac{2\pi i}{5}\right)}\right\}\exp\left\{\frac{\left(\frac{b_2}{a^2}-\frac{i\lambda_2}{a}-\frac{8a\pi i}{3}\right)^2}{4\left(\frac{1}{2a^2}-\frac{2\pi i}{5}\right)}\right\}. \qquad (26)
\end{aligned}$$

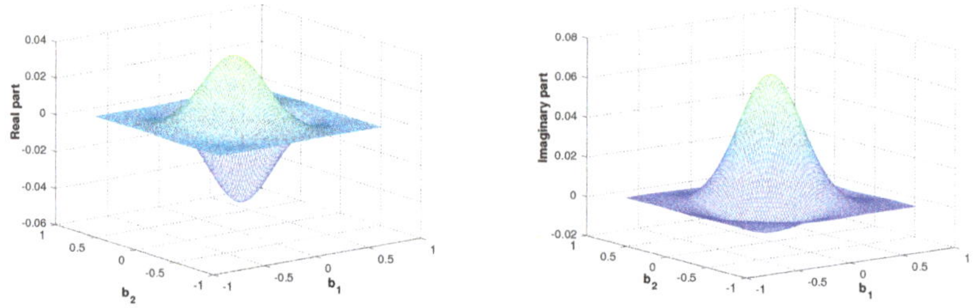

Figure 1. Real and imaginary parts of the non-separable linear canonical wavelet transform of f corresponding to a fixed scale $a = 1/4$.

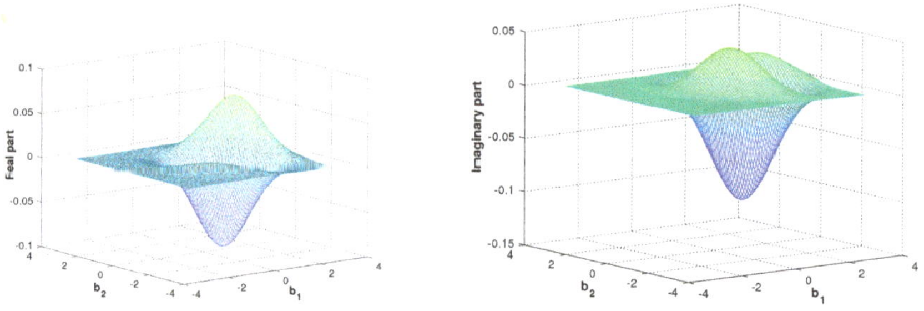

Figure 2. Real and imaginary parts of the non-separable linear canonical wavelet transform of f corresponding to a fixed scale $a = 1$.

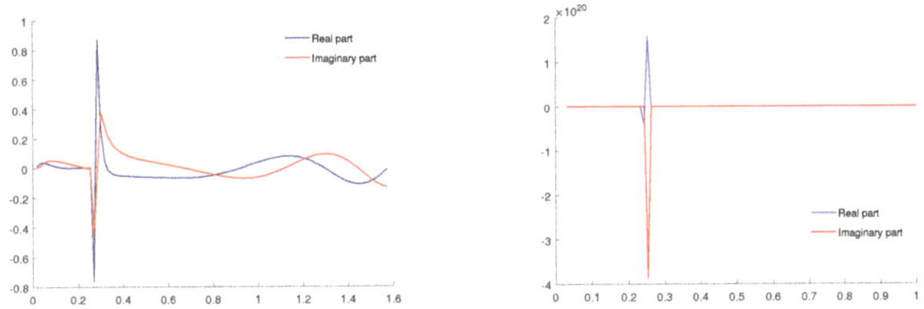

(a) (b)

Figure 3. (**a**) Frequency representation of f corresponding to a position $\mathbf{b} = (0,0)$. (**b**) Frequency representation of f corresponding to a position $\mathbf{b} = (1,1)$.

The non-separable linear canonical wavelet transforms shown in Equation (26) of f corresponding to $\Lambda = (1,1)$ are plotted in Figures 4–6.

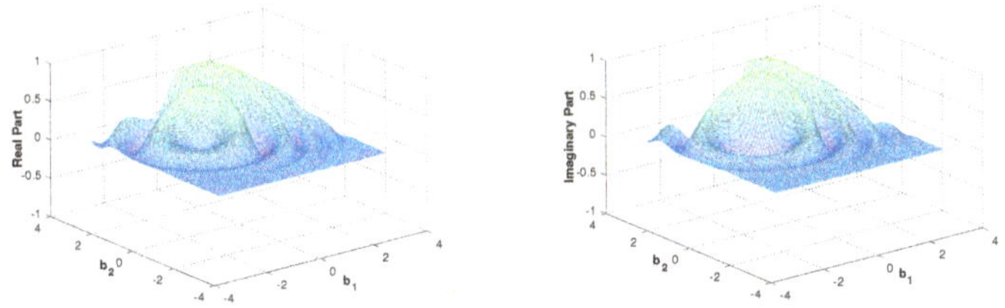

Figure 4. Real and imaginary parts of the non-separable linear canonical wavelet transform of f corresponding to a fixed scale $a = 1/4$.

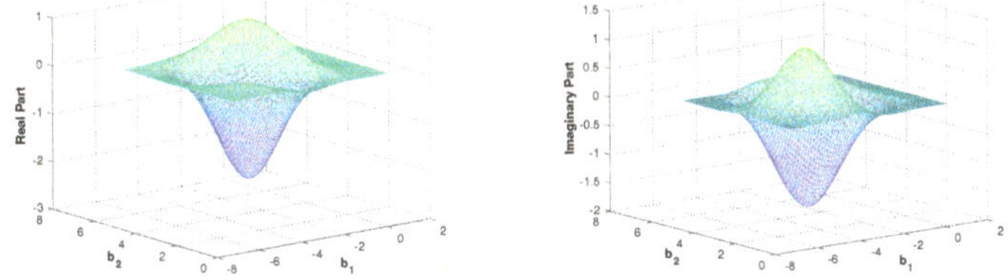

Figure 5. Real and imaginary parts of the non-separable linear canonical wavelet transform of f corresponding to a fixed scale $a = 1$.

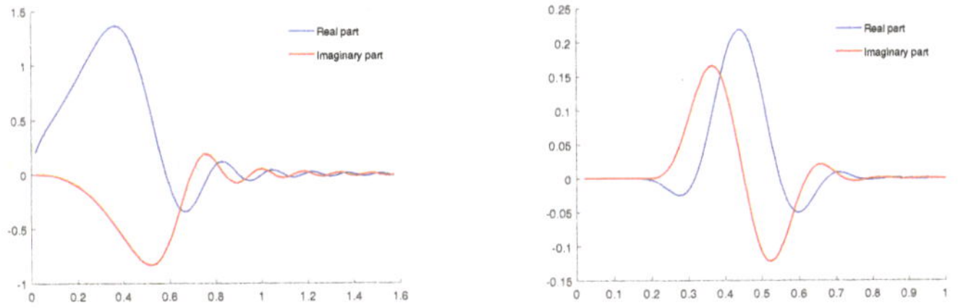

(a) (b)

Figure 6. (**a**) Frequency representation of f corresponding to a position $\mathbf{b} = (0,0)$. (**b**) Frequency representation of f corresponding to a position $\mathbf{b} = (1,1)$.

Next, we shall derive a fundamental relationship between the non-separable linear canonical wavelet transform (Equation (7)) and the non-separable linear canonical transform (Equation (1)). With the aid of this formula, we shall study the fundamental properties of the proposed transform.

Proposition 1. *Let $\mathcal{W}_{\psi}^{\mathbf{M}}[f](a,\mathbf{b})$ and $\mathcal{F}^{\mathbf{M}}[f](a)$ be the non-separable linear canonical wavelet transform and the non-separable linear canonical transform of any $f \in L^2(\mathbb{R}^n)$, respectively. Then, we have*

$$\mathcal{F}^{\mathbf{M}}\Big[\mathcal{W}_{\psi}^{\mathbf{M}}[f]\Big](\mathbf{w}) = \sqrt{a\,|\det B|}\,\mathcal{K}^{\mathbf{M}}(\mathbf{b},\Lambda_a)\,e^{\pi i\,a^2\mathbf{w}^T DB^{-1}\mathbf{w}}\mathcal{F}^{\mathbf{M}}[f](\mathbf{w})\,\overline{\mathcal{F}^{\mathbf{M}}[\Psi](a\mathbf{w})}, \quad (27)$$

where

$$\Psi(\mathbf{t},a) = e^{\pi i\left(2(a\Lambda_a)^T B^{T^{-1}}\mathbf{t}-\mathbf{t}^T B^{-1}A\mathbf{t}\right)}\,\psi(\mathbf{t}). \quad (28)$$

Proof. Applying the definition of the non-separable linear canonical transform, we have

$$\begin{aligned}
&\mathcal{F}^{\mathbf{M}}\Big[\psi_{a,\mathbf{b}}^{\mathbf{M}}(\mathbf{t})\Big](\mathbf{w})\\
&= \int_{\mathbb{R}^n}\frac{1}{\sqrt{a}}\psi\left(\frac{\mathbf{t}-\mathbf{b}}{a}\right)\frac{1}{\sqrt{|\det B|}}\exp\Big\{-\pi i\Big(\Lambda_a^T DB^{-1}\Lambda_a - 2\Lambda_a^T B^{T^{-1}}\mathbf{t}+\mathbf{t}^T B^{-1}A\mathbf{t}\Big)\Big\}\\
&\qquad\times\frac{1}{\sqrt{|\det B|}}\exp\Big\{\pi i\Big(\mathbf{w}^T DB^{-1}\mathbf{w}-2\mathbf{w}^T B^{T^{-1}}\mathbf{t}+\mathbf{t}^T B^{-1}A\mathbf{t}\Big)\Big\}d\mathbf{t}\\
&= \frac{\sqrt{a}}{|\det B|}\int_{\mathbb{R}^n}\psi(\mathbf{z})\exp\Big\{-\pi i\Big(\Lambda_a^T DB^{-1}\Lambda_a - 2\Lambda_a^T B^{T^{-1}}(\mathbf{b}+a\mathbf{z})\Big)\Big\}\\
&\qquad\times\exp\Big\{\pi i\Big(\mathbf{w}^T DB^{-1}\mathbf{w}-2\mathbf{w}^T B^{T^{-1}}(\mathbf{b}+a\mathbf{z})\Big)\Big\}d\mathbf{z}\\
&= \frac{\sqrt{a}}{|\det B|}\int_{\mathbb{R}^n}\psi(\mathbf{z})\exp\Big\{\pi i\Big(\mathbf{w}^T DB^{-1}\mathbf{w}-2(a\mathbf{w}^T)B^{T^{-1}}\mathbf{z}+\mathbf{z}^T B^{-1}A\mathbf{z}\Big)\Big\}\\
&\qquad\times\exp\Big\{-\pi i\Big(\Lambda_a^T DB^{-1}\Lambda_a-2(a\Lambda_a)^T B^{T^{-1}}\mathbf{z}+\mathbf{z}^T B^{-1}A\mathbf{z}\Big)\Big\}\\
&\qquad\times\exp\Big\{\pi i\Big(2\Lambda_a^T B^{T^{-1}}\mathbf{b}-2\mathbf{w}^T DB^{-1}\mathbf{w}\Big)\Big\}d\mathbf{z}\\
&= \frac{\sqrt{a}}{|\det B|}\int_{\mathbb{R}^n}\psi(\mathbf{z})\exp\Big\{\pi i\Big((a\mathbf{w})^T DB^{-1}(a\mathbf{w})-2(a\mathbf{w})^T B^{T^{-1}}\mathbf{z}+\mathbf{z}^T B^{-1}A\mathbf{z}\Big)\Big\}\\
&\qquad\times\exp\Big\{\pi i\Big(\mathbf{w}^T DB^{-1}\mathbf{w}-2\mathbf{w}^T B^{T^{-1}}\mathbf{b}+\mathbf{b}^T B^{-1}A\mathbf{b}\Big)\Big\}\\
&\qquad\times\exp\Big\{-\pi i\Big(\Lambda_a^T DB^{-1}\Lambda_a-2\Lambda_a^T B^{T^{-1}}\mathbf{b}+\mathbf{b}^T B^{-1}A\mathbf{b}\Big)\Big\}\\
&\qquad\times\exp\Big\{-\pi i\Big((a\mathbf{w})^T DB^{-1}(a\mathbf{w})-2(a\Lambda_a)^T B^{T^{-1}}\mathbf{z}+\mathbf{z}^T B^{-1}A\mathbf{z}\Big)\Big\}d\mathbf{z}\\
&= \sqrt{a\,|\det B|}\,e^{-\pi i\,a^2\mathbf{w}^T DB^{-1}\mathbf{w}}\,\mathcal{K}^{\mathbf{M}}(\mathbf{b},\mathbf{w})\,\overline{\mathcal{K}^{\mathbf{M}}(\mathbf{b},\Lambda_a)}\\
&\qquad\times\int_{\mathbb{R}^n}e^{\pi i\left(2(a\Lambda_a)^T B^{T^{-1}}\mathbf{z}-\mathbf{z}^T B^{-1}A\mathbf{z}\right)}\,\psi(\mathbf{z})\,\mathcal{K}^{\mathbf{M}}(\mathbf{z},a\mathbf{w})\,d\mathbf{z}\\
&= \sqrt{a\,|\det B|}\,e^{-\pi i\,a^2\mathbf{w}^T DB^{-1}\mathbf{w}}\,\mathcal{K}^{\mathbf{M}}(\mathbf{b},\mathbf{w})\,\overline{\mathcal{K}^{\mathbf{M}}(\mathbf{b},\Lambda_a)}\,\mathcal{F}^{\mathbf{M}}[\Psi](a\mathbf{w}),
\end{aligned} \quad (29)$$

where $\Psi(\mathbf{t},a) = e^{\pi i\left(2(a\Lambda_a)^T B^{T^{-1}}\mathbf{t}-\mathbf{z}^T B^{-1}A\mathbf{t}\right)}\,\psi(\mathbf{t})$.

Invoking the Plancheral theorem for the non-separable linear canonical transform and using Equation (29), we have

$$\mathcal{W}_\psi^{\mathbf{M}}[f](a,\mathbf{b}) = \sqrt{a\,|\det B|}\,\mathcal{K}^{\mathbf{M}}(\mathbf{b},\Lambda_a)\int_{\mathbb{R}^n} e^{\pi i\, a^2\mathbf{w}^T DB^{-1}\mathbf{w}}\,\mathcal{F}^{\mathbf{M}}[f](\mathbf{w})\,\overline{\mathcal{F}^{\mathbf{M}}[\Psi](a\mathbf{w})}\,\overline{\mathcal{K}^{\mathbf{M}}(\mathbf{b},\mathbf{w})}\,d\mathbf{w}$$
$$= \mathcal{F}^{\mathbf{M}^{-1}}\left[\sqrt{a\,|\det B|}\,\mathcal{K}^{\mathbf{M}}(\mathbf{b},\Lambda_a)\,e^{\pi i\, a^2\mathbf{w}^T DB^{-1}\mathbf{w}}\,\mathcal{F}^{\mathbf{M}}[f](\mathbf{w})\,\overline{\mathcal{F}^{\mathbf{M}}[\Psi](a\mathbf{w})}\right](\mathbf{b}).$$

Consequently,

$$\mathcal{F}^{\mathbf{M}}\left[\mathcal{W}_\psi^{\mathbf{M}}[f](a,\mathbf{b})\right](\mathbf{w}) = \sqrt{a\,|\det B|}\,\mathcal{K}^{\mathbf{M}}(\mathbf{b},\Lambda_a)\,e^{\pi i\, a^2\mathbf{w}^T DB^{-1}\mathbf{w}}\,\mathcal{F}^{\mathbf{M}}[f](\mathbf{w})\,\overline{\mathcal{F}^{\mathbf{M}}[\Psi](a\mathbf{w})}.$$

This completes the proof of Proposition 1. □

2.3. Basic Properties of the Non-Separable Linear Canonical Wavelet Transform

In this subsection, we shall study some mathematical properties of the proposed non-separable linear canonical wavelet transform (Equation (7)), including Rayleigh's theorem, inversion formula, and the range theorem. In this direction, we have the following theorem, which assembles some of the basic properties of the proposed transform.

Theorem 1. *For any $f, g \in L^2(\mathbb{R}^n)$ and $\alpha, \beta \in \mathbb{R}$, $\mathbf{k} \in \mathbb{R}^n$, and $\mu \in \mathbb{R}^+$, the non-separable linear canonical wavelet transform as defined by Equation (7) satisfies the following properties:*

(i) *Linearity:* $\mathcal{W}_\psi^{\mathbf{M}}[\alpha f + \beta g](a,\mathbf{b}) = \alpha\,\mathcal{W}_\psi^{\mathbf{M}}[f](a,\mathbf{b}) + \beta\,\mathcal{W}_\psi^{\mathbf{M}}[g](a,\mathbf{b})$

(ii) *Anti-linearity:* $\mathcal{W}_{\alpha\psi+\beta\phi}^{\mathbf{M}}[,f](a,\mathbf{b}) = \bar{\alpha}\,\mathcal{W}_\psi^{\mathbf{M}}[f](a,\mathbf{b}) + \bar{\beta}\,\mathcal{W}_\phi^{\mathbf{M}}[f](a,\mathbf{b})$

(iii) *Translation:*

$$\mathcal{W}_\psi^{\mathbf{M}}[f(\mathbf{t}-\mathbf{k})](a,\mathbf{b}) = e^{2\pi i\Lambda_a B^{T^{-1}}\mathbf{k}}\mathcal{W}_\psi^{\mathbf{M}}\left[f(\mathbf{t})\,e^{\pi i\left(\mathbf{k}^T B^{-1}A\mathbf{x}+\mathbf{x}^T B^{-1}A\mathbf{k}\right)}\right](a,\mathbf{b}-\mathbf{k})$$

(iv) *Scaling:* $\mathcal{W}_\psi^{\mathbf{M}}[f(\mu\mathbf{t})](a,\mathbf{b}) = |\mu|^{1-\frac{n}{2}}\,\mathcal{W}_\psi^{\mathbf{M}'}f(\mu a,\mu\mathbf{b}),\ \ \mathbf{M}' = (A/\mu, B/\mu : \mu C, \mu D)$

(v) *Conjugation:* $\mathcal{W}_\psi^{\mathbf{M}}[\bar{f}](a,\mathbf{b}) = \dfrac{1}{\sqrt{a}}\,\overline{\mathcal{W}_\psi^{\mathbf{M}'}[f](a,\mathbf{b})},\ \ \mathbf{M}' = (A, -B : -C, D).$

Proof. For the sake of brevity, we omit the proof of the theorem. □

Next, we shall define the admissibility condition for a function $\psi \in L^2(\mathbb{R}^n)$.

Definition 3. *A function $\psi \in L^2(\mathbb{R}^n)$ is said to be admissible with respect to a real free symplectic matrix $\mathbf{M} = (A, B : C, D)$ if*

$$C_\psi = \int_{\mathbb{R}^+} \frac{\left|\mathcal{F}^{\mathbf{M}}[\Psi](a\mathbf{w})\right|^2}{a}\,da < \infty, \quad a.e. \quad \mathbf{w} \in \mathbb{R}^n, \tag{30}$$

where $\Psi(\mathbf{t}, a)$ is given by Equation (28).

We are now in a position to derive the orthogonality relation for the proposed transform defined in Equation (7). As a consequence of the orthogonality relation, we will demonstrate that the non-separable wavelet transform is an isometry from the space of square-integrable functions $L^2(\mathbb{R}^n)$ to the space of transforms $L^2(\mathbb{R}^n \times \mathbb{R}^+)$.

Theorem 2. *Let $\mathcal{W}_\psi^{\mathbf{M}}[f](a,\mathbf{b})$ and $\mathcal{W}_\psi^{\mathbf{M}}[g](a,\mathbf{b})$ be the non-separable linear canonical wavelet transforms of f and g belonging to $L^2(\mathbb{R}^n)$, respectively. Then, we have*

$$\int_{\mathbb{R}^n\times\mathbb{R}^+} \mathcal{W}_\psi^{\mathbf{M}}[f](a,\mathbf{b})\,\overline{\mathcal{W}_\psi^{\mathbf{M}}[g](a,\mathbf{b})}\,\frac{d\mathbf{b}\,da}{a^2} = C_\psi\,\langle f, g\rangle, \tag{31}$$

where C_ψ is given by Equation (30).

Proof. For any pair of square integrable functions f and g, Proposition 1 implies that

$$\mathcal{W}_{\psi}^{\mathbf{M}}[f](a,\mathbf{b}) = \sqrt{a\,|\det B|}\int_{\mathbb{R}^n} e^{\pi i\, a^2 \mathbf{w}^T DB^{-1}\mathbf{w}}\,\mathcal{F}^{\mathbf{M}}[f](\mathbf{w})\,\overline{\mathcal{K}^{\mathbf{M}}(\mathbf{b},\mathbf{w})}\,\mathcal{K}^{\mathbf{M}}(\mathbf{b},\Lambda_a)\,\overline{\mathcal{F}^{\mathbf{M}}[\Psi](a\mathbf{w})}\,d\mathbf{w}$$

and

$$\mathcal{W}_{\psi}^{\mathbf{M}}[g](a,\mathbf{b}) = \sqrt{a\,|\det B|}\int_{\mathbb{R}^n} e^{\pi i\, a^2 \mathbf{x}^T DB^{-1}\mathbf{x}}\,\mathcal{F}^{\mathbf{M}}[g](\mathbf{x})\,\overline{\mathcal{K}^{\mathbf{M}}(\mathbf{b},\mathbf{x})}\,\mathcal{K}^{\mathbf{M}}(\mathbf{b},\Lambda_a)\,\overline{\mathcal{F}^{\mathbf{M}}[\Psi](a\mathbf{x})}\,d\mathbf{x},$$

where Ψ are given by Equation (28). Consequently, we have

$$\begin{aligned}
&\int_{\mathbb{R}^n\times\mathbb{R}^+} \mathcal{W}_{\psi}^{\mathbf{M}}[f](a,\mathbf{b})\,\overline{\mathcal{W}_{\psi}^{\mathbf{M}}[g](a,\mathbf{b})}\,\frac{d\mathbf{b}\,da}{a^2}\\
&\quad = |\det B|\int_{\mathbb{R}^n\times\mathbb{R}^n\times\mathbb{R}^n\times\mathbb{R}^+} e^{\pi i a^2\left(\mathbf{w}^T DB^{-1}\mathbf{w}-\mathbf{x}^T DB^{-1}\mathbf{x}\right)}\,\mathcal{F}^{\mathbf{M}}[f](\mathbf{w})\,\overline{\mathcal{F}^{\mathbf{M}}[g](\mathbf{x})}\\
&\qquad \times \mathcal{F}^{\mathbf{M}}[\Psi](a\mathbf{w})\,\overline{\mathcal{F}^{\mathbf{M}}[\Psi](a\mathbf{x})}\,\mathcal{K}^{\mathbf{M}}(\mathbf{b},\Lambda_a)\,\overline{\mathcal{K}^{\mathbf{M}}(\mathbf{b},\Lambda_a)}\,\mathcal{K}_{\mathbf{M}}(\mathbf{b},\mathbf{x})\,\overline{\mathcal{K}_{\mathbf{M}}(\mathbf{b},\mathbf{w})}\,\frac{d\mathbf{b}\,d\mathbf{w}\,d\mathbf{x}\,da}{a}\\
&\quad = \int_{\mathbb{R}^n\times\mathbb{R}^n\times\mathbb{R}^+} e^{\pi i a^2\left(\mathbf{w}^T DB^{-1}\mathbf{w}-\mathbf{x}^T DB^{-1}\mathbf{x}\right)}\,\mathcal{F}^{\mathbf{M}}[f](\mathbf{w})\,\overline{\mathcal{F}^{\mathbf{M}}[g](\mathbf{x})}\\
&\qquad \times \mathcal{F}^{\mathbf{M}}[\Psi](a\mathbf{w})\,\overline{\mathcal{F}^{\mathbf{M}}[\Psi](a\mathbf{x})}\left\{\int_{\mathbb{R}^n}\mathcal{K}_{\mathbf{M}}(\mathbf{b},\mathbf{x})\,\overline{\mathcal{K}_{\mathbf{M}}(\mathbf{b},\mathbf{w})}\,d\mathbf{b}\right\}\frac{d\mathbf{w}\,d\mathbf{x}\,da}{a}\\
&\quad = \int_{\mathbb{R}^n\times\mathbb{R}^n\times\mathbb{R}^+} e^{\pi i a^2\left(\mathbf{w}^T DB^{-1}\mathbf{w}-\mathbf{x}^T DB^{-1}\mathbf{x}\right)}\,\mathcal{F}^{\mathbf{M}}[f](\mathbf{w})\,\overline{\mathcal{F}^{\mathbf{M}}[g](\mathbf{v})}\\
&\qquad \times\, \mathcal{F}^{\mathbf{M}}[\Psi](a\mathbf{w})\,\overline{\mathcal{F}^{\mathbf{M}}[\Psi](a\mathbf{x})}\,\delta(\mathbf{w}-\mathbf{x})\,\frac{d\mathbf{w}\,d\mathbf{x}\,da}{a}\\
&\quad = \int_{\mathbb{R}^n\times\mathbb{R}^+}\mathcal{F}^{\mathbf{M}}[f](\mathbf{w})\,\overline{\mathcal{F}^{\mathbf{M}}[g](\mathbf{w})}\left|\mathcal{F}^{\mathbf{M}}[\Psi](a\mathbf{w})\right|^2\frac{d\mathbf{w}\,da}{a}\\
&\quad = \int_{\mathbb{R}^n}\mathcal{F}^{\mathbf{M}}[f](\mathbf{w})\,\overline{\mathcal{F}^{\mathbf{M}}[g](\mathbf{w})}\left\{\int_{\mathbb{R}^+}\frac{\left|\mathcal{F}^{\mathbf{M}}[\Psi](a\mathbf{w})\right|^2}{a}\,da\right\}d\mathbf{w}\\
&\quad = C_\psi\left\langle \mathcal{F}^{\mathbf{M}}[f](\mathbf{w}),\,\mathcal{F}^{\mathbf{M}}[g](\mathbf{w})\right\rangle_2\\
&\quad = C_\psi\,\langle f,g\rangle_2.
\end{aligned}$$

This completes the proof of Theorem 2. □

Remark 1. *(i). For $f = g$, Theorem 2 yields the energy preserving relation associated with the non-separable linear canonical wavelet transform (Equation (10)):*

$$\int_{\mathbb{R}^n\times\mathbb{R}^+}\left|\mathcal{W}_{\psi}^{\mathbf{M}}[f](a,\mathbf{b})\right|^2\frac{d\mathbf{b}\,da}{a^2} = C_\psi\,\|f\|_2^2. \tag{32}$$

(ii). The operator $\mathcal{W}_{\psi}^{\mathbf{M}}$ is a bounded-linear operator. Moreover, for $C_\psi = 1$ and $|\det B| = 1$, the operator $\mathcal{W}_{\psi}^{\mathbf{M}}$ becomes an isometry from $L^2(\mathbb{R}^n)$ to $L^2(\mathbb{R}^n\times\mathbb{R}^+)$.

In our next theorem, we demonstrate that the non-separable linear canonical wavelet transform $\mathcal{W}_{\psi}^{\mathbf{M}}[f](a,\mathbf{b})$ of any function $f\in L^2(\mathbb{R}^n)$ is reversible in the sense that f can be easily recovered from the transformed domain $L^2(\mathbb{R}^n\times\mathbb{R}^+)$.

Theorem 3. *Let $\mathcal{W}_{\psi}^{\mathbf{M}}[f](a,\mathbf{b})$ be the non-separable linear canonical wavelet transform of an arbitrary function $f\in L^2(\mathbb{R}^n)$. Then, f can be reconstructed via*

$$f(\mathbf{t}) = \frac{1}{C_\psi}\int_{\mathbb{R}^n\times\mathbb{R}^+}\mathcal{W}_{\psi}^{\mathbf{M}}[f](a,\mathbf{b})\,\psi_{a,\mathbf{b}}^{\mathbf{M}}(\mathbf{t})\,\frac{d\mathbf{b}\,da}{a^2}, \quad a.e. \tag{33}$$

Proof. According to Theorem 2, we can write

$$\begin{aligned}\langle f, g\rangle &= \frac{1}{C_\psi}\int_{\mathbb{R}^n\times\mathbb{R}^+}\mathcal{W}_\psi^{\mathbf{M}}[f](a,\mathbf{b})\,\overline{\mathcal{W}_\psi^{\mathbf{M}}[g](a,\mathbf{b})}\,\frac{d\mathbf{b}\,da}{a^2}\\
&= \frac{1}{C_\psi}\int_{\mathbb{R}^n\times\mathbb{R}^+}\mathcal{W}_\psi^{\mathbf{M}}[f](a,\mathbf{b})\left\{\int_{\mathbb{R}^n}\overline{g(\mathbf{t})}\,\psi_{a,\mathbf{b}}^{\mathbf{M}}(\mathbf{t})\,d\mathbf{t}\right\}\frac{d\mathbf{b}\,da}{a^2}\\
&= \frac{1}{C_\psi}\int_{\mathbb{R}^n\times\mathbb{R}^n\times\mathbb{R}^+}\mathcal{W}_\psi^{\mathbf{M}}[f](a,\mathbf{b})\,\psi_{a,\mathbf{b}}^{\mathbf{M}}(\mathbf{t})\,\overline{g(\mathbf{t})}\,\frac{d\mathbf{t}\,d\mathbf{b}\,da}{a^2}\\
&= \frac{1}{C_\psi}\left\langle\int_{\mathbb{R}^n\times\mathbb{R}^+}\mathcal{W}_\psi^{\mathbf{M}}[f](a,\mathbf{b})\,\psi_{a,\mathbf{b}}^{\mathbf{M}}(\mathbf{t})\,\frac{d\mathbf{b}\,da}{a^2},\, g(\mathbf{t})\right\rangle.\end{aligned}$$

Since g is chosen arbitrarily from $L^2(\mathbb{R}^n)$, using the elementary properties of inner products, one can obtain

$$f(\mathbf{t}) = \frac{1}{C_\psi}\int_{\mathbb{R}^n\times\mathbb{R}^+}\mathcal{W}_\psi^{\mathbf{M}}[f](a,\mathbf{b})\,\psi_{a,\mathbf{b}}^{\mathbf{M}}(\mathbf{t})\,\frac{d\mathbf{b}\,da}{a^2}\quad a.e.$$

This completes the proof of Theorem 3. □

Finally, we investigate the characterization of the range for the proposed transform (Equation (7)). As a consequence of the range theorem, we shall demonstrate that the range of the non-separable linear canonical wavelet transforms; that is, $\mathcal{W}_\psi^{\mathbf{M}}(L^2(\mathbb{R}^n))$ is a reproducing kernel Hilbert space.

Theorem 4. *If $f \in L^2(\mathbb{R}^n\times\mathbb{R}^+)$, then f belongs to the range $\mathcal{W}_\psi^{\mathbf{M}}(L^2(\mathbb{R}^n))$ if and only if*

$$f(a',\mathbf{b}') = \frac{1}{C_\psi}\int_{\mathbb{R}^n\times\mathbb{R}^+} f(a,\mathbf{b})\left\langle\psi_{a,\mathbf{b}}^{\mathbf{M}},\psi_{a',\mathbf{b}'}^{\mathbf{M}}\right\rangle_2\frac{d\mathbf{b}\,da}{a^2}, \tag{34}$$

where C_ψ satisfies Equation (27).

Proof. Assume that $f \in \mathcal{W}_\psi^{\mathbf{M}}(L^2(\mathbb{R}^n))$. Then, there exists a square integrable function g such that $\mathcal{W}_\psi^{\mathbf{M}} g = f$. In order to show that f satisfies Equation (34), we proceed as

$$\begin{aligned}f(a',\mathbf{b}') &= \mathcal{W}_\psi^{\mathbf{M}}[g](a',\mathbf{b}')\\
&= \int_{\mathbb{R}^n} g(\mathbf{t})\,\overline{\psi_{a',\mathbf{b}'}^{\mathbf{M}}(\mathbf{t})}\,d\mathbf{t}\\
&= \frac{1}{C_\psi}\int_{\mathbb{R}^n}\left\{\int_{\mathbb{R}^n\times\mathbb{R}^+}\mathcal{W}_\psi^{\mathbf{M}}[g](a,\mathbf{b})\,\psi_{a,\mathbf{b}}^{\mathbf{M}}(\mathbf{t})\,\frac{d\mathbf{b}\,da}{a^2}\right\}\overline{\psi_{a',\mathbf{b}'}^{\mathbf{M}}(\mathbf{t})}\,d\mathbf{t}\\
&= \frac{1}{C_\psi}\int_{\mathbb{R}^n\times\mathbb{R}^+}\mathcal{W}_\psi^{\mathbf{M}}[g](a,\mathbf{b})\left\{\int_{\mathbb{R}}\psi_{a,\mathbf{b}}^{\mathbf{M}}(\mathbf{t})\,\overline{\psi_{a',\mathbf{b}'}^{\mathbf{M}}(\mathbf{t})}\,d\mathbf{t}\right\}\frac{d\mathbf{b}\,da}{a^2}\\
&= \frac{1}{C_\psi}\int_{\mathbb{R}^n\times\mathbb{R}^+} f(a,\mathbf{b})\left\langle\psi_{a,\mathbf{b}}^{\mathbf{M}},\psi_{a',\mathbf{b}'}^{\mathbf{M}}\right\rangle_2\frac{d\mathbf{b}\,da}{a^2},\end{aligned}$$

which evidently verifies our claim. Conversely, suppose that the function f satisfies Equation (34). To verify that $f \in \mathcal{W}_\psi^{\mathbf{M}}(L^2(\mathbb{R}^n))$, it is sufficient to find out a function $g \in L^2(\mathbb{R}^n)$ such that $\mathcal{W}_\psi^{\mathbf{M}} g = f$. Therefore, the desired function g will be constructed as follows:

Let

$$g(\mathbf{t}) = \frac{1}{C_\psi} \int_{\mathbb{R}^n \times \mathbb{R}^+} f(a, \mathbf{b})\, \psi^{\mathbf{M}}_{a,\mathbf{b}}(\mathbf{t})\, \frac{d\mathbf{b}\, da}{a^2}. \tag{35}$$

Then, it is straightforward to obtain $\|g\|_2 \leq \|f\|_2 < \infty$; that is $g \in L^2(\mathbb{R}^n)$. Furthermore, by virtue of the Fubini theorem, we have

$$\begin{aligned}
\mathcal{W}^{\mathbf{M}}_\psi[g](a', \mathbf{b}') &= \int_{\mathbb{R}^n} g(\mathbf{x})\, \overline{\psi^{\mathbf{M}}_{a',\mathbf{b}'}(\mathbf{t})}\, d\mathbf{t} \\
&= \frac{1}{C_\psi} \int_{\mathbb{R}^n} \left\{ \int_{\mathbb{R}^n \times \mathbb{R}^+} f(a, \mathbf{b})\, \psi^{\mathbf{M}}_{a,\mathbf{b}}(\mathbf{t})\, \frac{d\mathbf{b}\, da}{a^2} \right\} \overline{\psi^{\mathbf{M}}_{a',\mathbf{b}'}(\mathbf{t})}\, d\mathbf{t} \\
&= \frac{1}{C_\psi} \int_{\mathbb{R}^n \times \mathbb{R}^+} f(a, \mathbf{b}) \left\langle \psi^{\mathbf{M}}_{a,\mathbf{b}}, \psi^{\mathbf{M}}_{a',\mathbf{b}'} \right\rangle_2 \frac{d\mathbf{b}\, da}{a^2} \\
&= f(a', \mathbf{b}').
\end{aligned}$$

This completes the proof of Theorem 4. □

Corollary 1. *For any admissible wavelet $\psi \in L^2(\mathbb{R}^n)$, the range of the proposed non-separable linear canonical wavelet transform; that is, $\mathcal{W}^{\mathbf{M}}_\psi(L^2(\mathbb{R}^n))$ is a reproducing kernel Hilbert space embedded as a subspace in $L^2(\mathbb{R}^n \times \mathbb{R}^+)$ with the kernel given by*

$$K^{\Lambda}_\psi(a, \mathbf{b}; a', \mathbf{b}') = \left\langle \psi^{\mathbf{M}}_{a,\mathbf{b}}, \psi^{\mathbf{M}}_{a',\mathbf{b}'} \right\rangle_2. \tag{36}$$

3. Uncertainty Principles for the Non-Separable Linear Canonical Wavelet Transform

The uncertainty principle lies at the heart of harmonic analysis, which asserts that "the position and the velocity of a particle cannot be both determined precisely at the same time" [23]. The harmonic analysis version of this principle states that "a non-trivial function cannot be properly localized in both the time and frequency domains at the same time" [24]. This standard inequality has been extensively studied in numerous domains and vistas [25–27]. Keeping in view the fact that the theory of uncertainty principles for the non-separable linear canonical wavelet transform is yet to be explored exclusively; therefore, it is both theoretically and practically fascinating to develop some new uncertainty principles, including the Heisenberg's, logarithmic, and Nazaros uncertainty principles for the non-separable linear canonical wavelet transform 7.

Theorem 5. *Let $\mathcal{W}^{\mathbf{M}}_\psi[f](a, \mathbf{b})$ be the non-separable linear canonical wavelet transform of any non-trivial function $f \in L^2(\mathbb{R}^n)$ with respect to a real free symplectic matrix $\mathbf{M} = (A, B : C, D)$, then the following uncertainty inequality holds:*

$$\left\{ \int_{\mathbb{R}^n \times \mathbb{R}^+} |\mathbf{b}|^2 \left| \mathcal{W}^{\mathbf{M}}_\psi[f](a, \mathbf{b}) \right|^2 \frac{da\, d\mathbf{b}}{a^2} \right\}^{1/2} \left\{ \int_{\mathbb{R}^n} |\mathbf{w}|^2 \left| \mathcal{F}^{\mathbf{M}}[f](\mathbf{w}) \right| d\mathbf{w} \right\}^{1/2} \geq \frac{n\, \sigma_{\min}(B)\, \sqrt{C_\psi}}{4\pi} \|f\|_2^2, \tag{37}$$

where $\sigma_{\min}(B)$ denotes the minimum singular value of matrix B.

Proof. The classical Heisenberg–Pauli–Weyl uncertainty inequality for any $f \in L^2(\mathbb{R}^n)$ in the non-separable linear canonical domain is given by [7]:

$$\left\{ \int_{\mathbb{R}^n} |\mathbf{b}|^2 |f(\mathbf{b})|^2 d\mathbf{b} \right\}^{1/2} \left\{ \int_{\mathbb{R}^n} |\mathbf{w}|^2 \left| \mathcal{F}^{\mathbf{M}}[f](\mathbf{w}) \right|^2 d\mathbf{w} \right\}^{1/2} \geq \frac{n\, \sigma_{\min}(B)}{4\pi} \left\{ \int_{\mathbb{R}^n} |f(\mathbf{b})|^2 d\mathbf{b} \right\}. \tag{38}$$

We shall identify $\mathcal{W}_\psi^{\mathbf{M}}[f](a,\mathbf{b})$ as a function of the time variable $\mathbf{b}$ and then invoke Equation (38) so that

$$\left\{\int_{\mathbb{R}^n}|\mathbf{b}|^2\left|\mathcal{W}_\psi^{\mathbf{M}}[f](a,\mathbf{b})\right|^2 d\mathbf{b}\right\}^{1/2}\left\{\int_{\mathbb{R}^n}|\mathbf{w}|^2\left|\mathcal{F}^{\mathbf{M}}\left[\mathcal{W}_\psi^{\mathbf{M}}[f](a,\mathbf{b})\right](\mathbf{w})\right|^2 d\mathbf{w}\right\}^{1/2} \geq \frac{n\,\sigma_{\min}(B)}{4\pi}\left\{\int_{\mathbb{R}^n}\left|\mathcal{W}_\psi^{\mathbf{M}}[f](a,\mathbf{b})\right|^2 d\mathbf{b}\right\}. \tag{39}$$

Integrating Equation (39) with respect to the da/a^2, we obtain

$$\int_{\mathbb{R}^+}\left\{\int_{\mathbb{R}^n}|\mathbf{b}|^2\left|\mathcal{W}_\psi^{\mathbf{M}}[f](a,\mathbf{b})\right|^2 d\mathbf{b}\right\}^{1/2}\left\{\int_{\mathbb{R}^n}|\mathbf{w}|^2\left|\mathcal{F}^{\mathbf{M}}\left[\mathcal{W}_\psi^{\mathbf{M}}[f](a,\mathbf{b})\right](\mathbf{w})\right|^2 d\mathbf{w}\right\}^{1/2}\frac{da}{a^2} \geq \frac{n\,\sigma_{\min}(B)}{4\pi}\left\{\int_{\mathbb{R}^n\times\mathbb{R}^+}\left|\mathcal{W}_\psi^{\mathbf{M}}[f](a,\mathbf{b})\right|^2\frac{d\mathbf{b}\,da}{a^2}\right\}. \tag{40}$$

As a consequence of the Cauchy–Schwartz's inequality, Fubini theorem, and Equation (30), we can express Equation (40) as

$$\left\{\int_{\mathbb{R}^n\times\mathbb{R}^+}|\mathbf{b}|^2\left|\mathcal{W}_\psi^{\mathbf{M}}[f](a,\mathbf{b})\right|^2\frac{da\,d\mathbf{b}}{a^2}\right\}^{1/2}\left\{\int_{\mathbb{R}^n\times\mathbb{R}^+}|\mathbf{w}|^2\left|\mathcal{F}^{\mathbf{M}}\left[\mathcal{W}_\psi^{\mathbf{M}}[f](a,\mathbf{b})\right](\mathbf{w})\right|^2\frac{d\mathbf{w}\,da}{a^2}\right\}^{1/2} \geq \frac{n\,\sigma_{\min}(B)\,C_\psi}{4\pi}\left\|f\right\|_2^2.$$

Using Proposition 1, we can rewrite the above inequality as follows

$$\left\{\int_{\mathbb{R}^n\times\mathbb{R}^+}|\mathbf{b}|^2\left|\mathcal{W}_\psi^{\mathbf{M}}[f](a,\mathbf{b})\right|^2\frac{da\,d\mathbf{b}}{a^2}\right\}^{1/2}\left\{\int_{\mathbb{R}^n\times\mathbb{R}^+}|\mathbf{w}|^2\left|\mathcal{F}^{\mathbf{M}}[f](\mathbf{w})\,\mathcal{F}^{\mathbf{M}}[\Psi](a\mathbf{w})\right|^2\frac{da\,d\mathbf{w}}{a}\right\}^{1/2} \geq \frac{n\,\sigma_{\min}(B)\,C_\psi}{4\pi}\left\|f\right\|_2^2,$$

or equivalently,

$$\left\{\int_{\mathbb{R}^n\times\mathbb{R}^+}|\mathbf{b}|^2\left|\mathcal{W}_\psi^{\mathbf{M}}[f](a,\mathbf{b})\right|^2\frac{da\,d\mathbf{b}}{a^2}\right\}^{1/2}\left\{\int_{\mathbb{R}^n}|\mathbf{w}|^2\left|\mathcal{F}^{\mathbf{M}}[f](\mathbf{w})\right|^2\left(\int_{\mathbb{R}^+}\frac{|\mathcal{F}^{\mathbf{M}}[\Psi](a\mathbf{w})|^2}{a}\,da\right)d\mathbf{w}\right\}^{1/2} \geq \frac{n\,\sigma_{\min}(B)\,C_\psi}{4\pi}\left\|f\right\|_2^2.$$

Finally, using Equation (27), we obtain the desired result:

$$\left\{\int_{\mathbb{R}^n\times\mathbb{R}^+}|\mathbf{b}|^2\left|\mathcal{W}_\psi^{\mathbf{M}}[f](a,\mathbf{b})\right|^2\frac{da\,d\mathbf{b}}{a^2}\right\}^{1/2}\left\{\int_{\mathbb{R}^n}|\mathbf{w}|^2\left|\mathcal{F}^{\mathbf{M}}[f](\mathbf{w})\right|d\mathbf{w}\right\}^{1/2} \geq \frac{n\,\sigma_{\min}(B)\,\sqrt{C_\psi}}{4\pi}\left\|f\right\|_2^2.$$

This completes the proof of Theorem 5. □

Remark 2. *The uncertainty inequality in Equation (37) embodies a wide class of uncertainty relations including the ones corresponding to the separable linear canonical wavelet transform, fractional wavelet transform, and classical wavelet transforms. The corresponding uncertainty principles can be obtained by choosing an appropriate matrix parameter* $\mathbf{M} = (A, B : C, D)$.

Example 2. *For the sake of computational convenience, we restrict ourselves to the two-dimensional space. From the inequality in Equation (37), we observe that the lower bound can be adjusted suitably by choosing a real, free symplectic matrix* $\mathbf{M} = (A, B : C, D)$ *and the analyzing function* ψ.

(i). *Consider the real, free symplectic matrix*

$$\mathbf{M}_1 = \begin{pmatrix} A_1 & B_1 \\ C_1 & D_1 \end{pmatrix} = \begin{pmatrix} 1/2 & -3/2 & 1 & -1 \\ 3/2 & 1/2 & -1 & -1 \\ 0 & -1 & 1 & -1 \\ -1 & 0 & 1 & 1 \end{pmatrix}$$

and the two-dimensional Morlet wavelet $\psi_1(\mathbf{t})$ *given by*

$$\psi_1(\mathbf{t}) = e^{i\Lambda\cdot\mathbf{t}-|\mathbf{t}|^2/2}, \quad \Lambda = (\lambda_1, \lambda_2) > 0.$$

Then, by virtue of Equation (28), *we obtain*

$$\begin{aligned} \Psi(\mathbf{t}, a) &= \exp\Big\{\pi i\Big(2a\Lambda_a^T B^{T^{-1}}\mathbf{t} - \mathbf{t}^T B^{-1} A\mathbf{t}\Big)\Big\}\psi(\mathbf{t}) \\ &= \exp\left\{\pi i\left(-2a^2 t_2 + \frac{t_1^2 + t_2^2}{2}\right)\right\}\exp\left\{i(\lambda_1 t_1 + \lambda_2 t_2) - \frac{t_1^2 + t_2^2}{2}\right\} \\ &= \exp\left\{-\left(\frac{1-\pi i}{2}\right)t_1^2 + \lambda_1 t_1\right\}\exp\left\{-\left(\frac{1-\pi i}{2}\right)t_2^2 + (\lambda_2 - 2a^2)t_2\right\}. \end{aligned}$$

Subsequently, we have

$$\begin{aligned} &\mathcal{F}^{\mathbf{M}}[\Psi](a\mathbf{w}) \\ &= \frac{1}{|\det B|^{1/2}}\int_{\mathbb{R}^2}\Psi(\mathbf{t}, a)\exp\Big\{\pi i\Big((a\mathbf{w})^T D B^{-1}(a\mathbf{w}) - 2(a\mathbf{w})^T B^{T^{-1}}\mathbf{t} + \mathbf{t}^T B^{-1} A\mathbf{t}\Big)\Big\}d\mathbf{t} \\ &= \frac{1}{|\det B|^{1/2}}\int_{\mathbb{R}^2}\exp\left\{-\left(\frac{1-\pi i}{2}\right)t_1^2 + \lambda_1 t_1\right\}\exp\left\{-\left(\frac{1-\pi i}{2}\right)t_2^2 + (\lambda_2 - 2a^2)t_2\right\} \\ &\quad \times \exp\left\{\pi i\left(a^2(\omega_1^2 - \omega_2^2) + a\omega_1(t_2 - t_1) + a\omega_2(t_1 + t_2) - \frac{t_1^2 + t_2^2}{2}\right)\right\}dt_1\, dt_2 \\ &= \sqrt{2}\, e^{\pi i a^2(\omega_1^2 - \omega_2^2)}\int_{\mathbb{R}}\exp\left\{-\frac{t_1^2}{2} + (\lambda_1 - a\pi i(\omega_1 - \omega_2))t_1\right\}dt_1 \\ &\quad \times \int_{\mathbb{R}}\exp\left\{-\frac{t_2^2}{2} + (\lambda_2 - 2a^2 + a\pi i(\omega_1 + \omega_2))t_1\right\}dt_2 \\ &= 2\pi\sqrt{2a}\, e^{\pi i a^2(\omega_1^2 - \omega_2^2)}\exp\left\{\frac{(\lambda_1 - a\pi i(\omega_1 - \omega_2))^2}{2}\right\}\exp\left\{\frac{(\lambda_2 - 2a^2 + a\pi i(\omega_1 + \omega_2))^2}{2}\right\}. \end{aligned}$$

Taking $\lambda_1 = a\pi i$ *and* $\lambda_2 = 2a^2$, *we obtain*

$$\Big|\mathcal{F}^{\mathbf{M}}[\Psi](a\mathbf{w})\Big|^2 = 8\pi^2 a\exp\Big\{-a^2\pi^2(1 + 2\omega_1^2 + 2\omega_2^2 + \omega_2)\Big\}. \tag{41}$$

Implementing Equation (41) *in Equation* (30) *yields*

$$C_\psi = 8\pi^2\int_{\mathbb{R}^+}\exp\Big\{-\pi^2(1 + 2\omega_1^2 + 2\omega_2^2 + \omega_2)a^2\Big\}\, da = \frac{4\pi^{3/2}}{\sqrt{1 + 2\omega_1^2 + 2\omega_2^2 + \omega_2}}.$$

In particular, for $(\omega_1, \omega_2) = (1, 1)$, *we obtain*

$$C_\psi = \frac{4\pi^{3/2}}{\sqrt{6}}. \tag{42}$$

Therefore, for any normalized function $f \in L^2(\mathbb{R}^2)$, an application of Equation (42) in Equation (37) yields the lower bound for the Heisenberg's inequality in Equation (37) of the form

$$\left\{\int_{\mathbb{R}^n\times\mathbb{R}^+}|\mathbf{b}|^2\left|\mathcal{W}_\psi^{\mathbf{M}}[f](a,\mathbf{b})\right|^2\frac{da\,d\mathbf{b}}{a^2}\right\}^{1/2}\left\{\int_{\mathbb{R}^n}|\mathbf{w}|^2\left|\mathcal{F}^{\mathbf{M}}[f](\mathbf{w})\right|d\mathbf{w}\right\}^{1/2}\geq\left(\frac{2}{3\pi}\right)^{1/4}. \tag{43}$$

(ii). *Consider the real, free symplectic matrix*

$$\mathbf{M}_2=\begin{pmatrix}A_2 & B_2\\ C_2 & D_2\end{pmatrix}=\begin{pmatrix}1/6 & 1 & -2 & 1/6\\ -5/6/2 & -1/6 & 1/6 & 5/3\\ 1 & 0 & 12/29 & -31/29\\ -6/29 & -36/29 & 36/29 & 6\end{pmatrix},$$

and the two-dimensional DOG wavelet ψ_2 given by

$$\psi_2(\mathbf{t})=\frac{1}{2\alpha^2}e^{-|\mathbf{t}|^2/(2\alpha^2)}-e^{-|\mathbf{t}|^2/2},\quad 0<\alpha<1.$$

Similar to computations carried out in (i), we can show that

$$C_{\psi_2}=\frac{6\pi}{11}\sqrt{\frac{1+3\alpha^2}{1+\alpha^2}},\quad \text{and,} \tag{44}$$

$$\left\{\int_{\mathbb{R}^n\times\mathbb{R}^+}|\mathbf{b}|^2\left|\mathcal{W}_\psi^{\mathbf{M}}[f](a,\mathbf{b})\right|^2\frac{da\,d\mathbf{b}}{a^2}\right\}^{1/2}\left\{\int_{\mathbb{R}^n}|\mathbf{w}|^2\left|\mathcal{F}^{\mathbf{M}}[f](\mathbf{w})\right|d\mathbf{w}\right\}^{1/2}\geq\sqrt{\frac{101(1+3\alpha^2)}{66\pi(1+\alpha^2)}}. \tag{45}$$

(iii). *Finally, for the real free symplectic matrix*

$$\mathbf{M}_3=\begin{pmatrix}A_3 & B_3\\ C_3 & D_3\end{pmatrix}=\begin{pmatrix}2 & -1/8 & 1/4 & -1\\ 1/8 & -2 & 1 & 1/4\\ -2/15 & -31/30 & 1 & 0\\ 1 & 2/3 & -4/5 & -14/15\end{pmatrix}$$

and the two-dimensional Maxican-hat wavelet ψ_3

$$\psi_3(\mathbf{t})=(2-|\mathbf{t}|^2)e^{-|\mathbf{t}|^2/2}.$$

The admissibility constant C_{ψ_3} and inequality in Equation (37) turn out to be

$$C_{\psi_3}=\frac{\pi^{5/2}}{2}\sqrt{\frac{16}{17}},\quad \text{and,} \tag{46}$$

$$\left\{\int_{\mathbb{R}^n\times\mathbb{R}^+}|\mathbf{b}|^2\left|\mathcal{W}_\psi^{\mathbf{M}}[f](a,\mathbf{b})\right|^2\frac{da\,d\mathbf{b}}{a^2}\right\}^{1/2}\left\{\int_{\mathbb{R}^n}|\mathbf{w}|^2\left|\mathcal{F}^{\mathbf{M}}[f](\mathbf{w})\right|d\mathbf{w}\right\}^{1/2}\geq\left(\frac{16\pi}{17}\right)^{1/4}\sqrt{\frac{17}{32}}. \tag{47}$$

The lower bounds of the Heisenberg's uncertainty inequality in Equation (37) corresponding to the aforementioned parametric symplectic matricies and analyzing functions are summarized in Table 2.

Table 2. Lower bounds associated with the Heisenberg's inequality in Equation (37).

Symplectic Matrix	Admissibility Constant C_ψ	Lower Bound
$\mathbf{M}_1 = (A_1, B_1 : C_1, D_1)$	$C_{\psi_1} = 4\pi^3\sqrt{\frac{\pi}{6}}$	$\left(\frac{2}{3\pi}\right)^{1/4}$
	$C_{\psi_2} = \frac{6\pi}{11}\sqrt{\frac{1+3\alpha^2}{1+\alpha^2}}$	$\frac{1}{2^{1/4}}\sqrt{\frac{1+3\alpha^2}{1+\alpha^2}}$
	$C_{\psi_3} = \frac{\pi^{5/2}}{2}\sqrt{\frac{16}{17}}$	$\left(\frac{\pi}{32}\right)^{1/4}$
$\mathbf{M}_2 = (A_2, B_2 : C_2, D_2)$	$C_{\psi_1} = 4\pi^3\sqrt{\frac{\pi}{6}}$	$\left(\frac{11\pi^3}{3}\right)^{1/4}\frac{\sqrt{101}}{(12\pi)}$
	$C_{\psi_2} = \frac{6\pi}{11}\sqrt{\frac{1+3\alpha^2}{1+\alpha^2}}$	$\sqrt{\frac{101(1+3\alpha^2)}{66\pi(1+\alpha^2)}}$
	$C_{\psi_3} = \frac{\pi^{5/2}}{2}\sqrt{\frac{16}{17}}$	$\pi^{1/4}\sqrt{\frac{101}{88}}$
$\mathbf{M}_3 = (A_3, B_3 : C_3, D_3)$	$C_{\psi_1} = 4\pi^3\sqrt{\frac{\pi}{6}}$	$\left(\frac{17\pi^3}{8}\right)^{1/4}\frac{\sqrt{17}}{8\pi}$
	$C_{\psi_2} = \frac{6\pi}{11}\sqrt{\frac{1+3\alpha^2}{1+\alpha^2}}$	$\frac{1}{8}\left(\frac{16}{17}\right)^{1/4}\sqrt{\frac{17(1+3\alpha^2)}{\pi(1+\alpha^2)}}$
	$C_{\psi_3} = \frac{\pi^{5/2}}{2}\sqrt{\frac{16}{17}}$	$\left(\frac{16\pi}{17}\right)^{1/4}\sqrt{\frac{17}{32}}$

In our next theorem, we shall establish the logarithmic uncertainty principle for the non-separable linear canonical wavelet transform in Equation (14).

Theorem 6. *Let ψ be an admissible function and suppose that $\mathcal{W}_\psi^{\mathbf{M}}[f](\cdot, \mathbf{b}) \in \mathbb{S}(\mathbb{R}^n)$, then the non-separable linear canonical wavelet transform (Equation (14)) of any $f \in \mathbb{S}(\mathbb{R}^n)$ satisfies the following logarithmic estimate of the uncertainty inequality:*

$$\int_{\mathbb{R}^n\times\mathbb{R}^+} \ln|\mathbf{b}|\left|\mathcal{W}_\psi^{\mathbf{M}}[f](a,\mathbf{b})\right|^2\frac{da\,d\mathbf{b}}{a^2} + C_\psi\int_{\mathbb{R}^n}\ln\left|\mathbf{w}B^{T^{-1}}\right|\left|\mathcal{F}^{\mathbf{M}}[f](\mathbf{w})\right|^2 d\mathbf{w} \geq \left[\frac{\Gamma'(n/2)}{\Gamma(n/2)} - \ln\pi\right]C_\psi\|f\|_2^2. \quad (48)$$

whenever the L.H.S of Equation (48) is defined.

Proof. For any $f \in \mathbb{S}(\mathbb{R}) \subseteq L^2(\mathbb{R}^n)$, the logarithmic uncertainty principle for the non-separable linear canonical transform (Equation (7)) is given by

$$\int_{\mathbb{R}^n}\ln|\mathbf{t}||f(\mathbf{t})|^2 d\mathbf{t} + \int_{\mathbb{R}^n}\ln\left|\mathbf{w}B^{T^{-1}}\right|\left|\mathcal{F}^{\mathbf{M}}[f](\mathbf{w})\right|^2 d\mathbf{w} \geq \left(\frac{\Gamma'(n/2)}{\Gamma(n/2)} - \ln\pi\right)\int_{\mathbb{R}^n}|f(\mathbf{t})|^2 d\mathbf{t}. \quad (49)$$

Identifying $\mathcal{W}_\psi^{\mathbf{M}}[f](a,\mathbf{b})$ as a function of the translation parameter $\mathbf{b}$ and then replace $f \in \mathbb{S}(\mathbb{R}^n)$ with $\mathcal{W}_\psi^{\mathbf{M}}[f](a,\mathbf{b})$, we have

$$\int_{\mathbb{R}^n}\ln|\mathbf{b}|\left|\mathcal{W}_\psi^{\mathbf{M}}[f](a,\mathbf{b})\right|^2 d\mathbf{b} + \int_{\mathbb{R}^n}\ln\left|\mathbf{w}B^{T^{-1}}\right|\left|\mathcal{F}^{\mathbf{M}}[\mathcal{W}_\psi^{\mathbf{M}}[f](a,\mathbf{b})](\mathbf{w})\right|^2 d\mathbf{w} \geq \left(\frac{\Gamma'(n/2)}{\Gamma(n/2)} - \ln\pi\right)\int_{\mathbb{R}^n}\left|\mathcal{W}_\psi^{\mathbf{M}}[f](a,\mathbf{b})\right|^2 d\mathbf{b}. \quad (50)$$

Integrating Equation (50) with respect to the measure da/a^2, we obtain

$$\int_{\mathbb{R}^n\times\mathbb{R}^+}\ln|\mathbf{b}|\left|\mathcal{W}_\psi^{\mathbf{M}}[f](a,\mathbf{b})\right|^2\frac{da\,d\mathbf{b}}{a^2}+\int_{\mathbb{R}^n\times\mathbb{R}^+}\ln\left|\mathbf{w}B^{T^{-1}}\right|\left|\mathcal{F}^{\mathbf{M}}[\mathcal{W}_\psi^{\mathbf{M}}[f](a,\mathbf{b})](\mathbf{w})\right|^2\frac{da\,d\mathbf{w}}{a^2}$$
$$\geq\left(\frac{\Gamma'(n/2)}{\Gamma(n/2)}-\ln\pi\right)\int_{\mathbb{R}^n\times\mathbb{R}^+}\left|\mathcal{W}_\psi^{\mathbf{M}}[f](a,\mathbf{b})\right|^2\frac{da\,d\mathbf{b}}{a^2}. \tag{51}$$

As a consequence of Proposition 1, we can simplify Equation (51) as:

$$\int_{\mathbb{R}^n\times\mathbb{R}^+}\ln|\mathbf{b}|\left|\mathcal{W}_\psi^{\mathbf{M}}[f](a,\mathbf{b})\right|^2\frac{da\,d\mathbf{b}}{a^2}+\int_{\mathbb{R}^n}\ln\left|\mathbf{w}B^{T^{-1}}\right|\left|\mathcal{F}^{\mathbf{M}}[f](\mathbf{w})\right|^2\left\{\int_{\mathbb{R}^+}\frac{\left|\mathcal{F}^{\mathbf{M}}[\Psi](a\mathbf{w})\right|^2}{a}\,da\right\}d\mathbf{w}$$
$$\geq\left(\frac{\Gamma'(n/2)}{\Gamma(n/2)}-\ln\pi\right)C_\psi\,\|f\|_2^2. \tag{52}$$

Equivalently,

$$\int_{\mathbb{R}^n\times\mathbb{R}^+}\ln|\mathbf{b}|\left|\mathcal{W}_\psi^{\mathbf{M}}[f](a,\mathbf{b})\right|^2\frac{da\,d\mathbf{b}}{a^2}+C_\psi\int_{\mathbb{R}^n}\ln\left|\mathbf{w}B^{T^{-1}}\right|\left|\mathcal{F}^{\mathbf{M}}[f](\mathbf{w})\right|^2d\mathbf{w}$$
$$\geq\left(\frac{\Gamma'(n/2)}{\Gamma(n/2)}-\ln\pi\right)C_\psi\,\|f\|_2^2. \tag{53}$$

This completes the proof of Theorem 6. □

Nazarov's uncertainty principle measures the localization of a non-trivial function f by taking into consideration the notion of support of the function instead of the dispersion as used in the Heisenberg–Pauli–Weyl inequality (38). In this direction, we have the following theorem.

Theorem 7. *Let $\mathcal{W}_\psi^{\mathbf{M}}[f](a,\mathbf{b})$ be the non-separable linear canonical wavelet transform of any function $f\in L^2(\mathbb{R}^n)$. Then, the following inequality holds:*

$$Ce^{C(T_1,T_2)}\left(\int_{\mathbb{R}^n\setminus T_1\times\mathbb{R}^+}\left|\mathcal{W}_\psi^{\mathbf{M}}[f](a,\mathbf{b})\right|^2\frac{da\,d\mathbf{b}}{a^2}+C_\psi\int_{\mathbb{R}^n\setminus(T_2B^T)}\left|\mathcal{F}^M[f](\mathbf{w})\right|^2d\mathbf{w}\right)$$
$$\geq C_\psi\int_{\mathbb{R}^n}\left|f(\mathbf{t})\right|^2d\mathbf{t}, \tag{54}$$

where $C(T_1,T_2)=C\min\left(|T_1||T_2|,|T_1|^{1/n}\mathfrak{W}(T_2),\mathfrak{W}(T_1)T_2^{1/n}\right)$, $\mathfrak{W}(T_1)$ is the mean width of T_1, and $|T_1|$ denotes the Lebesgue measure of T_1.

Proof. For an arbitrary function $f\in L^2(\mathbb{R}^n)$ and a pair of finite measurable subsets T_1 and T_2 of $\mathbb{R}^n$, Nazarov's uncertainty principle in the linear canonical domain reads [5]

$$Ce^{C(T_1,T_2)}\left(\int_{\mathbb{R}^n\setminus T_1}\left|f(\mathbf{t})\right|^2d\mathbf{t}+\int_{\mathbb{R}^n\setminus(T_2B^T)}\left|\mathcal{F}^M[f](\mathbf{w})\right|^2d\mathbf{w}\right)\geq\int_{\mathbb{R}^n}\left|f(\mathbf{t})\right|^2d\mathbf{t}, \tag{55}$$

where $C(T_1,T_2)=C\min\left(|T_1||T_2|,|T_1|^{1/n}\mathfrak{W}(T_2),\mathfrak{W}(T_1)T_2^{1/n}\right)$, $\mathfrak{W}(\cdot)$ is the mean width of the measurable subset, and $|\cdot|$ denotes the Lebesgue measure.

By identifying $\mathcal{W}_{\psi}^{\mathbf{M}}[f](a,\mathbf{b})$ as a function of $\mathbf{b}$ followed by invoking Equation (55), we obtain

$$Ce^{C(T_1,T_2)}\left(\int_{\mathbb{R}^n\setminus T_1}\left|\mathcal{W}_{\psi}^{\mathbf{M}}[f](a,\mathbf{b})\right|^2 d\mathbf{b}+\int_{\mathbb{R}^n\setminus(T_2B^T)}\left|\mathcal{F}^M\left[\mathcal{W}_{\psi}^{\mathbf{M}}[f](a,\mathbf{b})\right](\mathbf{w})\right|^2 d\mathbf{w}\right) \geq \int_{\mathbb{R}^n}\left|\mathcal{W}_{\psi}^{\mathbf{M}}[f](a,\mathbf{b})\right|^2 d\mathbf{b}, \quad (56)$$

Upon integrating Equation (56) with respect to the measure da/a^2, we have

$$Ce^{C(T_1,T_2)}\left(\int_{\mathbb{R}^n\setminus T_1\times\mathbb{R}^+}\left|\mathcal{W}_{\psi}^{\mathbf{M}}[f](a,\mathbf{b})\right|^2 \frac{da\,d\mathbf{b}}{a^2}+\int_{\mathbb{R}^n\setminus(T_2B^T)\times\mathbb{R}^+}\left|\mathcal{F}^M\left[\mathcal{W}_{\psi}^{\mathbf{M}}[f](a,\mathbf{b})\right](\mathbf{w})\right|^2 \frac{da\,d\mathbf{w}}{a^2}\right) \geq \int_{\mathbb{R}^n\times\mathbb{R}^+}\left|\mathcal{W}_{\psi}^{\mathbf{M}}[f](a,\mathbf{b})\right|^2 \frac{da\,d\mathbf{b}}{a^2}.$$

Finally, as a consequence of orthogonality relation in Equation (31) and Proposition 1, we obtain the desired result

$$Ce^{C(T_1,T_2)}\left(\int_{\mathbb{R}^n\setminus T_1\times\mathbb{R}^+}\left|\mathcal{W}_{\psi}^{\mathbf{M}}[f](a,\mathbf{b})\right|^2 \frac{da\,d\mathbf{b}}{a^2}+C_{\psi}\int_{\mathbb{R}^n\setminus(T_2B^T)}\left|\mathcal{F}^M[f](\mathbf{w})\right|^2 d\mathbf{w}\right) \geq C_{\psi}\int_{\mathbb{R}^n}|f(\mathbf{t})|^2 d\mathbf{t}.$$

This completes the proof of Theorem 7. □

4. Conclusions

In the present article, we introduced the notion of a kernel-based non-separable linear canonical wavelet transform in $L^2(\mathbb{R}^n)$ for obtaining an efficient time-frequency representation of higher-dimensional non-transient signals that has more degrees of freedom. Besides studying all the fundamental properties, such as Rayleigh's theorem, inversion formula, and range theorem, we have also formulated several uncertainty inequalities for the proposed transform containing Heisenberg's, logarithmic, and Nazarov's inequalities in the non-separable linear canonical domain.

Author Contributions: Writing original draft preparation, H.M.S.; Conceptualization, methodology, F.A.S.; Software and editing, T.K.G.; Methodology and software, W.Z.L. and H.L.Q.; Funding acquisition and research support, T.K.G. All authors have read and agreed to the published version of the manuscript.

Funding: This research received no external funding.

Institutional Review Board Statement: Not applicable.

Informed Consent Statement: Not applicable.

Acknowledgments: The authors are deeply indebted to the anonymous referees for meticulously reading the manuscript, pointing out many inaccuracies, and giving several valuable suggestions to improve the initial version of the manuscript to the present stage. The second named author is supported by SERB (DST), Government of India under Grant No. EMR/2016/007951.

Conflicts of Interest: The authors declare no conflict of interest.

References

1. Moshinsky, M.; Quesne, C. Linear canonical transformations and their unitary representations. *J. Math. Phys.* **1971**, *12*, 1772–1780. [CrossRef]
2. Xu, T.Z.; Li, B.Z. *Linear Canonical Transform and Its Applications*; Science Press: Beijing, China, 2013.
3. Healy, J.J.; Kutay, M.A.; Ozaktas, H.M.; Sheridan, J.T. *Linear Canonical Transforms: Theory and Applications*; Springer: New York, NY, USA, 2016.

4. Zhang, Z. Uncertainty principle of complex-valued functions in specic free metaplectic transformation domains. *J. Fourier Anal. Appl.* **2021**, *27*. [CrossRef]
5. Zhang, Z. Uncertainty principle for real functions in free metaplectic transformation domains. *J. Fourier Anal. Appl.* **2019**, *25*, 2899–2922. [CrossRef]
6. Gosson, M. *Symplectic Geometry and Quantum Mechanics*; Birkhäuser: Basel, Switzerland, 2006.
7. Jing, R.; Liu, B.; Li, R.; Liu, R. The N-dimensional uncertainty principle for the free metaplectic transformation. *Mathematics* **2020**, *8*, 1685. [CrossRef]
8. Debnath, L.; Shah, F.A. *Lectuer Notes on Wavelet Transforms*; Birkhäuser, Boston, MA, USA, 2017.
9. Debnath, L.; Shah, F.A. *Wavelet Transforms and Their Applications*; Birkhäuser: New York, NY, USA, 2015.
10. Dai, H.; Zheng, Z.; Wang, W. A new fractional wavelet transform. *Commun. Nonlinear Sci. Numer. Simulat.* **2017**, *44*, 19–36. [CrossRef]
11. Wei, D.; Li, Y.M. Generalized wavelet transform based on the convolution operator in the linear canonical transform domain, *Optik* **2014**, *125*, 4491–4496. [CrossRef]
12. Wang, J.; Wang, Y.; Wang, W.; Ren, S. Discrete linear canonical wavelet transform and its applications. *EURASIP J. Adv. Sig. Process.* **2018**, *29*, 1–18. [CrossRef]
13. Shah, F.A.; Teali, A.A.; Tantary, A.Y. Special affine wavelet transform and the corresponding Poisson summation formula. *Int. J. Wavelets Multiresolut. Inf. Process.* **2021**, *19*. [CrossRef]
14. Shah, F.A.; Tantary, A.Y.; Zayed, A.I. A convolution-based special affine wavelet transforms. *Integ. Trans. Special Funct.* **2020**, 1–21. [CrossRef]
15. Shah, F.A.; Teali, A.A.; Tantary, A.Y. Linear canonical wavelet transforms in quaternion domains. *Adv. Appl. Clifford Algebr.* **2021**, *31*, 42. [CrossRef]
16. Shah, F.A.; Lone, W.Z. Quadratic-phase wavelet transform with applications to generalized differential equations. *Math. Methods Appl. Sci.* **2021**, accepted. [CrossRef]
17. Antoine, J.P.; Murenzi, R.; Vandergheynst, P.; Ali, S.T. *Two-Dimensional Wavelets and Their Relatives*; Cambridge University Press: Cambridge, UK, 2004.
18. Pandey, J.N.; Pandey, J.S.; Upadhyay, S.K.; Srivastava, H.M. Continuous wavelet transform of Schwartz tempered distributions in $S'(\mathbb{R}^n)$. *Symmetry* **2019**, *11*, 235. [CrossRef]
19. Srivastava, H.M.; Shah, F.A.; Tantary, A.Y. A family of convolution-based generalized Stockwell transforms, *J. Pseudo-Differ. Oper. Appl.* **2020**, *11*, 1505–1536. [CrossRef]
20. Srivastava, H.M.; Singh, A.; Rawat, A.; Singh, S. A family of Mexican hat wavelet transforms associated with an isometry in the heat equation, *Math. Methods Appl. Sci.* **2021**, *44*, 11340–11349. [CrossRef]
21. Ali, S.T.; Antoine, J.P.; Gazeau, J.P. *Coherent States, Wavelets, and Their Generalizations*; Springer: New York, NY, USA, 2014.
22. Pandey, J.N.; Jha, N.K.; Singh, O.P. The continuous wavelet transform in n-dimensions. *Int. J. Wavelets Multiresolut. Inf. Process.* **2016**, *14*, 1650037. [CrossRef]
23. Folland, G.B.; Sitaram, A. The uncertainty principle: A mathematical survey. *J. Fourier Anal. Appl.* **1997**, *3*, 207–238. [CrossRef]
24. Cowling, M.G.; Price, J.F. Bandwidth verses time concentration: The Heisenberg-Pauli-Weyl inequality. *SIAM J. Math. Anal.* **1994**, *15*, 151–165. [CrossRef]
25. Beckner, W. Pitt's inequality and the uncertainty principle. *Proc. Am. Math. Soc.* **1995**, *123*, 1897–1905.
26. Wilczok, E. New uncertainty principles for the continuous Gabor transform and the continuous wavelet transform. *Doc. Math.* **2000**, *5*, 201–226.
27. Shah, F.A.; Nisar, K.S.; Lone, W.Z.; Tantary, A.Y. Uncertainty principles for the quadratic-phase Fourier transforms. *Math. Methods Appl. Sci.* **2021**, *44*, 10416–10431. [CrossRef]

symmetry

MDPI

Article

A Double Logarithmic Transform Involving the Exponential and Polynomial Functions Expressed in Terms of the Hurwitz–Lerch Zeta Function

Robert Reynolds * **and Allan Stauffer**

Department of Mathematics and Statistics, York University, Toronto, ON M3J1P3, Canada; stauffer@yorku.ca
* Correspondence: milver@my.yorku.ca

Abstract: The object of this paper is to derive a double integral in terms of the Hurwitz–Lerch zeta function. Almost all Hurwitz–Lerch zeta functions have an asymmetrical zero-distribution. Special cases are evaluated in terms of fundamental constants. All the results in this work are new.

Keywords: Lerch function; double integral; Catalan's constant; Aprey's constant

Citation: Reynolds, R.; Stauffer, A. A Double Logarithmic Transform Involving the Exponential and Polynomial Functions Expressed in Terms of the Hurwitz–Lerch Zeta Function. *Symmetry* **2021**, *13*, 1983. https://doi.org/10.3390/sym13111983

Academic Editor: Hari Mohan Srivastava

Received: 26 September 2021
Accepted: 18 October 2021
Published: 20 October 2021

1. Significance Statement

P.S. Laplace (1749–1827) introduced the Laplace transform as part of their famous study of probability theory and celestial mechanics [1]. R.H Mellin (1854–1933) first gave a systematic formulation of the Mellin transformation and its inverse [2]. He used their transform to develop applications of the theory of special functions to the solution of hypergeometric differential equations.

The double Laplace transform is studied in the work of Debnath [3] to solve initial and boundary value problems in applied mathematics, and mathematical physics. The double Laplace transform has been used to study European vulnerable options under constant as well as stochastic (the Hull–White) interest rates [4].

In this work, we derive a double integral whose kernel involves generalized exponential, logarithmic and polynomial functions. The form of this kernel can be viewed as a double Laplace transform of logarithmic and polynomial functions when the powers of the variable x in the exponential function is equal to one. On the other hand, since a polynomial is involved in this kernel, we can also consider this double integral as a double Mellin transform of the logarithm and exponential functions when $c = e = 1$.

These integral types are also used to derive geometric probability constants by calculating the expected Euclidean distance $\delta(n)$ to the center of an n-dimensional cube [5], and have some similarities with the generalized hyperterminants in globally valid remainder terms for asymptotic expansions about saddles and contour endpoints of arbitrary order degeneracy derived from the method of steepest descents [6].

2. Introduction

In this paper, we derive the double integral given by

$$\int_0^\infty \int_0^\infty t^{cm-1} x^{-em+e-1} e^{-bt^c - dx^e} \log^k\left(at^c x^{-e}\right) dx dt \tag{1}$$

where the parameters $k, a, b, d, m \in \mathbb{C}, Re(c) > 0, Re(e) > 0$. The derivations follow the method used by us in [7]. This method involves using a form of the generalized Cauchy's integral formula given by

$$\frac{y^k}{\Gamma(k+1)} = \frac{1}{2\pi i} \int_C \frac{e^{wy}}{w^{k+1}} dw. \tag{2}$$

where C is in general an open contour in the complex plane, which has the same value at the end points of the contour. We then multiply both sides by a function of x and t, then take a definite double integral of both sides. This yields a definite integral in terms of a contour integral. Then, we multiply both sides of Equation (2) by another function of y and take the infinite sums of both sides such that the contour integral of both equations are the same.

3. Definite Integral of the Contour Integral

We use the method shown in [7]. The variable of integration in the contour integral is $\alpha = w + m$. The cut and contour are in the first quadrant of the complex α-plane. The cut approaches the origin from the interior of the first quadrant and the contour goes round the origin with zero radius and is on opposite sides of the cut. Using a generalization of Cauchy's integral formula, we replace equations by replacing y by $\log^k(at^c x^{-e})$ and multiplying by $t^{cm-1}x^{e(-m)+e-1}e^{-bt^c-dx^e}$ then taking the definite integral with respect $x \in [0,\infty)$ and $t \in [0,\infty)$ to obtain

$$\begin{aligned}
&\frac{1}{\Gamma(k+1)}\int_0^\infty\int_0^\infty t^{cm-1}x^{-em+e-1}e^{-bt^c-dx^e}\log^k(at^c x^{-e})dxdt \\
&= \frac{1}{2\pi i}\int_0^\infty\int_0^\infty\int_C a^w w^{-k-1}t^{c(m+w)-1}x^{e(-(m+w))+e-1}e^{-bt^c-dx^e}dwdxdt \\
&= \frac{1}{2\pi i}\int_C\int_0^\infty\int_0^\infty a^w w^{-k-1}t^{c(m+w)-1}x^{e(-(m+w))+e-1}e^{-bt^c-dx^e}dxdtdw \\
&= \frac{1}{2\pi i}\int_C \frac{\pi a^w w^{-k-1}b^{-m-w}d^{m+w-1}\csc(\pi(m+w))}{ce}dw
\end{aligned} \tag{3}$$

from Equation (3.381.10) in [8], where $Re(w+m) > 0$, $Re(c) > 0, Re(e) > 0, Re(b) > 0, Re(d) > 0, Re(m) > 0$ and using the reflection formula for the Gamma function (25.4.1) in [9]. We are able to switch the order of integration over α, x and t using Fubini's theorem since the integrand is of bounded measure over the space $\mathbb{C} \times [0,\infty) \times [0,\infty)$.

4. The Hurwitz–Lerch Zeta Function and Infinite Sum of the Contour Integral

4.1. The Hurwitz–Lerch Zeta Function

The Hurwitz–Lerch zeta function, shown in Section (25.14) in [9–11] has a series representation given by

$$\Phi(z,s,v) = \sum_{n=0}^{\infty}(v+n)^{-s}z^n \tag{4}$$

where $|z| < 1, vs. \neq 0, -1, ..$ and is continued analytically by its integral representation given by

$$\Phi(z,s,v) = \frac{1}{\Gamma(s)}\int_0^\infty \frac{t^{s-1}e^{-vt}}{1-ze^{-t}}dt = \frac{1}{\Gamma(s)}\int_0^\infty \frac{t^{s-1}e^{-(v-1)t}}{e^t - z}dt \tag{5}$$

where $Re(v) > 0$, and either $|z| \leq 1, z \neq 1, Re(s) > 0$, or $z = 1, Re(s) > 1$.

4.2. Infinite Sum of the Contour Integral

Using Equation (2) and replacing y by $\log(a) - \log(b) + \log(d) + i\pi(2y+1)$ then multiplying both sides by

$$-\frac{2i\pi b^{-m}d^{m-1}e^{i\pi m(2y+1)}}{ce} \tag{6}$$

then taking the infinite sum over $y \in [0,\infty)$ and simplifying in terms of the Hurwitz–Lerch zeta function to obtain

$$
\begin{aligned}
&-\frac{(2i\pi)^{k+1}e^{i\pi m}b^{-m}d^{m-1}\Phi\left(e^{2im\pi},-k,\frac{-i\log(a)+i\log(b)-i\log(d)+\pi}{2\pi}\right)}{ce\Gamma(k+1)} \\
&= -\frac{1}{2\pi i}\sum_{y=0}^{\infty}\int_C \frac{1}{ce}2i\pi b^{-m}d^{m-1}w^{-k-1}\exp(w(\log(a)-\log(b)+\log(d)) \\
&\qquad +i\pi(2y+1)(m+w))dw \\
&= -\frac{1}{2\pi i}\int_C\sum_{y=0}^{\infty}\frac{1}{ce}2i\pi b^{-m}d^{m-1}w^{-k-1}\exp(w(\log(a)-\log(b)+\log(d)) \\
&\qquad +i\pi(2y+1)(m+w))dw \\
&= \frac{1}{2\pi i}\int_C \frac{\pi a^w w^{-k-1}b^{-m-w}d^{m+w-1}\csc(\pi(m+w))}{ce}dw
\end{aligned} \tag{7}
$$

from Equation (1.232.3) in [8], where $Im(w+m) > 0$ in order for the sum to converge.

5. Definite Integral in Terms of the Hurwitz–Lerch Zeta Function

Theorem 1. *For all* $k, a, b, d, m \in \mathbb{C}, Re(c) > 0, Re(e) > 0,$

$$
\begin{aligned}
&\int_0^{\infty}\int_0^{\infty} t^{cm-1}x^{-em+e-1}e^{-bt^c-dx^e}\log^k(at^cx^{-e})dxdt \\
&= -\frac{1}{ce}(2i\pi)^{k+1}e^{i\pi m}b^{-m}d^{m-1}\Phi\left(e^{2im\pi},-k,\frac{-i\log(a)+i\log(b)-i\log(d)+\pi}{2\pi}\right)
\end{aligned} \tag{8}
$$

Proof. Observe the right-hand side of Equations (3) and (7) are equal so we may equate the left-hand sides and simplify the factorial to yield the stated result. □

Lemma 1.

$$
\begin{aligned}
&\int_0^{\infty}\int_0^{\infty} t^{\frac{c}{2}-1}x^{\frac{e}{2}-1}e^{-bt^c-dx^e}\log^k(-t^cx^{-e})dxdt \\
&= -\frac{1}{\sqrt{bc}\sqrt{de}}i^{k+2}(2\pi)^{k+1}\left(2^k\zeta\left(-k,\frac{i\log(b)-i\log(d)+2\pi}{4\pi}\right)\right. \\
&\qquad \left.-2^k\zeta\left(-k,\frac{1}{2}\left(\frac{i\log(b)-i\log(d)+2\pi}{2\pi}+1\right)\right)\right)
\end{aligned} \tag{9}
$$

Proof. Use Equation (8) and set $m = 1/2, a = -1$ and simplify using entry (4) in the table below (64:12:7) in [12]. □

6. Special Cases

In this section, we evaluate Equation (8) for various values of the parameters in terms of special functions and fundamental constants. In this section, we use the following functions and fundamental constants; Euler's constant γ, Catalan's constant C, Glaisher's constant A, Aprey's constant $\zeta(3)$, Hurwitz zeta function $\zeta(s,a)$, hypergeometric function ${}_2F_1(a,b;c;z)$, Polylogarithm function $Li_n(z)$ and Riemann zeta function $\zeta(s)$.

Example 1.

$$
\int_0^{\infty}\int_0^{\infty} t^{cm-1}x^{-em+e-1}e^{-bt^c-dx^e}dxdt = \frac{\pi b^{-m}d^{m-1}\csc(\pi m)}{ce} \tag{10}
$$

Proof. Use Equation (8) and set $k = 0$ and simplify using entry (2) in the table below (64:12:7) in [12]. □

Example 2.

$$\int_0^\infty \int_0^\infty \frac{e^{-2\left(t^2+x^3\right)}\left(t^{2p}x^{2-3p}-t^{2m}x^{2-3m}\right)}{t\log\left(\frac{t^2}{x^3}\right)}dxdt = \frac{1}{6}\left(\tanh^{-1}\left(e^{i\pi m}\right)-\tanh^{-1}\left(e^{i\pi p}\right)\right) \tag{11}$$

Proof. Use Equation (8) and form a second equation by replacing $m \to p$ and taking their difference and setting $k=-1, a=1, b=c=d=2, e=3$ and simplifying using entry (3) in the table below (64:12:7) in [12]. □

Example 3.

$$\int_0^\infty \int_0^\infty \sqrt{x}e^{-2t^2-3x^3}\log\left(\log\left(-\frac{t^2}{x^3}\right)\right)dxdt = \frac{\pi}{12\sqrt{6}}\left(4\log\Gamma\left(-\frac{i\log\left(\frac{3}{2}\right)}{4\pi}\right)-4\log\Gamma\left(-\frac{1}{2}-\frac{i\log\left(\frac{3}{2}\right)}{4\pi}\right)+3i\pi+\log(16)\right. \\ \left.+2\log(\pi)+4\log\left(\log\left(\frac{3}{2}\right)\right)+\log\left(\frac{1}{\left(\log\left(\frac{3}{2}\right)-2i\pi\right)^4}\right)\right) \tag{12}$$

Proof. Use Equation (9) and take the first partial derivative with respect to k and set $k=0, b=c=2, d=e=3$ and simplify using Equation (25.11.18) in [9]. □

Example 4.

$$\int_0^\infty \int_0^\infty \frac{\sqrt{x}e^{-5x^3-it^2}}{\log^2\left(\frac{it^2}{5x^3}\right)}dxdt = \frac{(-1)^{3/4}C}{3\sqrt{5}\pi} \tag{13}$$

Proof. Use Equation (9) and set $k=-2, a=i/5, b=i, c=2, d=5, e=3$ and simplify using Equation (3) in [13]. □

Example 5.

$$\int_0^\infty \int_0^\infty \frac{t^{\frac{1}{6}(\sqrt{2}-6)}x^{\frac{1}{4}(\sqrt{2}-4)}e^{-t^{\frac{1}{\sqrt{2}}}-x^{\frac{1}{\sqrt{2}}}}\left(x^{\frac{1}{6\sqrt{2}}}-t^{\frac{1}{6\sqrt{2}}}\right)}{\log\left(t^{\frac{1}{\sqrt{2}}}x^{-\frac{1}{\sqrt{2}}}\right)}dxdt = -\log(3) \tag{14}$$

Proof. Use Equation (8) and form a second equation by replacing $m \to p$ and taking their difference and setting $k=-1, a=1, b=1, d=1, m=1/2, p=1/3, c=1/\sqrt{2}, e=1/\sqrt{2}$ and simplify using Equation (7) in [14]. □

Example 6.

$$\int_0^\infty \int_0^\infty \frac{e^{-t^{\frac{1}{\sqrt{2}}}-x^{\frac{1}{\sqrt{2}}}}x^{-\frac{m+p-1}{\sqrt{2}}-1}\left(x^{\frac{m}{\sqrt{2}}}t^{\frac{p}{\sqrt{2}}}-t^{\frac{m}{\sqrt{2}}}x^{\frac{p}{\sqrt{2}}}\right)}{t\log\left(t^{\frac{1}{\sqrt{2}}}x^{-\frac{1}{\sqrt{2}}}\right)}dxdt = 2\log\left(\cot\left(\frac{\pi m}{2}\right)\tan\left(\frac{\pi p}{2}\right)\right) \tag{15}$$

Proof. Use Equation (8) and form a second equation by replacing $m \to p$ and taking their difference and setting $k=-1, a=1, b=1, d=1, c=1/\sqrt{2}, e=1/\sqrt{2}$ and simplifying using entry (1) in the table below (64:12:7) in [12]. □

Example 7.

$$\int_0^\infty \int_0^\infty \frac{t^{\frac{1}{8}(\sqrt{2}-8)} x^{\frac{1}{8}(\sqrt{2}-8)} e^{-t^{\frac{1}{\sqrt{2}}}-x^{\frac{1}{\sqrt{2}}}} \left(t^{\frac{1}{2\sqrt{2}}} - x^{\frac{1}{2\sqrt{2}}}\right)}{\log\left(t^{\frac{1}{\sqrt{2}}} x^{-\frac{1}{\sqrt{2}}}\right)} dxdt = 4\sinh^{-1}(1) \tag{16}$$

Proof. Use Equation (15) and set $m = \frac{2\cot^{-1}(1+\sqrt{2})}{\pi}, p = \frac{2\tan^{-1}(1+\sqrt{2})}{\pi}$ and simplify. Note this is a double integral representation for the Universal Parabolic constant, P, given in [15]. This integral has similarities with the generalized hyperterminants in [6]. □

Example 8.

$$\int_0^\infty \int_0^\infty \frac{x^{3/2} e^{-t^2-x^5}}{4\log^2\left(\frac{t^2}{x^5}\right) + \pi^2} dxdt = \frac{\pi + \log\left(3-2\sqrt{2}\right)}{20\sqrt{2}\pi} \tag{17}$$

and

$$\int_0^\infty \int_0^\infty \frac{x^{3/2} e^{-t^2-x^5} \log\left(\frac{t^2}{x^5}\right)}{\log^2\left(\frac{t^2}{x^5}\right) + \frac{\pi^2}{4}} dxdt = 0 \tag{18}$$

Proof. Use Equation (8) and set $k = -1, a = i, c = 2, e = 5, b = 1, d = 1, m = 1/2$ and simplify using entry (1) in table below (64:12:7) in [12]; then, rationalize the denominator and equate real and imaginary parts. □

Example 9.

$$\int_0^\infty \int_0^\infty \frac{\sqrt{t} x^{3/2} e^{-t^3-x^5} \left(\log^2\left(\frac{t^3}{x^5}\right) - \pi^2\right)}{\left(\log^2\left(\frac{t^3}{x^5}\right) + \pi^2\right)^2} dxdt = -\frac{\pi}{360} \tag{19}$$

and

$$\int_0^\infty \int_0^\infty \frac{\sqrt{t} x^{3/2} e^{-t^3-x^5} \log\left(\frac{t^3}{x^5}\right)}{\left(\log^2\left(\frac{t^3}{x^5}\right) + \pi^2\right)^2} dxdt = 0 \tag{20}$$

Proof. Use Equation (8) and set $k = -1, a = -1, c = 3, e = 5, b = 1, d = 1, m = 1/2$ and simplify using entry (1) in table below (64:12:7) in [12]; then, rationalize the denominator and equate real and imaginary parts. □

Example 10.

$$\begin{aligned} &\int_0^\infty \int_0^\infty \sqrt{x} e^{-5t^2-5x^3} \log\left(-\tfrac{t^2}{x^3}\right) \log\left(\log\left(-\tfrac{t^2}{x^3}\right)\right) dxdt \\ &= \tfrac{1}{30} i\pi^2 \log(2i\pi) - \tfrac{2}{15} i\pi^2 \log\left(\frac{A^3}{\sqrt[3]{2}\sqrt[4]{e}}\right) \end{aligned} \tag{21}$$

Proof. Use Equation (8) and take the first partial derivative with respect to k and set $m = 1/2, k = 1, b = d = 5, a = -1, c = 2, e = 3$ and simplify using Equation (8) in [14]. □

Example 11.

$$\int_0^\infty \int_0^\infty \sqrt{x} e^{-5t^2-5x^3} \log^2\left(-\frac{t^2}{x^3}\right) \log\left(\log\left(-\frac{t^2}{x^3}\right)\right) dxdt = \frac{7\pi\zeta(3)}{15} \tag{22}$$

Proof. Use Equation (8) and take the first partial derivative with respect to k and set $m = 1/2, k = 2, b = d = 5, a = -1, c = 2, e = 3$ and simplify using Equation (9) in [14]. □

Example 12.

$$\int_0^\infty \int_0^\infty \frac{\sqrt{t}e^{-t^3-x^2}}{\sqrt{\log\left(\frac{it^3}{x^2}\right)}}dxdt = \frac{1}{6}(-1)^{3/4}\sqrt{\pi}\left(\zeta\left(\frac{1}{2},\frac{7}{8}\right)-\zeta\left(\frac{1}{2},\frac{3}{8}\right)\right) \tag{23}$$

Proof. Use Equation (8) and set $m = 1/2, k = -1/2, b = 1, a = i, c = 3, d = 1, e = 2$ and simplify using entry (4) in the table below (64:12:7) in [12]. □

Example 13.

$$\begin{aligned} &\int_0^\infty \int_0^\infty \sqrt{t}e^{-x^2+(-1-i)t^3}\log^{\frac{i}{2}}\left(\frac{it^3}{x^2}\right)dxdt \\ &= \frac{1}{3}(-1)^{\frac{7}{8}+\frac{i}{4}}2^{-\frac{1}{4}+i}\pi^{1+\frac{i}{2}}\left(\zeta\left(-\frac{i}{2},\frac{13}{16}+\frac{i\log(2)}{8\pi}\right)-\zeta\left(-\frac{i}{2},\frac{5}{16}+\frac{i\log(2)}{8\pi}\right)\right) \end{aligned} \tag{24}$$

Proof. Use Equation (8) and set $m = 1/2, k = -i/2, b = 1+i, a = i, c = 3, d = 1, e = 2$ and simplify using entry (4) in the table below (64:12:7) in [12]. □

Example 14.

$$\int_0^\infty \int_0^\infty \frac{te^{-t^3-x^2}}{\sqrt[3]{x}\log\left(\frac{it^3}{x^2}\right)}dxdt = -\frac{2}{9}\left(-\frac{1}{2}+\frac{i\sqrt{3}}{2}\right){}_2F_1\left(\frac{3}{4},1;\frac{7}{4};-\frac{1}{2}-\frac{i\sqrt{3}}{2}\right) \tag{25}$$

Proof. Use Equation (8) and set $m = 2/3, k = -1, b = 1, a = i, c = 3, d = 1, e = 2$ and simplify using Equation (9.559) in [8]. □

Example 15.

$$\begin{aligned} &\int_0^\infty \int_0^\infty t^{3m-1}x^{1-2m}e^{-d\left(t^3+x^2\right)}\log^k\left(-\frac{t^3}{x^2}\right)dxdt = \\ &-\frac{i(2i)^k\pi^{k+1}e^{-i\pi m}Li_{-k}\left(e^{2im\pi}\right)}{3d} \end{aligned} \tag{26}$$

Proof. Use Equation (8) and set $b = d, a = -1, c = 3, e = 2$ and simplify using Equation (25.14.3) in [9]. □

Example 16.

$$\int_0^\infty \int_0^\infty \sqrt{t}e^{-d\left(t^3+x^2\right)}\log^k\left(-\frac{t^3}{x^2}\right)dxdt = -\frac{(2i)^k\left(2^{k+1}-1\right)\pi^{k+1}\zeta(-k)}{3d} \tag{27}$$

Proof. Use Equation (26) and set $m = 1/2$ and simplify using □

Example 17.

$$\int_0^\infty \int_0^\infty \frac{\sqrt{t}e^{-t^3-x^2}\left(\pi^2-3\log^2\left(\frac{t^3}{x^2}\right)\right)}{\left(\log^2\left(\frac{t^3}{x^2}\right)+\pi^2\right)^3}dxdt = \frac{\zeta(3)}{32\pi^3} \tag{28}$$

and

$$\int_0^\infty \int_0^\infty \frac{\sqrt{t}e^{-t^3-x^2}\log\left(\frac{t^3}{x^2}\right)\left(\log^2\left(\frac{t^3}{x^2}\right)-3\pi^2\right)}{\left(\log^2\left(\frac{t^3}{x^2}\right)+\pi^2\right)^3}dxdt = 0 \tag{29}$$

Proof. Use Equation (27) and set $k = -3, d = 1$, rationalize the denominator and compare real and imaginary parts and simplify. □

Example 18.

$$\int_0^\infty \int_0^\infty \frac{\sqrt{t}e^{-t^3-x^2}\left(5\log^4\left(\frac{t^3}{x^2}\right) - 10\pi^2\log^2\left(\frac{t^3}{x^2}\right) + \pi^4\right)}{\left(\log^2\left(\frac{t^3}{x^2}\right) + \pi^2\right)^5} dxdt = \frac{5\zeta(5)}{512\pi^5} \tag{30}$$

and

$$\int_0^\infty \int_0^\infty \frac{\sqrt{t}e^{-t^3-x^2}\log\left(\frac{t^3}{x^2}\right)\left(\log^4\left(\frac{t^3}{x^2}\right) - 10\pi^2\log^2\left(\frac{t^3}{x^2}\right) + 5\pi^4\right)}{\left(\log^2\left(\frac{t^3}{x^2}\right) + \pi^2\right)^5} dxdt = 0 \tag{31}$$

Proof. Use Equation (27) and set $k = -5, d = 1$, rationalize the denominator and compare real and imaginary parts and simplify. □

Example 19.

$$\int_0^\infty \int_0^\infty \sqrt{t}e^{-t^3-x^2}\sqrt{\log\left(-\frac{t^3}{x^2}\right)}dxdt = \left(\frac{1}{12} + \frac{i}{12}\right)\left(2\sqrt{2}-1\right)\sqrt{\pi}\zeta\left(\frac{3}{2}\right) \tag{32}$$

Proof. Use Equation (27) and set $k = 1/2, d = 1$ and simplify. □

Example 20.

$$\begin{aligned} &\int_0^\infty \int_0^\infty \frac{\sqrt{t}e^{-t^3-x^2}\log\left(\log\left(-\frac{t^3}{x^2}\right)\right)}{\log\left(-\frac{t^3}{x^2}\right)} dxdt = \\ &\frac{1}{12}(\pi\log(2) + i(\gamma\log(4) - \log(2)(\log(8) + 2\log(\pi)))) \end{aligned} \tag{33}$$

Proof. Use Equation (27) to take the first partial derivative with respect to k and apply l'Hopital's rule as $k \to -1$ and simplify using Equation (37) in [16]. □

7. Discussion

In this work, we derived a double integral in terms of the Hurwitz–Lerch zeta function. We then used this integral to derive special cases in terms of other special functions and fundamental constants. We were also able to derive double integral representations for geometric constants [15] and double integrals associated with with generalized hyperterminants in [6]. We will be using our method to derive more double integrals in terms of special functions. We checked our results numerically using Mathematica by Wolfram.

Author Contributions: Conceptualization, R.R.; funding acquisition, A.S.; supervision, A.S. All authors have read and agreed to the published version of the manuscript.

Funding: Natural Sciences and Engineering Research Council of Canada, Grant No. 504070.

Institutional Review Board Statement: Not applicable.

Informed Consent Statement: Not applicable.

Data Availability Statement: Not applicable.

Conflicts of Interest: The authors declare no conflict of interest.

References

1. Simon, P.; De Laplace, M. *Théorie Analytique des Probabilités*; Courcier: Paris, France, 1820.
2. Poularikas, A.D. *The Transforms and Applications Handbook*, 2nd ed.; CRC Press LLC: Boca Raton, FL, USA, 2000.
3. Debnath, L. The Double Laplace Transforms and Their Properties with Applications to Functional, Integral and Partial Differential Equations. *Int. J. Appl. Comput. Math.* **2016**, *2*, 223–241. [CrossRef]
4. Yoon, J.-H.; Kim, J.-H. The pricing of vulnerable options with double Mellin transforms. *J. Math. Anal. Appl.* **2015**, *422*, 838–857. [CrossRef]

5. Finch, S.R. f *Mathematical Constants*; Cambridge University Press: Cambridge, UK, 2003.
6. Bennett, T.; Howls, C.J.; Nemes, G.; Daalhuis, A.B.O. Globally Exact Asymptotics for Integrals with Arbitrary Order Saddles. *SIAM J. Math. Anal.* **2018**, *50*, 2144–2177. [CrossRef]
7. Reynolds, R.; Stauffer, A. A Method for Evaluating Definite Integrals in Terms of Special Functions with Examples. *Int. Math. Forum* **2020**, *15*, 235–244. [CrossRef]
8. Gradshteyn, I.S.; Ryzhik, I.M. *Tables of Integrals, Series and Products*, 6th ed.; Academic Press: Cambridge, MA, USA, 2000.
9. Olver, F.W.J.; Lozier, D.W.; Boisvert, R.F.; Clark, C.W. (Eds.) *NIST Digital Library of Mathematical Functions*; U.S. Department of Commerce, National Institute of Standards and Technology: Washington, DC, USA; Cambridge University Press: Cambridge, UK, 2010; With 1 CD-ROM (Windows, Macintosh and UNIX). MR 2723248 (2012a:33001).
10. Srivastava, H.M. Some general families of the Hurwitz-Lerch Zeta functions and their applications: Recent developments and directions for further researches. *Proc. Inst. Math. Mech. Nat. Acad. Sci. Azerbaijan* **2019**, *45*, 234–269. [CrossRef]
11. Srivastava, H.M. The Zeta and Related Functions: Recent Developments. *J. Adv. Eng. Comput.* **2019**, *3*, 329–354. [CrossRef]
12. Oldham, K.B.; Myland, J.C.; Spanier, J. *An Atlas of Functions: With Equator, the Atlas Function Calculator*, 2nd ed.; Springer: New York, NY, USA, 2009.
13. Marichev, O.; Sondow, J.; Weisstein, E.W. Catalan's Constant, From MathWorld–A Wolfram Web Resource. Available online: https://mathworld.wolfram.com/CatalansConstant.html (accessed on 1 September 2021).
14. Weisstein, E.W. Lerch Transcendent, From MathWorld–A Wolfram Web Resource. Available online: https://mathworld.wolfram.com/LerchTranscendent.html (accessed on 1 September 2021).
15. Reese, S.; Sondow, J. Universal Parabolic Constant, From MathWorld–A Wolfram Web Resource, created by Eric W. Weisstein. Available online: https://mathworld.wolfram.com/UniversalParabolicConstant.html (accessed on 1 September 2021).
16. Sondow, J.; Weisstein, E.W. Riemann Zeta Function, From MathWorld–A Wolfram Web Resource. Available online: https://mathworld.wolfram.com/RiemannZetaFunction.html (accessed on 1 September 2021).

symmetry MDPI

Article

Double Integral of the Product of the Exponential of an Exponential Function and a Polynomial Expressed in Terms of the Lerch Function

Robert Reynolds * and Allan Stauffer

Department of Mathematics and Statistics, York University, 4700 Keele Street, Toronto, ON M3J 1P3, Canada; stauffer@yorku.ca
* Correspondence: milver@my.yorku.ca; Tel.: +416-319-8383

Abstract: In this work, the authors use their contour integral method to derive an application of the Fourier integral theorem given by $\int_{-\infty}^{\infty}\int_{-\infty}^{\infty} e^{mx-my-e^x-e^y+y}(\log(a)+x-y)^k dxdy$ in terms of the Lerch function. This integral formula is then used to derive closed solutions in terms of fundamental constants and special functions. Almost all Lerch functions have an asymmetrical zero distribution. There are some useful results relating double integrals of certain kinds of functions to ordinary integrals for which we know no general reference. Thus, a table of integral pairs is given for interested readers. All of the results in this work are new.

Keywords: Fourier integral theorem; double integral; Lerch function; contour integral; exponential function

Citation: Reynolds, R.; Stauffer, A. Double Integral of the Product of the Exponential of an Exponential Function and a Polynomial Expressed in Terms of the Lerch Function. *Symmetry* **2021**, *13*, 1962. https://doi.org/10.3390/sym13101962

Academic Editors: José Carlos R. Alcantud and Paolo Emilio Ricci

Received: 10 September 2021
Accepted: 13 October 2021
Published: 18 October 2021

1. Statement of Significance

In 1906, Niels Nielsen [1] produced his famous book on the Gamma function. In this work, the authors use their contour integral method and apply it to an interesting integral in the book of Nielsen [1] to yield a double integral and express its closed form in terms of the Lerch function. This derived integral formula is then used to provide formal derivations and new formulae in the form of a summary table of integrals, Table 1. The Lerch function being a special function has the fundamental property of analytic continuation, which enables enables the expansion of the range of evaluation for the parameters involved in the definite integral.

Double integrals over a real line are used in very interesting areas in mathematics. Some areas of high interest are namely in the use of the Fourier integral theorem in Electromagnetic Theory of Propagation, Interference, and Diffraction of Light [2], evaluation of two-dimensional Gaussian integrals in the constructions of representation theory and related topics of differential geometry and analysis [3], and the implementation of Cahn's scheme for simulating the morphology of isotropic spinodal decomposition [4].

2. Introduction

In 1882, Joseph Fourier (1768–1830) discovered a double integral representation [5] of a non-periodic function $f(x)$ for all real x, which is universally known as the Fourier Integral Theorem in the form

$$f(x) = \frac{1}{2\pi}\int_{-\infty}^{\infty} e^{ikx}\left(\int_{-\infty}^{\infty} f(\alpha)e^{-ik\alpha}d\alpha\right)dk \tag{1}$$

Throughout the nineteenth and twentieth centuries, mathematicians and mathematical physicists recognized the significance of this theorem. It is regarded as one of the most fundamental representation theorems of modern mathematical analysis according to Lord Kelvin (1824–1907) and Peter Guthrie Tait (1831–1901).

In this work, we will derive the Fourier integral theorem applied to a function involving the product of the exponential of an exponential function and a polynomial and express this integral in terms of the Lerch function. The application of the Fourier integral theorem is in the form of a double integral over a real line. The definite integral derived in this manuscript is given by

$$\int_{-\infty}^{\infty}\int_{-\infty}^{\infty} e^{mx-my-e^x-e^y+y}(\log(a)+x-y)^k dxdy \tag{2}$$

where the parameters k and a are general complex numbers and $0 < Re(m) < 1$. In the book of Titchmarsh [6], examples on the Fourier integral theorem are applied to a vast number of functions and real-world applications are showcased. This work is important because the authors were unable to find similar derivations in current literature. The derivation of the definite integral follows the method used by us in [7], which involves Cauchy's integral formula. The generalized Cauchy's integral formula is given by

$$\frac{y^k}{\Gamma(k+1)} = \frac{1}{2\pi i}\int_C \frac{e^{wy}}{w^{k+1}}dw. \tag{3}$$

where C is, in general, an open contour in the complex plane, where the bilinear concomitant has the same value at the end points of the contour. This method involves using a form of Equation (3), then multiplying both sides by a function, and then taking a definite integral of both sides. This yields a definite integral in terms of a contour integral. A second contour integral is derived by multiplying Equation (3) by a function and performing some substitutions so that the contour integrals are the same.

3. Definite Integral of the Contour Integral

We use the method in [7]. The variable of integration in the contour integral is $t = m + w$. The cut and contour are in the second quadrant of the complex t-plane. The cut approaches the origin from the interior of the second quadrant and the contour goes around the origin with zero radius and is on opposite sides of the cut. Using Equation (3), we replace y by $x - y + \log(a)$ and then multiply it by $e^{mx-my-e^x-e^y+y}$. Next, we take the double infinite integral over $x \in (-\infty, \infty)$ and $y \in (-\infty, \infty)$ to obtain

$$\begin{aligned}
&\frac{1}{\Gamma(k+1)}\int_{-\infty}^{\infty}\int_{-\infty}^{\infty} e^{mx-my-e^x-e^y+y}(\log(a)+x-y)^k dxdy \\
&= \frac{1}{2\pi i}\int_{-\infty}^{\infty}\int_{-\infty}^{\infty}\int_C a^w w^{-k-1} e^{mx-my+w(x-y)-e^x-e^y+y} dwdxdy \\
&= \frac{1}{2\pi i}\int_C\int_{-\infty}^{\infty}\int_{-\infty}^{\infty} a^w w^{-k-1} e^{mx-my+w(x-y)-e^x-e^y+y} dxdydw \\
&= \frac{1}{2\pi i}\int_C \pi a^w w^{-k-1}\csc(\pi(m+w))dw
\end{aligned} \tag{4}$$

from Equation (3.328) in [8], where $-1 < Re(w+m) < 0$, and using the reflection formula for the Gamma function. We are able to switch the order of integration over t, x, and y using Fubini's theorem since the integrand is of bounded measure over the space $C \times \mathbb{R} \times \mathbb{R}$.

4. The Lerch Function

We use Equation (1.11.3) in [9], where $\Phi(z,s,v)$ is the Lerch function, which is a generalization of the Hurwitz zeta $\zeta(s,v)$ and Polylogarithm functions $Li_n(z)$. The Lerch function has a series representation given by

$$\Phi(z,s,v) = \sum_{n=0}^{\infty}(v+n)^{-s}z^n \tag{5}$$

where $|z| < 1, v \neq 0, -1, -2, -3, ..$, and is continued analytically by its integral representation given by

$$\Phi(z,s,v) = \frac{1}{\Gamma(s)}\int_0^\infty \frac{t^{s-1}e^{-vt}}{1-ze^{-t}}dt = \frac{1}{\Gamma(s)}\int_0^\infty \frac{t^{s-1}e^{-(v-1)t}}{e^t - z}dt \tag{6}$$

where $Re(v) > 0$ and either $|z| \leq 1, z \neq 1, Re(s) > 0$, or $z = 1, Re(s) > 1$.

5. Infinite Sum of the Contour Integral

In this section, we again use Cauchy's integral formula from Equation (3) and take the infinite sum to derive equivalent sum representations for the contour integrals. We proceed by using Equation (3), replacing y by $\log(a) + i\pi(2y+1)$, multiplying both sides by $-2i\pi e^{i\pi m(2y+1)}$, and simplifying to obtain

$$-\frac{2i\pi e^{\frac{1}{2}i\pi(k+4my+2m)}(-i\log(a) + 2\pi y + \pi)^k}{\Gamma(k+1)} = -\frac{1}{2\pi i}\int_C 2i\pi a^w w^{-k-1} e^{i\pi(2y+1)(m+w)}dw \tag{7}$$

Next, we take the infinite over $y \in [0,\infty)$ and simplify it using the Lerch function to obtain

$$\begin{aligned}
-\frac{(2i\pi)^{k+1}e^{i\pi m}\Phi\left(e^{2im\pi}, -k, \frac{\pi - i\log(a)}{2\pi}\right)}{\Gamma(k+1)} &= -\frac{1}{2\pi i}\sum_{y=0}^{\infty}\int_C 2i\pi a^w w^{-k-1}e^{i\pi(2y+1)(m+w)}dw \\
&= -\frac{1}{2\pi i}\int_C \sum_{y=0}^{\infty} 2i\pi a^w w^{-k-1}e^{i\pi(2y+1)(m+w)}dw \\
&= \frac{1}{2\pi i}\int_C \pi a^w w^{-k-1}\csc(\pi(m+w))dw
\end{aligned} \tag{8}$$

from (1.232.3) in [8] and $Im(m+w) > 0$ for convergence of the sum.

6. Definite Integral in Terms of the Lerch Function

Theorem 1. *For all* $k, a \in \mathbb{C}, 0 < Re(m) < 1$,

$$\int_{-\infty}^{\infty}\int_{-\infty}^{\infty} e^{mx-my-e^x-e^y+y}(\log(a) + x - y)^k dxdy = (2i\pi)^{k+1}\left(-e^{i\pi m}\right)\Phi\left(e^{2im\pi}, -k, \frac{\pi - i\log(a)}{2\pi}\right) \tag{9}$$

Proof. Since the right-hand sides of Equations (4) and (8) are equal, we can equate the left-hand sides and simplify the factorial to achieve the stated result. □

Main Results and Table of Integrals

In this section, we evaluate Equation (9) for various values of the parameters in terms of special functions and fundamental constants and create a table of integrals. Some of the fundamental constants evaluated are Aprey's constant $\zeta(3)$ from Section 1.6 in [10]; Catalan's constant C, Equation (9.73) in [8]; Euler's constant γ, Equation (9.73) in [8]; and the Glaisher–Kinkelin constant A, Section 2.15 in [10]. Some special functions used are the polylogarithm function $Li_n(z)$ from Equation (64:12:2) in [11] and the hypergeometric function ${}_2F_1(a,b;c;z)$ from Equation (9.559) in [8].

7. Derivation of the Degenerate Case

7.1. Derivation of Entry (1)

Lemma 1. *For* $0 < Re(m) < 1$,

$$\int_{-\infty}^{\infty}\int_{-\infty}^{\infty} e^{mx-my-e^x-e^y+y}dxdy = \pi\csc(\pi m) \tag{10}$$

Proof. Use Equation (9), set $k = 0$, and simplify using entry (2) from the Table below (64:12:7) in [11]. □

7.2. Derivation of Entry (2)

Lemma 2.

$$\int_{-\infty}^{\infty}\int_{-\infty}^{\infty} \frac{e^{\frac{1}{2}(-2(e^x+e^y)+x+y)}}{(x-y)^2+\pi^2}dxdy = \frac{\log(2)}{\pi} \tag{11}$$

and

$$\int_{-\infty}^{\infty}\int_{-\infty}^{\infty} \frac{e^{\frac{1}{2}(-2(e^x+e^y)+x+y)}(x-y)}{(x-y)^2+\pi^2}dxdy = 0 \tag{12}$$

Proof. Use Equation (9), set $k = -1, a = -1, m = 1/2$, rationalize the denominator, compare the real and imaginary parts, and simplify using entry (3) from the table below (64:12:7) in [11]. □

7.3. Derivation of Entry (3)

Theorem 2. *For all* $k, a \in \mathbb{C}, 0 < Re(m) < 1, 0 < Re(n) < 1$,

$$\begin{aligned}&\int_{-\infty}^{\infty}\int_{-\infty}^{\infty} e^{y(-(m+n-1))-e^x-e^y}\left(e^{my+nx}-e^{mx+ny}\right)(\log(a)+x-y)^k dxdy \\ &= (2i\pi)^{k+1}\left(e^{i\pi m}\Phi\left(e^{2im\pi},-k,\frac{\pi-i\log(a)}{2\pi}\right)-e^{i\pi n}\Phi\left(e^{2in\pi},-k,\frac{\pi-i\log(a)}{2\pi}\right)\right)\end{aligned} \tag{13}$$

Proof. Use Equation (9) to form a second equation by replacing m by n, and take their difference. □

7.4. Derivation of Entry (4)

Lemma 3. *For all* $0 < Re(m) < 1, 0 < Re(n) < 1$,

$$\int_{-\infty}^{\infty}\int_{-\infty}^{\infty} \frac{e^{y(-(m+n-1))-e^x-e^y}\left(e^{my+nx}-e^{mx+ny}\right)}{x-y}dxdy = \log\left(\cot\left(\frac{\pi m}{2}\right)\tan\left(\frac{\pi n}{2}\right)\right) \tag{14}$$

Proof. Use Equation (13), set $k = -1, a = 1$, and simplify using entry (5) from the table below (64:12:7). □

7.5. Derivation of Entry (5)

Theorem 3. *For* $k \in \mathbb{C}, 0 < Re(m) < 1$,

$$\int_{-\infty}^{\infty}\int_{-\infty}^{\infty}(x-y+i\pi)^k e^{mx-my-e^x-e^y+y}dxdy = (2i\pi)^{k+1}\left(-e^{-i\pi m}\right)Li_{-k}\left(e^{2im\pi}\right) \tag{15}$$

Proof. Use Equation (9), set $a = -1$, and simplify using Equation (64:12:2) in [11]. □

7.6. Derivation of Entry (6)

Lemma 4. *For $a \in \mathbb{C}$,*

$$\int_{-\infty}^{\infty}\int_{-\infty}^{\infty} e^{\frac{1}{2}(-2(e^x+e^y)+x+y)} \log(ia+x-y)dxdy = \pi \log\left(\frac{4i\pi\Gamma\left(\frac{a}{4\pi}+\frac{3}{4}\right)^2}{\Gamma\left(\frac{a+\pi}{4\pi}\right)^2}\right) \tag{16}$$

Proof. Use Equation (9), and set $m = 1/2$. Next, take the first partial derivative with respect to k, set $k = 0$, and simplify using entry (4) from the table below (64:12:7) and from Equation (64:10:2) in [11]. □

7.7. Derivation of Entry (7)

Lemma 5.

$$\int_{-\infty}^{\infty}\int_{-\infty}^{\infty} e^{\frac{1}{2}(-2(e^x+e^y)+x+y)} \log\left(x-y-\frac{1}{2}\right)dxdy = \pi \log\left(\frac{4i\pi\Gamma\left(\frac{3}{4}+\frac{i}{8\pi}\right)^2}{\Gamma\left(\frac{1}{4}+\frac{i}{8\pi}\right)^2}\right) \tag{17}$$

Proof. Use Equation (16), set $a = i/2$, and simplify. □

7.8. Derivation of Entry (8)

Lemma 6.

$$\begin{aligned}\int_{-\infty}^{\infty}\int_{-\infty}^{\infty} & e^{\frac{1}{6}(-6(e^x+e^y)+2x+3y)}\left(e^{y/6}-e^{x/6}\right)\log(x-y)dxdy \\ &= 2(-1)^{5/6}\pi\Phi'\left((-1)^{2/3},0,\frac{1}{2}\right) - \frac{i\pi^2}{2} + \frac{i\pi^2}{\sqrt{3}} + \frac{\pi\log(4)}{\sqrt{3}} \\ &\quad - \pi\log(\pi) + \frac{2\pi\log(\pi)}{\sqrt{3}} + \pi\log\left(\frac{4\Gamma\left(\frac{1}{4}\right)^2}{\Gamma\left(-\frac{1}{4}\right)^2}\right)\end{aligned} \tag{18}$$

Proof. Use Equation (13), set $m = 1/2, n = 1/2, a = 1$, take the first partial derivative with respect to k, set $k = 0$, and simplify using entry (4) from the table below (64:12:7) in [11]. □

7.9. Derivation of Entry (9)

Lemma 7.

$$\int_{-\infty}^{\infty}\int_{-\infty}^{\infty} \frac{e^{\frac{1}{6}(-6(e^x+e^y)+2x+3y)}\left(e^{y/6}-e^{x/6}\right)}{x-y}dxdy = -\frac{\log(3)}{2} \tag{19}$$

Proof. Use Equation (13), set $k = -1, a = 1, m = 1/2, n = 1/3$, and simplify using entry (1) from the table below (64:12:7) in [11]. □

7.10. Derivation of Entry (10)

Lemma 8.

$$\begin{aligned}\int_{-\infty}^{\infty}\int_{-\infty}^{\infty} & e^{\frac{1}{6}(-6(e^x+e^y)+2x+3y)}\left(e^{y/6}-e^{x/6}\right)(x-y)\log(x-y)dxdy \\ &= \frac{1}{3}\pi\left((-1)^{5/6}\pi\left(12i\Phi'\left(\frac{1}{2}i\left(\sqrt{3}+i\right),-1,\frac{1}{2}\right) + \left(\sqrt{3}+i\right)\log(2\pi)\right) + 12iC - i\pi^2\right)\end{aligned} \tag{20}$$

Proof. Use Equation (13), set $a = 1, m = 1/2, n = 1/3$, and take the first partial derivative with respect to k, set $k = 1$, and simplify using Equation (21) in [12]. □

7.11. Derivation of Entry (11)

Lemma 9.

$$\int_{-\infty}^{\infty}\int_{-\infty}^{\infty} e^{\frac{1}{6}(-6(e^x+e^y)+2x+3y)}\left(e^{x/6}-e^{y/6}\right)(\pi-i(x-y))^2\log(x-y+i\pi)dxdy \\ =\frac{8\pi^3\Phi'\left((-1)^{2/3},-2,1\right)}{\sqrt{3}} \\ -\frac{8(-1)^{2/3}\pi^3\Phi'\left((-1)^{2/3},-2,1\right)}{\sqrt{3}} \\ -14\pi\zeta(3)+\frac{8\pi^3\log(2i\pi)}{3\sqrt{3}}-\frac{8\sqrt[3]{-1}\pi^3\log(2i\pi)}{3\sqrt{3}} \tag{21}$$

Proof. Use Equation (13), set $a=-1, m=1/2, n=1/3$, and take the first partial derivative with respect to k, then set $k=2$, and simplify using Equation (19) in [12]. □

7.12. Derivation of Entry (12)

Lemma 10.

$$\int_{-\infty}^{\infty}\int_{-\infty}^{\infty} e^{\frac{1}{6}(-6(e^x+e^y)+2x+3y)}\left(e^{y/6}-e^{x/6}\right)(x-y+i\pi)\log(x-y+i\pi)dxdy \\ =4(-1)^{2/3}\pi^2 Li'_{-1}\left((-1)^{2/3}\right)+12i\pi^2\log(A) \\ -i\pi^2-\frac{1}{3}i\pi^2\log(16)-\left(\frac{2}{3}+i\right)\pi^2\log(2i\pi)+\frac{2i\pi^2\log(2i\pi)}{\sqrt{3}} \tag{22}$$

Proof. Use Equation (13), set $a=-1, m=1/2, n=1/3$, take the first partial derivative with respect to k, then set $k=1$, and simplify using Equation (18) in [12]. □

7.13. Derivation of Entry (13)

Lemma 11.

$$\int_{-\infty}^{\infty}\int_{-\infty}^{\infty} e^{\frac{1}{8}(-8(e^x+e^y)+3x+5y)}(x-y)\sinh\left(\frac{x-y}{8}\right)dxdy=\frac{\pi^2}{\sqrt{2}} \tag{23}$$

Proof. Use Equation (13), set $k=1, a=1, m=1/4, n=1/2$, and simplify using entry (1) from the table below (64:12:7) in [11]. □

7.14. Derivation of Entry (14)

Lemma 12.

$$\int_{-\infty}^{\infty}\int_{-\infty}^{\infty} e^{\frac{1}{8}(-8(e^x+e^y)+3x+5y)}\sinh\left(\frac{x-y}{8}\right)dxdy=-\frac{1}{2}\left(\sqrt{2}-1\right)\pi \tag{24}$$

Proof. Use Equation (13), set $k=0, a=1, m=1/4, n=1/2$, and simplify using entry (2) from the table below (64:12:7) in [11]. □

7.15. Derivation of Entry (15)

Lemma 13.

$$\int_{-\infty}^{\infty}\int_{-\infty}^{\infty}\frac{e^{\frac{1}{6}(-6(e^x+e^y)+2x+3y)}\left(e^{x/6}-e^{y/6}\right)\log(x-y)}{x-y}dxdy$$
$$=-\sqrt[3]{-1}\Phi'\left((-1)^{2/3},1,\frac{1}{2}\right)+\frac{i\gamma\pi}{2}+\frac{1}{4}\log(9)\log(2\pi)+\frac{1}{4}i\pi\log\left(\frac{192\pi^6}{\Gamma\left(\frac{1}{4}\right)^8}\right) \quad (25)$$

Proof. Use Equation (13), set $a=1, n=1/2, m=1/3$, take the first partial derivative with respect to k, then set $k=-1$, and simplify using entry (1) from the table below (64:12:7) in [11]. □

7.16. Derivation of Entry (16)

Theorem 4. *For $k \in \mathbb{C}$,*

$$\int_{-\infty}^{\infty}\int_{-\infty}^{\infty}e^{\frac{1}{2}(-2(e^x+e^y)+x+y)}(x-y+i\pi)^k dxdy=-i^k\left(2^{k+1}-1\right)(2\pi)^{k+1}\zeta(-k) \quad (26)$$

Proof. Use Equation (9), set $a=-1, m=1/2$, and simplify using entry (4) from the table below (64:12:7) and entry (2) from the table below (64:7) in [11]. □

7.17. Derivation of Entry (17)

Lemma 14.

$$\int_{-\infty}^{\infty}\int_{-\infty}^{\infty}\frac{e^{\frac{1}{6}(-6(e^x+e^y)+2x+3y)}\left(e^{x/6}-e^{y/6}\right)}{x-y}dxdy=\frac{\log(3)}{2} \quad (27)$$

Proof. Use Equation (13), set $k=-1, a=1, n=1/2, m=1/3$, and simplify using entry (1) from the table below (64:12:7) in [11]. □

7.18. Derivation of Entry (18)

Theorem 5. *For $0<Re(m)<1, 0<Re(n)<1$,*

$$\int_{-\infty}^{\infty}\int_{-\infty}^{\infty}\frac{e^{y(-(m+n-1))-e^x-e^y}\left(e^{my+nx}-e^{mx+ny}\right)}{2x-2y+i\pi}dxdy$$
$$=\frac{2}{3}\left(e^{i\pi m}\,{}_2F_1\left(\frac{3}{4},1;\frac{7}{4};e^{2im\pi}\right)-e^{i\pi n}\,{}_2F_1\left(\frac{3}{4},1;\frac{7}{4};e^{2in\pi}\right)\right) \quad (28)$$

Proof. Use Equation (13), set $k=-1, a=i$, and simplify using Equation (9.559) in [8]. □

7.19. Derivation of Entry (19)

Theorem 6. *For $0<Re(m)<1, 0<Re(n)<1$,*

$$\int_{-\infty}^{\infty}\int_{-\infty}^{\infty}\frac{e^{y(-(m+n-1))-e^x-e^y}\left(e^{my+nx}-e^{mx+ny}\right)}{x-y+i\pi}dxdy$$
$$=e^{-i\pi n}\log\left(1-e^{2i\pi n}\right)-e^{-i\pi m}\log\left(1-e^{2i\pi m}\right) \quad (29)$$

Proof. Use Equation (13), set $k=-1, a=-1$, and simplify using Equation (9.559) in [8]. □

7.20. Derivation of Entry (20)

Lemma 15.

$$\int_{-\infty}^{\infty}\int_{-\infty}^{\infty} e^{\frac{1}{2}(-2(e^x+e^y)+x+y)}(x-y)\log(x-y)dxdy = -4i\pi C \tag{30}$$

Proof. Use Equation (9), set $m = 1/2$, and simplify using entry (4) below Table (64:12:7) in [11]. Then, take the first partial derivative with respect to k, set $k = 1, a = 1$, and simplify using Equation (9.73) in [8]. □

7.21. Derivation of Entry (21)

Lemma 16.

$$\int_{-\infty}^{\infty}\int_{-\infty}^{\infty} e^{\frac{1}{2}(-2(e^x+e^y)+x+y)}(x-y+i\pi)\log(x-y+i\pi)dxdy = \frac{1}{6}\pi^2\left(-3\pi+2i\log\left(\frac{128e^3\pi^3}{A^{36}}\right)\right) \tag{31}$$

Proof. Use Equation (9), set $m = 1/2$, and simplify using entry (4) below Table (64:12:7) in [11]. Then, take the first partial derivative with respect to k, set $k = 1, a = -1$, and simplify using 2.15 in [10]. □

7.22. Derivation of Entry (22)

Lemma 17.

$$\int_{-\infty}^{\infty}\int_{-\infty}^{\infty} \frac{e^{\frac{1}{2}(-2(e^x+e^y)+x+y)}\log(x-y+i\pi)}{x-y+i\pi}dxdy = \frac{1}{2}\log(2)\left(2i\gamma+\pi-i\log\left(8\pi^2\right)\right) \tag{32}$$

Proof. Use Equation (9), set $m = 1/2$, and simplify using entry (4) below Table (64:12:7) in [11]. Then, take the first partial derivative with respect to k; set $a = -1$; apply L'Hopital's rule to the right-hand side as $k \to -1$; and simplify using Equations (1:7:7), (44:7:7), and (64:4:1) in [11]. □

7.23. Derivation of Entry (23)

Lemma 18.

$$\int_{-\infty}^{\infty}\int_{-\infty}^{\infty} e^{\frac{1}{2}(-2(e^x+e^y)+x+y)}\log(x-y+i\pi)dxdy = \pi\log(4)+\frac{i\pi^2}{2} \tag{33}$$

Proof. Use Equation (9) set $m = 1/2$ and simplify using entry (4) below Table (64:12:7) in [11]. Then, take the first partial derivative with respect to k, set $k = 0, a = -1$, and simplify. □

7.24. Derivation of Entry (24)

Lemma 19.

$$\int_{-\infty}^{\infty}\int_{-\infty}^{\infty} \frac{e^{\frac{x}{2}-e^x-e^y+\frac{y}{2}}}{\sqrt{x-y+i\pi}}dxdy = \left(\sqrt{2}-1\right)e^{\frac{3i\pi}{4}}\sqrt{2\pi}\zeta\left(\frac{1}{2}\right) \tag{34}$$

Proof. Use Equation (9), set $k = -1/2, m = 1/2, a = -1$, and simplify using Equation (64:12:1) and entry (2) from the table below (64:7) in [11] and 1.5 in [13]. □

7.25. Derivation of Entry (25)

Lemma 20.

$$\int_{-\infty}^{\infty}\int_{-\infty}^{\infty} e^{\frac{x}{2}-e^x-e^y+\frac{y}{2}}\sqrt{x-y+i\pi}dxdy = 2\sqrt{2}\left(1-2\sqrt{2}\right)e^{\frac{i\pi}{4}}\pi^{3/2}\zeta\left(-\frac{1}{2}\right) \tag{35}$$

Proof. Use Equation (9), set $k = 1/2, m = 1/2, a = -1$, and simplify using Equation (64:12:1) and entry (2) from the table below (64:7) in [11]. □

7.26. Derivation of Entry (26)

Lemma 21.

$$\int_{-\infty}^{\infty}\int_{-\infty}^{\infty}\frac{e^{\frac{1}{2}(-2(e^x+e^y)+x+y)}\left((x-y)^2-\pi^2\right)}{\left((x-y)^2+\pi^2\right)^2}dxdy=-\frac{\pi}{24} \tag{36}$$

and

$$\int_{-\infty}^{\infty}\int_{-\infty}^{\infty}\frac{e^{\frac{1}{2}(-2(e^x+e^y)+x+y)}(x-y)}{\left((x-y)^2+\pi^2\right)^2}dxdy=0 \tag{37}$$

Proof. Use Equation (9), set $k = -1$, then take the first partial derivative with respect to a, set $a = -1, m = 1/2$, rationalize the denominator, compare the real and imaginary parts, and simplify using entry (3) from the table below (64:12:7) in [11]. □

7.27. Derivation of Entry (27)

Lemma 22.

$$\int_{-\infty}^{\infty}\int_{-\infty}^{\infty}\frac{e^{\frac{1}{2}(-2(e^x+e^y)+x+y)}\left((x-y+2)(x-y)^3+6\pi^2(y-x)-\pi^4\right)}{\left((x-y)^2+\pi^2\right)^3}dxdy$$
$$=-\frac{\pi}{24} \tag{38}$$

and

$$\int_{-\infty}^{\infty}\int_{-\infty}^{\infty}\frac{\pi e^{\frac{1}{2}(-2(e^x+e^y)+x+y)}\left((x-y+3)(x-y)^2+\pi^2(x-y-1)\right)}{\left((x-y)^2+\pi^2\right)^3}dxdy$$
$$=-\frac{3\zeta(3)}{16\pi^2} \tag{39}$$

Proof. Use Equation (9), set $k = -1$, then take the second partial derivative with respect to a, set $a = -1, m = 1/2$, rationalize the denominator, compare the real and imaginary parts, and simplify using entries (1) and (3) from the table below (64:12:7) and entry (2) from the table below (64:7) in [11]. □

7.28. Derivation of Entry (28)

Lemma 23.

$$\int_{-\infty}^{\infty}\int_{-\infty}^{\infty}\frac{e^{\frac{1}{4}(-4(e^x+e^y)+x+3y)}(x-y)}{\left((x-y)^2+\pi^2\right)^2}dxdy=\frac{\pi^2-48C}{192\sqrt{2}\pi^2} \tag{40}$$

and

$$\int_{-\infty}^{\infty}\int_{-\infty}^{\infty}\frac{e^{\frac{1}{4}(-4(e^x+e^y)+x+3y)}\left((x-y)^2-\pi^2\right)}{\left((x-y)^2+\pi^2\right)^2}dxdy=-\frac{48C+\pi^2}{96\sqrt{2}\pi} \tag{41}$$

Proof. Use Equation (9), set $k = -1$, then take the first partial derivative with respect to a, set $a = -1, m = 1/4$, rationalize the denominator, compare the real and imaginary parts, and simplify using entry (3) from the table below (64:12:7) in [11]. □

8. Summary Table of Results

In this section we will summarize the evaluation of Equation (9) from the previous section.

Table 1. Summary Table of Results.

$f(x,y)$	$\int_{-\infty}^{\infty}\int_{-\infty}^{\infty} f(x,y)\,dxdy$
$e^{mx-my-e^x-e^y+y}$	$\pi\csc(\pi m)$
$\frac{e^{\frac{1}{2}(-2(e^x+e^y)+x+y)}}{(x-y)^2+\pi^2}$	$\frac{\log(2)}{\pi}$
$\frac{e^{y(-(m+n-1))-e^x-e^y}\left(e^{my+nx}-e^{mx+ny}\right)}{x-y}$	$\log\left(\cot\left(\frac{\pi m}{2}\right)\tan\left(\frac{\pi n}{2}\right)\right)$
$(x-y+i\pi)^k e^{mx-my-e^x-e^y+y}$	$(2i\pi)^{k+1}\left(-e^{-i\pi m}\right)\mathrm{Li}_{-k}\left(e^{2im\pi}\right)$
$e^{\frac{1}{2}(-2(e^x+e^y)+x+y)}\log(ia+x-y)$	$\pi\log\left(\frac{4i\pi\Gamma\left(\frac{a}{4\pi}+\frac{3}{4}\right)^2}{\Gamma\left(\frac{a+\pi}{4\pi}\right)^2}\right)$
$e^{\frac{1}{2}(-2(e^x+e^y)+x+y)}\log\left(x-y-\frac{1}{2}\right)$	$\pi\log\left(\frac{4i\pi\Gamma\left(\frac{3}{4}+\frac{i}{8\pi}\right)^2}{\Gamma\left(\frac{1}{4}+\frac{i}{8\pi}\right)^2}\right)$
$\frac{e^{\frac{1}{6}(-6(e^x+e^y)+2x+3y)}\left(e^{y/6}-e^{x/6}\right)}{x-y}$	$-\frac{\log(3)}{2}$
$e^{\frac{1}{8}(-8(e^x+e^y)+3x+5y)}(x-y)\sinh\left(\frac{x-y}{8}\right)$	$\frac{\pi^2}{\sqrt{2}}$
$e^{\frac{1}{8}(-8(e^x+e^y)+3x+5y)}\sinh\left(\frac{x-y}{8}\right)$	$\frac{1}{2}\left(\sqrt{2}-1\right)\pi$
$e^{\frac{1}{2}(-2(e^x+e^y)+x+y)}(x-y+i\pi)^k$	$-i^k\left(2^{k+1}-1\right)(2\pi)^{k+1}\zeta(-k)$
$\frac{e^{\frac{1}{6}(-6(e^x+e^y)+2x+3y)}\left(e^{x/6}-e^{y/6}\right)}{x-y}$	$\frac{\log(3)}{2}$
$\frac{e^{y(-(m+n-1))-e^x-e^y}\left(e^{my+nx}-e^{mx+ny}\right)}{2x-2y+i\pi}$	$\frac{2}{3}\left(e^{i\pi m}\,{}_2F_1\left(\frac{3}{4},1;\frac{7}{4};e^{2im\pi}\right)-e^{i\pi n}\,{}_2F_1\left(\frac{3}{4},1;\frac{7}{4};e^{2in\pi}\right)\right)$
$\frac{e^{y(-(m+n-1))-e^x-e^y}\left(e^{my+nx}-e^{mx+ny}\right)}{x-y+i\pi}$	$e^{-i\pi n}\log\left(1-e^{2i\pi n}\right)-e^{-i\pi m}\log\left(1-e^{2i\pi m}\right)$
$e^{\frac{1}{2}(-2(e^x+e^y)+x+y)}(x-y)\log(x-y)$	$-4i\pi C$
$\frac{e^{\frac{1}{2}(-2(e^x+e^y)+x+y)}\log(x-y+i\pi)}{x-y+i\pi}$	$\frac{1}{2}\log(2)\left(2i\gamma+\pi-i\log\left(8\pi^2\right)\right)$
$e^{\frac{1}{2}(-2(e^x+e^y)+x+y)}\log(x-y+i\pi)$	$\pi\log(4)+\frac{i\pi^2}{2}$
$\frac{e^{\frac{x}{2}-e^x-e^y+\frac{y}{2}}}{\sqrt{x-y+i\pi}}$	$\left(\sqrt{2}-1\right)e^{\frac{3i\pi}{4}}\sqrt{2\pi}\zeta\left(\frac{1}{2}\right)$
$e^{\frac{x}{2}-e^x-e^y+\frac{y}{2}}\sqrt{x-y+i\pi}$	$2\sqrt{2}\left(1-2\sqrt{2}\right)e^{\frac{i\pi}{4}}\pi^{3/2}\zeta\left(-\frac{1}{2}\right)$

9. Discussion

In this work, the authors derived a double integral formula in terms of the Lerch function. This integral formula was then used to derive special cases in terms of fundamental constants and special functions. A table of integrals featuring some of the integral results was presented for the benefit of interested readers. We used Wolfram Mathematica to numerically verify the formulas for various ranges of the parameters for real and imaginary values. We will use our contour integral method to derive other double integrals and produce more tables of integrals in future work.

Author Contributions: Conceptualization, R.R.; methodology, R.R.; writing—original draft preparation, R.R.; writing—review and editing, R.R. and A.S.; funding acquisition, A.S. All authors have read and agreed to the published version of the manuscript.

Funding: This research is supported by NSERC Canada under grant 504070.

Institutional Review Board Statement: Not applicable.

Informed Consent Statement: Not applicable.

Data Availability Statement: Not applicable.

Conflicts of Interest: The authors declare no conflicts of interest.

References

1. Nielsen, N. *Handbuch der Theorie der Gammafunktion*; Teubner: Leipzig, Germany, 1906.
2. Max, B.; Emil, W. *Principles of Optics: Electromagnetic Theory of Propagation, Interference and Diffraction of Light*, 6th ed.; Pergamon Press: Oxford, UK, 1980.
3. Neretin, Y.A. *Lectures on Gaussian Integral Operators and Classical Groups*; European Mathematical Society: Zurich, Switzerland, 2011; pp. 1–559.
4. Berk, N.F. Scattering Properties of a Model Bicontinuous Structure with a Well Defined Length Scale. *Phys. Rev. Lett.* **1987**, *58*, 25. [CrossRef] [PubMed]
5. Fourier, J. *Théorie Analytique de la Chaleur*; F. Didot: Paris, France, 1822.
6. Titchmarsh, E.C. *Introduction to the Theory of Fourier Integrals*; Later Printing of Second Correct Edition; Oxford at the Clarendon Press: Oxford, UK, 1962.
7. Reynolds, R.; Stauffer, A. A Method for Evaluating Definite Integrals in Terms of Special Functions with Examples. *Int. Math. Forum* **2020**, *15*, 235–244. [CrossRef]
8. Gradshteyn, I.S.; Ryzhik, I.M. *Tables of Integrals, Series and Products*, 6th ed.; Academic Press: Cambridge, MA, USA, 2000.
9. Erdéyli, A.; Magnus, W.; Oberhettinger, F.; Tricomi, F.G. *Higher Transcendental Functions*; McGraw-Hill Book Company, Inc.: New York, NY, USA; Toronto, ON, Canada; London, UK, 1953; Volume I.
10. Finch, S.R. *Encyclopedia of Mathematics and Its Applications: Mathematical Constants*; Cambridge University Press: Cambridge, UK, 2003.
11. Oldham, K.B.; Myland, J.C.; Spanier, J. *An Atlas of Functions: With Equator, the Atlas Function Calculator*, 2nd ed.; Springer: New York, NY, USA, 2009.
12. Guillera, J.; Sondow, J. Double integrals and infinite products for some classical constants via analytic continuations of Lerch's transcendent. *Ramanujan J.* **2008**, *16*, 247. [CrossRef]
13. Borwein, J.; Bailey, D.; Girgensohn, R. *Experimentation in Mathematics: Computational Paths to Discovery*; A K Peters: Wellesley, MA, USA, 2004.

symmetry

Article

Fuzzy Mixed Variational-like and Integral Inequalities for Strongly Preinvex Fuzzy Mappings

Muhammad Bilal Khan [1], Hari Mohan Srivastava [2,3,4,5], Pshtiwan Othman Mohammed [6,*] and Juan L. G. Guirao [7,8,*]

1 Department of Mathematics, COMSATS University Islamabad, Islamabad 44000, Pakistan; bilal42742@gmail.com
2 Department of Mathematics and Statistics, University of Victoria, Victoria, BC V8W 3R4, Canada; harimsri@math.uvic.ca
3 Department of Medical Research, China Medical University Hospital, China Medical University, Taichung 40402, Taiwan
4 Department of Mathematics and Informatics, Azerbaijan University, 71 Jeyhun Hajibeyli Street, Baku AZ1007, Azerbaijan
5 Section of Mathematics, International Telematic University Uninettuno, I-00186 Rome, Italy
6 Department of Mathematics, College of Education, University of Sulaimani, Sulaimani 46001, Iraq
7 Department of Applied Mathematics and Statistics, Technical University of Cartagena, Hospital de Marina, 30203 Cartagena, Spain
8 Nonlinear Analysis and Applied Mathematics (NAAM)-Research Group, Department of Mathematics, Faculty of Science, King Abdulaziz University, Jeddah 21589, Saudi Arabia
* Correspondence: pshtiwan.muhammad@univsul.edu.iq (P.O.M.); Juan.Garcia@upct.es (J.L.G.G.)

Abstract: It is a familiar fact that convex and non-convex fuzzy mappings play a critical role in the study of fuzzy optimization. Due to the behavior of its definition, the idea of convexity plays a significant role in the subject of inequalities. The concepts of convexity and symmetry have a tight connection. We may use whatever we learn from one to the other, thanks to the significant correlation that has developed between both in recent years. Our aim is to consider a new class of fuzzy mappings (FMs) known as strongly preinvex fuzzy mappings (strongly preinvex-FMs) on the invex set. These FMs are more general than convex fuzzy mappings (convex-FMs) and preinvex fuzzy mappings (preinvex-FMs), and when generalized differentiable (briefly, G-differentiable), strongly preinvex-FMs are strongly invex fuzzy mappings (strongly invex-FMs). Some new relationships among various concepts of strongly preinvex-FMs are established and verified with the support of some useful examples. We have also shown that optimality conditions of G-differentiable strongly preinvex-FMs and the fuzzy functional, which is the sum of G-differentiable preinvex-FMs and non G-differentiable strongly preinvex-FMs, can be distinguished by strongly fuzzy variational-like inequalities and strongly fuzzy mixed variational-like inequalities, respectively. In the end, we have established and verified a strong relationship between the Hermite–Hadamard inequality and strongly preinvex-FM. Several exceptional cases are also discussed. These inequalities are a very interesting outcome of our main results and appear to be new ones. The results in this research can be seen as refinements and improvements to previously published findings.

Keywords: preinvex fuzzy mappings; strongly preinvex fuzzy mappings; strongly invex fuzzy mappings; strongly fuzzy monotonicity; strongly fuzzy mixed variational-like inequalities

Citation: Khan, M.B.; Srivastava, H.M.; Mohammed, P.O.; Guirao, J.L.G. Fuzzy Mixed Variational-like and Integral Inequalities for Strongly Preinvex Fuzzy Mappings. *Symmetry* **2021**, *13*, 1816. https://doi.org/10.3390/sym13101816

Academic Editor: José Carlos R. Alcantud

Received: 2 September 2021
Accepted: 24 September 2021
Published: 29 September 2021

1. Introduction

Recently, many generalizations and extensions have been studied for classical convexity. Polyak [1] introduced and studied the idea of strongly convex functions on the convex set, which have a significant impact on optimization theory and related fields. Karmardian [2] discussed how strongly convex functions can be used to solve nonlinear complementarity problems for the first time. Qu and Li [3] and Nikodem and Pales [4] developed the convergence analysis for addressing equilibrium issues and variational inequalities, using strongly convex functions. For further study, we refer the reader to applications and properties of the strongly convex functions of [5–10], and the references therein. For differentiable functions, invex functions were introduced by Hanson [11], which played a significant role in mathematical programing. The concept of invex sets and preinvex functions were introduced and studied by Israel and Mond [12]. It is well known that differential preinvex function are invex functions. The converse also holds under Condition C [13]. Furthermore, Noor [14], studied the optimality conditions of differentiable preinex functions and proved that the minimum can be characterized by variational-like inequalities. Noor et al. [15,16] studied the properties of the strongly preinvex function and investigated its applications. For more applications and properties of strongly preinvex functions, see [17–19] and the references therein.

In [20], a large amount of research work on fuzzy sets and systems was devoted to the advancement of various fields, playing an important role in the analysis of broad class problems emerging in pure and applied sciences, such as operation research, computer science, decision sciences, control engineering, artificial intelligence, and management sciences. Convex analysis has made significant contributions to the improvement of several practical and pure science domains. In the same way, fuzzy convex analysis is a fundamental principle in fuzzy optimization and it is worthwhile to explore some basic principles of convex sets in fuzzy set theory. Many scholars have addressed fuzzy convex sets. Liu [21] investigated some properties of convex fuzzy sets and updated the definition of shadow of fuzzy sets with the support of useful examples. Lowen [22] gathered some well-known convex sets' results and proved the separation theorem for convex fuzzy sets. Ammar and Metz [23,24] investigated forms of convexity and established the generalized convexity of fuzzy sets. Furthermore, they used the principle of convexity to formulate a general fuzzy nonlinear programming problem.

A fuzzy number is a generalized version of an interval that can be discussed (in crisp set theory). Zadeh [20] defined fuzzy numbers, while Dubois and Prade [25] built on Zadeh's work by adding new fuzzy number conditions. Furthermore, Goetschel and Voxman [26] adjusted many conditions on fuzzy numbers to make them easier to handle. For example, in [25], one of the conditions for a fuzzy number is that it is a continuous function, whereas in [26], the fuzzy number is upper semi-continuous. The purpose is to establish a metric for a collection of fuzzy numbers, using the relaxation of requirements on fuzzy numbers, and then use this metric to examine some basic features of topological space. Nanda and Kar [27], Syau [28] and Furukawa [29] introduced the concept of convex-FMs from $\mathbb{R}^n$ to the set of fuzzy numbers. Furthermore, they also defined different type of convex-FMs, such as logarithmic convex-FMs and quasi-convex-FMs, as well studying Lipschitz continuity of fuzzy valued mappings. Yan and Xu [30] provided the notions of epigraphs and the convexity of FMs, as well as the characteristics of convex-FMs and quasi-convex-FMs, based on Goetschel and Voxman's concept of ordering [31]. The concept of fuzzy preinvex mapping on the invex set was introduced and studied by Noor [32]. He also demonstrated that variational inequalities may be used to specify the fuzzy optimality conditions of differentiable fuzzy preinex mappings. Syau [33], introduced notions of $(\phi_1, \phi_2)-$convexity, ϕ_1-B-vexity and ϕ_1-convexity-FMs through the so-called fuzzy max order among the fuzzy numbers, and proved that the ϕ_1-B-vexity and ϕ_1-convexity, B-vexity, convexity and preinvexity of FMs are the subclasses. Syau and Lee [34] examined various aspects of fuzzy optimization and discussed continuity and convexity through linear ordering and metrics defined on fuzzy integers. They also extended the Weirstrass theorem from real-valued functions to FMs. For recent applications, see [35–39] and the references therein.

On the other hand, integral inequalities have various applications in linear programing, combinatory, orthogonal polynomials, quantum theory, number theory, optimization theory, dynamics, and the theory of relativity; see [40,41] and the references therein. The HH-inequality is a familiar, supreme and broadly useful inequality. This inequality has fundamental significance [42,43], due to other classical inequalities, such as the Olsen, Gagliardo–Nirenberg, Hardy, Opial, Young, Linger, arithmetic–geometric, Ostrowski, Levinson, Minkowski, Beckenbach–Dresher, Ky Fan and Holer inequalities [44–49], which are closely linked to the classical HH-inequality. It can be stated as follows:

Let $\mathcal{H}: K \to \mathbb{R}$ be a convex function on a convex set K and , $\nu \in K$ with $u \leq \nu$. Then,

$$\mathcal{H}\left(\frac{u+\nu}{2}\right) \leq \frac{1}{\nu-u}\int_u^\nu \mathcal{H}(z)dz \leq \frac{\mathcal{H}(u)+\mathcal{H}(\nu)}{2}. \quad (1)$$

If $\mathcal{H}$ is a concave function, then inequality (1) is reversed.

There are several integrals that deal with FMs and have FMs as integrands. For FMs, Oseuna-Gomez et al. [50] and Costa et al. [51] constructed Jensen's integral inequality. Costa and Floures [52] used the same method to present Minkowski and Beckenbach's inequalities, where the integrands are fuzzy mappings. Costa et al. established a relationship between elements of fuzzy-interval space and interval space and introduced level-wise fuzzy order relation on fuzzy-interval space through Kulisch–Miranker order relation defined on an interval space. This was motivated by [48–53] and particularly [54], because Costa et al. established a relationship between elements of fuzzy-interval space and interval space and introduced a level-wise fuzzy order relation on fuzzy-interval space through the Kulisch–Miranker order relation defined on interval space. By using this relation on the fuzzy-interval space, we generalize integral inequality (1) by constructing fuzzy integral inequalities for strongly preinvex-FMs, where the integrands are strongly preinvex-FMs. Recently, Khan et al. [55] introduced the new class of convex-FMs, which is known as (h_1, h_2)-convex-FMs by means of the fuzzy order relation and presented the following new version of HH-type inequality for (h_1, h_2)-convex-FM involving fuzzy-interval Riemann integrals:

Theorem 1. *Let $\mathcal{H}: [u, \nu] \to \mathbb{F}_0$ be a (h_1, h_2) -convex-FM with $h_1, h_2: [0,1] \to \mathbb{R}^+$ and $h_1\left(\frac{1}{2}\right)h_2\left(\frac{1}{2}\right) \neq 0$. If $\mathcal{H}$ is fuzzy Riemann integrable (in sort, FR -integrable), then the following holds:*

$$\begin{aligned} \frac{1}{2h_1\left(\frac{1}{2}\right)h_2\left(\frac{1}{2}\right)}\mathcal{H}\left(\tfrac{u+\nu}{2}\right) &\preccurlyeq \tfrac{1}{\nu-u}\int_u^\nu \mathcal{H}(z)dz \\ &\preccurlyeq \left[\mathcal{H}(u) \tilde{+} \mathcal{H}(\nu)\right]\int_0^1 h_1(\tau)h_2(1-\tau)d\tau. \end{aligned} \quad (2)$$

Theorem 1 reduces to the result for convex fuzzy-IVF:

$$\mathcal{H}\left(\frac{u+\nu}{2}\right) \preccurlyeq \frac{1}{\nu-u}\int_u^\nu \mathcal{H}(z)dz \preccurlyeq \frac{H(u)\tilde{+}H(\nu)}{2}. \quad (3)$$

For further information related to fuzzy integral inequalities, see [56–68].

Motivated by ongoing studies as well as the relevance of the concepts of invexity and preinvexity of FMs, in Section 2, we provide an overview of some fundamental concepts, preliminary notations, and findings that will be useful in further research. In the parts that follow, the key results are considered and discussed. Section 3 introduces the concepts of strongly preinvex-FMs and discusses some of their properties. Moreover, new relationships among various concepts of strongly preinvex-FMs are also investigated in Section 3. In Section 4, we introduce fuzzy variational-like and Hermite–Hadamard inequalities for strong preinvex-FMs.

2. Preliminaries

In this section, we first provide some definitions, preliminary notations and results, which will be helpful for further study.

A fuzzy set of $\mathbb{R}$ is a mapping $\Psi: \mathbb{R} \to [0,1]$, for each fuzzy set and $\gamma \in (0, 1]$; then, γ-level sets of Ψ are denoted and defined as follows: $\Psi_\gamma = \{u \in \mathbb{R} | \Psi(u) \geq \gamma\}$. The support of Ψ is denoted by supp(Ψ) and is defined as supp(Ψ) $= \{u \in \mathbb{R} | \Psi(u) \geq \gamma\}$. A fuzzy set is normal if there exist $u \in \mathbb{R}$ such that $\Psi(u) = 1$. A fuzzy set is convex and concave if $\Psi((1-\tau)u + \tau v) \geq min(\Psi(u), \Psi(v))$ and $\Psi((1-\tau)u + \tau v) \leq max(\Psi(u), \Psi(v))$ for $u, v \in \mathbb{R}$, $\tau \in [0,1]$, respectively. A fuzzy convex set is a generalization of the classical convex set.

A fuzzy set is said to be fuzzy number with the following properties.

(a) Ψ is normal. (b) Ψ is a convex fuzzy set. (c) Ψ is upper semi-continuous. (d) Ψ_0 is compact.

$\mathbb{F}_0$ denotes the set of all fuzzy numbers. For a fuzzy number, it is convenient to distinguish the following γ-levels:

$$\Psi_\gamma = \{u \in \mathbb{R} | \Psi(u) \geq \gamma\},$$

From these definitions, we have the following:

$$\Psi_\gamma = [\Psi_*(\gamma),\ \Psi^*(\gamma)]$$

where

$$\Psi_*(\gamma) = inf\{u \in \mathbb{R} | \Psi(u) \geq \gamma\}, \Psi^*(\gamma) = sup\{u \in \mathbb{R} | \Psi(u) \geq \gamma\}.$$

Since each $r \in \mathbb{R}$ is also a fuzzy number, it is defined as follows:

$$\widetilde{r}(u) = \begin{cases} 1 \text{ if } u = r \\ 0 \text{ if } u \neq r \end{cases}.$$

It is also well known that for any $\Psi, \phi \in \mathbb{F}_0$ and $r \in \mathbb{R}$, the following holds:

$$\Psi \widetilde{+} \phi = \{(\Psi_*(\gamma) + \phi_*(\gamma), \Psi^*(\gamma) + \phi^*(\gamma), \gamma) : \gamma \in [0,\ 1]\}, \quad (4)$$

$$r\Psi = \{(r\Psi_*(\gamma), r\Psi^*(\gamma), \gamma) : \gamma \in [0,\ 1]\}. \quad (5)$$

Obviously, $\mathbb{F}_0$ is closed under addition and nonnegative scaler multiplication. Furthermore, for each scaler number $r \in \mathbb{R}$, the following holds:

$$\Psi \widetilde{+} r = \{(\Psi_*(\gamma) + r, \Psi^*(\gamma) + r, \gamma) : \gamma \in [0,\ 1]\}. \quad (6)$$

For any $\Psi, \phi \in \mathbb{F}_0$, we say that $\Psi \preccurlyeq \phi$ ("$\preccurlyeq$" relation between fuzzy numbers Ψ and ϕ if for all $\gamma \in (0,\ 1]$, $\Psi^*(\gamma) \leq \phi^*(\gamma)$ ("$\leq$" relation $\Psi^*(\gamma)$ and $\phi^*(\gamma)$) and $\Psi_*(\gamma) \leq \phi_*(\gamma)$. We say it is comparable if for any $\Psi, \phi \in \mathbb{F}_0$, we have $\Psi \preccurlyeq \phi$ or $\Psi \succcurlyeq \phi$; otherwise, they are non-comparable.

We can state that $\mathbb{F}_0$ is a partial ordered set under the relation $\preccurlyeq$ if we write $\Psi \preccurlyeq \phi$ instead of $\phi \succcurlyeq \Psi$. If $\Psi, \phi \in \mathbb{F}_0$, there exist $\omega \in \mathbb{F}_0$ such that $\Psi = \phi \widetilde{+} \omega$; then, we have the existence of the Hukuhara difference (in short, H-difference) of Ψ and ϕ, and we say that ω is the H-difference of Ψ and ϕ, denoted by $\Psi \widetilde{-} \phi$; see [37]. If this fuzzy operation exists, then we have the following:

$$(\omega)^*(\gamma) = \left(\Psi \widetilde{-} \phi\right)^*(\gamma) = \Psi^*(\gamma) - \phi^*(\gamma), (\omega)_*(\gamma) = \left(\Psi \widetilde{-} \phi\right)_*(\gamma) = \Psi_*(\gamma) - \phi_*(\gamma).$$

A mapping $\mathcal{H} : K \to \mathbb{F}_0$ is called fuzzy mapping (FM). For each $\gamma \in [0,\ 1]$, denote $[\mathcal{H}(u)]^\gamma = [\mathcal{H}_*(u,\gamma),\ \mathcal{H}^*(u,\gamma)]$ and in parameterized form, denote $\mathcal{H}(u) = \{(\mathcal{H}_*(u,\ \gamma), \mathcal{H}^*(u,\ \gamma), \gamma) : \gamma \in [0,\ 1]\}$.

Definition 1. *Let us say $I = (m,\ n)$ and $\sqcap \in (m,\ n)$ [35]. Then, FM $\mathcal{H} : (m,\ n) \to \mathbb{F}_0$ is said to be a generalized differentiable (briefly, G-differentiable) at $\sqcap$ if there exists an element $\mathcal{H}'(u) \in \mathbb{F}_0$*

such that for any $0 < \tau$, *sufficiently small, there exist* $\mathcal{H}(u+\tau)\tilde{-}\mathcal{H}(u)$, $\mathcal{H}(u)\tilde{-}\mathcal{H}(u-\tau)$, *and the limits are (in the metric D) as follows:*

$$\begin{aligned} \lim_{\tau\to 0^+} \frac{\mathcal{H}(u+\tau)\tilde{-}\mathcal{H}(u)}{\tau} &= \lim_{\tau\to 0^+} \frac{\mathcal{H}(u)\tilde{-}\mathcal{H}(u-\tau)}{\tau} = \mathcal{H}'(u) \\ \text{or } \lim_{\tau\to 0^+} \frac{\mathcal{H}(u)\tilde{-}\mathcal{H}(u+\tau)}{-\tau} &= \lim_{\tau\to 0^+} \frac{\mathcal{H}(u-\tau)\tilde{-}\mathcal{H}(u)}{-\tau} = \mathcal{H}'(u) \\ \text{or } \lim_{\tau\to 0^+} \frac{\mathcal{H}(u+\tau)\tilde{-}\mathcal{H}(u)}{\tau} &= \lim_{\tau\to 0^+} \frac{\mathcal{H}(u-\tau)\tilde{-}\mathcal{H}(u)}{-\tau} = \mathcal{H}'(u) \\ \text{or } \lim_{\tau\to 0^+} \frac{\mathcal{H}(u)\tilde{-}\mathcal{H}(u+\tau)}{-\tau} &= \lim_{\tau\to 0^+} \frac{\mathcal{H}(u)\tilde{-}\mathcal{H}(u-\tau)}{\tau} = \mathcal{H}'(u), \end{aligned}$$

where the limits are taken in the metric space (E, D), *for* $\Psi,\ \phi \in \mathbb{F}_0$ *as follows:*

$$D(\Psi,\ \phi) = \sup_{0\le\gamma\le 1} H(\Psi_\gamma,\ \phi_\gamma),$$

and H denotes the well-known Hausdorff metric on the space of intervals.

Definition 2. *A FM* $\mathcal{H} : K \to \mathbb{F}_0$ *is said to be convex on the convex set K if the following holds [27]:*

$$\mathcal{H}((1-\tau)u + \tau\nu) \preccurlyeq (1-\tau)\mathcal{H}(u)\tilde{+}\tau\mathcal{H}(\nu),\ \forall\, u,\nu \in K,\ \tau \in [0,\ 1]. \tag{7}$$

Similarly, $\mathcal{H}$ *is said to be concave-FM on K if inequality (7) is reversed.*

Definition 3. *The set* K_ξ *in* $\mathbb{R}$ *is said to be invex set with respect to (w.r.t.) arbitrary bifunction* $\xi(.,.)$, *if the following holds [12]:*

$$u + \tau\xi(\nu, u) \in K_\xi,\ \forall\, u,\nu \in K_\xi,\ \tau \in [0,\ 1].$$

The invex set K_ξ *is also known as a* ξ *-connected set. Note that each convex set with* $\nu - u = \xi(\nu, u)$ *is an invex set in the classical sense, but the reverse is not true. For instance, the following set* $K_\xi = [-7, -2] \cup [2, 10]$ *is an invex set w.r.t. non-trivial bi-function* $\xi : \mathbb{R} \times \mathbb{R} \to \mathbb{R}$ *given as follows:*

$$\begin{gathered} \xi(\nu, u) = \nu - u,\ \nu \ge 0,\ u \ge 0, \\ \xi(\nu, u) = \nu - u,\ 0 \ge \nu,\ 0 \ge u, \\ \xi(\nu, u) = -7 - u,\ \nu \ge 0 \ge u, \\ \xi(\nu, u) = 2 - u,\ u \ge 0 \ge \nu. \end{gathered}$$

Definition 4. *A FM* $\mathcal{H} : K_\xi \to \mathbb{F}_0$ *is said to be preinvex on the invex set* K_ξ *w.r.t. bi-function* ξ *if the following holds [32]:*

$$\mathcal{H}(u + \tau\xi(\nu, u)) \preccurlyeq (1-\tau)\mathcal{H}(u)\tilde{+}\tau\mathcal{H}(\nu), \tag{8}$$

for all $u,\nu \in K_\xi$, $\tau \in [0,\ 1]$, *where* $\xi : K_\xi \times K_\xi \to \mathbb{R}$. $\mathcal{H}$ *is said to be preconcave-FM on* K_ξ *if inequality (8) is reversed.*

Lemma 1. *Let* K_ξ *be an invex set w.r.t.* ξ *and let* $\mathcal{H} : K_\xi \to \mathbb{F}_0$ *be a FM, parameterized by the following [21]:*

$$\mathcal{H}(u) = \{(\mathcal{H}_*(u,\ \gamma), \mathcal{H}^*(u,\ \gamma), \gamma) : \gamma \in [0,\ 1]\},\ \forall\, u \in K_\xi$$

Then, $\mathcal{H}$ *is preinvex on* K_ξ *if, and only if, for all* $\gamma \in [0,\ 1]$,
$\mathcal{H}_*(u,\ \gamma)$ *and* $\mathcal{H}^*(u,\ \gamma)$ *are preinvex w.r.t.* ξ *on* K_ξ.
If $\xi(\nu, u) = \nu - u$, *then Lemma 1 reduces to the following result:*

"Let K_ξ be a convex set and let $\mathcal{H} : K_\xi \to \mathbb{F}_0$ be a FM parameterized by the following:

$$\mathcal{H}(u) = \{(\mathcal{H}_*(u,\ \gamma), \mathcal{H}^*(u,\ \gamma), \gamma) : \gamma \in [0,\ 1]\},\ \forall\, u \in K_\xi$$

Then, $\mathcal{H}$ is convex on K_ξ if, and only if, for all $\gamma \in [0,\ 1]$, $\mathcal{H}_*(u,\ \gamma)$ and $\mathcal{H}^*(u,\ \gamma)$ are convex w.r.t. ξ on K_ξ.

Theorem 2. *If $\mathcal{H} : [c,d] \subset \mathbb{R} \to \mathcal{K}_C$ is an interval valued function on : $[c,d]$ such that $[\mathcal{H}_*,\ \mathcal{H}^*]$[54]. Then $\mathcal{H}$ is Riemann integrable over : $[c,d]$ if and only if, $\mathcal{H}_*$ and $\mathcal{H}^*$ both are Riemann integrable over: $[c,d]$ such that the following holds:*

$$(IR)\int_c^d \mathcal{H}(z)dz = \left[(R)\int_c^d \mathcal{H}_*(u)dz,\ (R)\int_c^d \mathcal{H}^*(u)dz\right] \tag{9}$$

From the above literature review, the following results can be concluded; see [31,32,53,54].

Definition 5. *Let $\mathcal{H} : [c,\ d] \subset \mathbb{R} \to \mathbb{F}_0$ be a FM [47]. The fuzzy Riemann integral of $\mathcal{H}$ over $[c,\ d]$, denoted by $(FR)\int_c^d \mathcal{H}(z)dz$, is defined by the following:*

$$\left[(FR)\int_c^d \mathcal{H}(z)dz\right]^\gamma = (IR)\int_c^d \mathcal{H}_\gamma(z)dz = \left\{\int_c^d \mathcal{H}(z,\gamma)dz : \mathcal{H}(z,\gamma) \in \mathcal{R}_{[c,\ d]}\right\}, \tag{10}$$

for all $\gamma \in [0,\ 1]$, where $\mathcal{R}_{[c,\ d]}$ is the collection of end-point functions of IVFs. $\mathcal{H}$ is (FR)-integrable over $[c,\ d]$ if $(FR)\int_c^d \mathcal{H}(z)dz \in \mathbb{F}_0$. Note that, if both end-point functions are Lebesgue-integrable, then $\mathcal{H}$ is fuzzy Aumann-integrable.

Let K_ξ be a nonempty invex set in $\mathbb{R}$ for future investigation. Let $\xi : K_\xi \times K_\xi \to \mathbb{R}$ be an arbitrary bifunction and $\mathcal{H} : K_\xi \to \mathbb{F}_0$ be an FM. We denote $\|.\|$ and $\langle .,\ . \rangle$ as the norm and inner product, respectively. Furthermore, throughout this article, FMs are discussed through the so-called "fuzzy-max" order among fuzzy numbers. As is well known, the fuzzy-max order is a partial order relation " $\preccurlyeq$ " on the set of fuzzy numbers.

3. Strongly Preinvex Fuzzy Mappings

In this section, we propose and study the class of strongly preinvex-FMs. We also establish the relationship between strongly preinvex-FMs, strongly monotone operators and strongly invex-FMs. Firstly, we define the following notion of strongly preinvex-FM.

Definition 6. *Let K_ξ be an invex set and ω be a positive number. Then, FM $\mathcal{H} : K_\xi \to \mathbb{F}_0$ is said to be strongly preinvex-FM on K_ξ w.r.t. bi-function $\xi(.,.)$ if the following holds:*

$$\mathcal{H}(u + \tau\xi(\nu,u)) \preccurlyeq (1-\tau)\mathcal{H}(u)\tilde{+}\tau\mathcal{H}(\nu)\tilde{-}\omega\tau(1-\tau)\|\xi(\nu,u)\|^2, \tag{11}$$

for all $u,\ \nu \in K_\xi$, $\tau \in [0,\ 1]$. $\mathcal{H}$ is said to be strongly preconcave-FM on K_ξ if inequality (11) is reversed. $\mathcal{H}$ is said to be strongly affine preinvex-FM on K_ξ if the following holds:

$$\mathcal{H}(u + \tau\xi(\nu,u)) = (1-\tau)\mathcal{H}(u)\tilde{+}\tau\mathcal{H}(\nu)\tilde{-}\omega\tau(1-\tau)\|\xi(\nu,u)\|^2, \tag{12}$$

for all $u,\nu \in K_\xi$, $\tau \in [0,\ 1]$.

Remark 1. *Strongly preinvex-FMs, such as preinvex-FMs, have the following highly desirable features:*

(1) *$Y\mathcal{H}$ is also strongly preinvex for $Y \geq 0$, if $\mathcal{H}$ is strongly preinvex-FM.*
(2) *$max(\mathcal{H}(u), \varpi(u))$ is also strongly preinvex-FM if $\mathcal{H}$ and ϖ both are strongly preinvex-FMs.*

Now, we discuss some special cases of strongly preinvex-FMs:

If $\xi(v,u) = v - u$, then strongly preinvex-FM becomes strongly convex-FM, that is

$$\mathcal{H}((1-\tau)u+\tau v) \preccurlyeq (1-\tau)\mathcal{H}(u)\tilde{+}\tau\mathcal{H}(v)\tilde{-}\omega\tau(1-\tau)\|v-u\|^2,\ \forall\, u,\, v \in K_\xi,\ \tau \in [0,\ 1].$$

If $\omega = 0$, then inequality (11) reduces to inequality (8).

If $\omega = 0$ and $\xi(v,u) = v - u$, then inequality (11) reduces to inequality (7).

The following result characterizes the definition of strongly preinvex-FMs and establishes the relationship between strongly preinvex-FMs and end-point functions. With the help of this theorem, we can easily handle the upcoming results.

Theorem 3. *Let $\mathcal{H} : K_\xi \to \mathbb{F}_0$ be a FM parametrized by the following:*

$$\mathcal{H}(u) = \{(\mathcal{H}_*(u,\ \gamma), \mathcal{H}^*(u,\ \gamma), \gamma) : \gamma \in [0,\ 1]\},\ \forall\, u \in K_\xi. \tag{13}$$

Then, $\mathcal{H}$ is strongly preinvex on K w.r.t. ξ, with modulus ω if and only if, for all $\gamma \in [0,\ 1]$,

$\mathcal{H}_*(u,\ \gamma)$ and $\mathcal{H}^*(u,\ \gamma)$ are strongly preinvex w.r.t. ξ and modulus ω.

Proof. Assume that for each $\gamma \in [0,\ 1]$, $\mathcal{H}_*(u,\ \gamma)$ and $\mathcal{H}^*(u,\ \gamma)$ are strongly preinvex w.r.t. ξ and modulus ω on K_ξ. Then, from (11), for all $u, v \in K_\xi$, $\tau \in [0,\ 1]$, we have the following:

$$\mathcal{H}_*(u+\tau\xi(v,u),\ \gamma) \leq (1-\tau)\mathcal{H}_*(u,\ \gamma) + \tau\mathcal{H}_*(v,\ \gamma) - \omega\tau(1-\tau)\|\xi(v,u)\|^2$$

and

$$\mathcal{H}^*(u+\tau\xi(v,u),\ \gamma) \leq (1-\tau)\mathcal{H}^*(u,\ \gamma) + \tau\mathcal{H}^*(v,\ \gamma) - \omega\tau(1-\tau)\|\xi(v,u)\|^2.$$

Then, by (13), (4), (5) and (6), we obtain the following:

$$\begin{aligned}\mathcal{H}(u+\tau\xi(v,u)) &= \{(\mathcal{H}_*(u+\tau\xi(v,u),\ \gamma), \mathcal{H}^*(u+\tau\xi(v,u),\ \gamma), \gamma) : \gamma \in [0,\ 1]\},\\ &\preccurlyeq \{((1-\tau)\mathcal{H}_*(u,\ \gamma), (1-\tau)\mathcal{H}^*(u,\ \gamma), \gamma) : \gamma \in [0,\ 1]\}\tilde{+}\{(\tau\mathcal{H}_*(v,\ \gamma), \tau\mathcal{H}^*(v,\ \gamma), \gamma) : \gamma \in [0,\ 1]\}\\ &\quad \tilde{-}\omega\tau(1-\tau)\|\xi(v,u)\|^2,\\ &= (1-\tau)\mathcal{H}(u)\tilde{+}\mathcal{H}(v)\tilde{-}\omega\tau(1-\tau)\|\xi(v,u)\|^2.\end{aligned}$$

Hence, $\mathcal{H}$ is strongly preinvex-FM on K_ξ with modulus ω. □

Conversely, let $\mathcal{H}$ be a strongly preinvex-FM on K_ξ with modulus ω. Then, for all $u, v \in K_\xi$ and $\tau \in [0,\ 1]$, we have $\mathcal{H}(u+\tau\xi(v,u)) \preccurlyeq (1-\tau)\mathcal{H}(u)\tilde{+}\tau\mathcal{H}(v)\tilde{-}\omega\tau(1-\tau)\|\xi(v,u)\|^2$. From (13), we have the following:

$$\mathcal{H}(u+\tau\xi(v,u)) = \{(\mathcal{H}_*(u+\tau\xi(v,u),\ \gamma), \mathcal{H}^*(u+\tau\xi(v,u),\ \gamma), \gamma) : \gamma \in [0,\ 1]\}.$$

Again, from (13), (4), (5) and (6), we obtain the following:

$$\begin{aligned}&(1-\tau)\mathcal{H}(u)\tilde{+}\tau\mathcal{H}(u)\tilde{-}\omega\tau(1-\tau)\|\xi(v,u)\|^2\\ &= \{((1-\tau)\mathcal{H}_*(u,\ \gamma), (1-\tau)\mathcal{H}^*(u,\ \gamma), \gamma) : \gamma \in [0,\ 1]\}\\ &\tilde{+}\{(\tau\mathcal{H}_*(v,\ \gamma),\ \tau\mathcal{H}^*(v,\ \gamma), \gamma) : \gamma \in [0,\ 1]\}\tilde{-}\omega\tau(1-\tau)\|\xi(v,u)\|^2,\end{aligned}$$

for all $u, v \in K_\xi$ and $\tau \in [0,\ 1]$. Then, by strongly preinvexity of $\mathcal{H}$, we have for all $u, v \in K_\xi$ and $\tau \in [0,\ 1]$ such that the following holds:

$$\mathcal{H}_*(u+\tau\xi(v,u),\ \gamma) \leq (1-\tau)\mathcal{H}_*(u,\ \gamma) + \tau\mathcal{H}_*(v,\ \gamma) - \omega\tau(1-\tau)\|\xi(v,u)\|^2,$$

and

$$\mathcal{H}^*(u+\tau\xi(\nu,u),\ \gamma) \leq (1-\tau)\mathcal{H}^*(u,\ \gamma)+\tau\mathcal{H}^*(\nu,\ \gamma)-\omega\tau(1-\tau)\|\xi(\nu,u)\|^2,$$

for each $\gamma \in [0,\ 1]$. Hence, the result follows.

Example 1. *We consider the FM $\mathcal{H} : [0,\ 1] \to \mathbb{F}_0$ defined by the following:*

$$\mathcal{H}(u)(\sigma) = \begin{cases} \frac{\sigma}{2u^2}\ \sigma \in [0,\ 2u^2] \\ \frac{4u^2-\sigma}{2u^2}\ \sigma \in (2u^2,\ 4u^2] \\ 0\ otherwise, \end{cases} \tag{14}$$

Then, for each $\gamma \in [0,\ 1]$, we have $\mathcal{H}_\gamma(u) = [2\gamma u^2, (4-2\gamma)u^2\ ]$. Since $\mathcal{H}_*(u,\gamma)$, $\mathcal{H}^*(u,\gamma)$ are strongly preinvex functions for each $\gamma \in [0,1]$, $\mathcal{H}(u)$ is strongly preinvex-FM w.r.t. the following:

$$\xi(\nu,u) = \nu - u,$$

with $0 < \omega = \gamma \leq 1$. It can be easily seen that for each $\omega \in (0,\ 1]$, there exists a strongly preinvex-FM, and $\mathcal{H}(u)$ is neither a convex FM nor a preinvex-FM w.r.t. bifunction $\xi(\nu,u) = \nu - u$ with $0 < \omega \leq 1$.

Now, we show that the difference between a strongly preinvex-FM and a strongly affine preinvex-FM is, again, a preinvex-FM for a strongly preinvex-FM.

Theorem 4. *Let FM $f : K_\xi \to \mathbb{F}_0$ be a strongly affine preinvex w.r.t. ξ and $0 \leq \omega$. Then, $\mathcal{H}$ is strongly preinvex-FM w.r.t. the same bi-function ξ if, and only if, $\varpi = \mathcal{H} - f$ is a preinvex-FM.*

Proof. The "If" part is obvious. To prove the "only if", assume that $f : K_\xi \to \mathbb{F}_0$ is a strongly fuzzy affine preinvex w.r.t. the non-negative bi-function ξ and $0 \leq \omega$. Then, the following holds:

$$f(u+\tau\xi(\nu,u)) = (1-\tau)f(u)\tilde{+}\tau f(\nu)\tilde{-}\omega\tau(1-\tau)\|\xi(\nu,u)\|^2 \tag{15}$$

Therefore, for each $\gamma \in [0,\ 1]$, we have the following:

$$f_*(u+\tau\xi(\nu,u),\gamma) = (1-\tau)f_*(u,\gamma)+\tau f_*(\nu,\gamma)-\omega\tau(1-\tau)\|\xi(\nu,u)\|^2,$$
$$f^*(u+\tau\xi(\nu,u),\gamma) = (1-\tau)f^*(u,\gamma)+\tau f^*(\nu,\gamma)-\omega\tau(1-\tau)\|\xi(\nu,u)\|^2.$$

Since $\mathcal{H}$ is strongly preinvex-FM w.r.t. the same bi function ξ, then, for each $\gamma \in [0,\ 1]$, we have the following:

$$\begin{aligned} \mathcal{H}_*(u+\tau\xi(\nu,u),\gamma) &\leq (1-\tau)\mathcal{H}_*(u,\gamma)+\tau\mathcal{H}_*(\nu,\gamma)-\omega\tau(1-\tau)\|\xi(\nu,u)\|^2, \\ \mathcal{H}^*(u+\tau\xi(\nu,u),\gamma) &\leq (1-\tau)\mathcal{H}^*(u,\gamma)+\tau\mathcal{H}^*(\nu,\gamma)-\omega\tau(1-\tau)\|\xi(\nu,u)\|^2. \end{aligned} \tag{16}$$

From (15) and (16), we have the following:

$$\begin{aligned} \mathcal{H}_*(u+\tau\xi(\nu,u),\gamma) - f_*(u+\tau\xi(\nu,u),\gamma) &\leq (1-\tau)\mathcal{H}_*(u,\gamma)+\tau\mathcal{H}_*(\nu,\gamma) \\ &\quad -(1-\tau)f_*(u,\gamma)-\tau f_*(\nu,\gamma), \\ \mathcal{H}^*(u+\tau\xi(\nu,u),\gamma) - f^*(u+\tau\xi(\nu,u),\gamma) &\leq (1-\tau)\mathcal{H}^*(u,\gamma)+\tau\mathcal{H}^*(\nu,\gamma) \\ &\quad -(1-\tau)f^*(u,\gamma)-\tau f^*(\nu,\gamma), \end{aligned}$$

$$\begin{aligned} \mathcal{H}_*(u+\tau\xi(\nu,u),\gamma) - f_*(u+\tau\xi(\nu,u),\gamma) &\leq (1-\tau)(\mathcal{H}_*(u,\gamma)-f_*(u,\gamma)) \\ &\quad +\tau(\mathcal{H}_*(\nu,\gamma)-f_*(\nu,\gamma)), \\ \mathcal{H}^*(u+\tau\xi(\nu,u),\gamma) - f^*(u+\tau\xi(\nu,u),\gamma) &\leq (1-\tau)(\mathcal{H}^*(u,\gamma)-f^*(u,\gamma)) \\ &\quad +\tau(\mathcal{H}^*(\nu,\gamma)-f^*(\nu,\gamma)), \end{aligned}$$

from which it follows that

$$\begin{aligned} \varpi_*(u+\tau\xi(v,u),\gamma) &= \mathcal{H}_*(u+\tau\xi(v,u),\gamma) - f_*(u+\tau\xi(v,u),\gamma), \\ \varpi^*(u+\tau\xi(v,u),\gamma) &= \mathcal{H}^*(u+\tau\xi(v,u),\gamma) - f^*(u+\tau\xi(v,u),\gamma), \\ \varpi_*(u+\tau\xi(v,u),\gamma) &\le (1-\tau)\varpi_*(u,\gamma) + \tau\varpi_*(v,\gamma), \\ \varpi_*(u+\tau\xi(v,u),\gamma) &\le (1-\tau)\varpi^*(u,\gamma) + \tau\varpi^*(v,\gamma), \end{aligned}$$

that is

$$\varpi(u+\tau\xi(v,u)) \preccurlyeq (1-\tau)\varpi(u)\tilde{+}\tau\varpi(u),$$

showing that $\varpi = \mathcal{H} - f$ is preinvex-FM. □

We know that under certain condition invex-FMs, we obtain a solution of the fuzzy optimization problem because with the help of these FMs, we can obtain the relationship between the fuzzy variational inequalities and optimization problems.

Definition 7. *The G-differentiable FM $\mathcal{H} : K_\xi \to \mathbb{F}_0$ on K_ξ is said to be strongly invex-FM w.r.t. bi-function ξ if there exist a constant $0 \le \omega$ such that the following holds:*

$$\mathcal{H}(v)\tilde{-}\mathcal{H}(u) \succcurlyeq F'(u),\ \xi(v,u)\tilde{+}\omega\|\xi(v,u)\|^2,\ for\ all\ u,v \in K_\xi. \tag{17}$$

Example 2. *We consider the FMs $\mathcal{H} : (0,\ 1) \to \mathbb{F}_0$ defined by, $\mathcal{H}_\gamma(u) = \left[2\gamma u^2, (4-2\gamma)u^2\right]$, as in Example 1; then, $\mathcal{H}(u)$ is strongly invex-FM w.r.t. bifunction $\xi(v,u) = v - u$, with $0 < \omega = \gamma \le 1$, where $u \le v$. We have $\mathcal{H}_*(u,\ \gamma) = \gamma u^2$ and $\mathcal{H}^*(u,\ \gamma) = (2-\gamma)u^2$. Now, we compute the following:*

$$\mathcal{H}_*(v,\ \gamma) - \mathcal{H}_*(u,\ \gamma) = \gamma v^2 - \gamma u^2,$$

while

$$\langle \mathcal{H}_*{}'(u,\gamma),\ \xi(v,u)\rangle + \omega\|\xi(v,u)\|^2 = 2\gamma(v-u) + \omega\|v-u\|^2.$$

and $\gamma v^2 - \gamma u^2 \ge 2\gamma(v-u) + \omega v - u^2$, with $0 < \omega \le 1$, where $u \le v$.
Similarly, it can be easily shown that

$$\mathcal{H}^*(v,\ \gamma) - \mathcal{H}^*(u,\ \gamma) \ge \langle \mathcal{H}^{*\prime}(u,\gamma),\ \xi(v,u)\rangle + \omega\|\xi(v,u)\|^2$$

Hence, $\mathcal{H}(u)$ is strongly invex-FM w.r.t. bifunction $\xi(v,u) = v-u$, with $0 < \omega \le 1$. It can be easily seen that $\mathcal{H}(u)$ is not invex-FM w.r.t. bifunction $\xi(v,u) = v - u$.

Definition 8. *The G-differentiable FM $\mathcal{H} : K_\xi \to \mathbb{F}_0$ on K_ξ is said to be strongly pseudo invex-FM w.r.t. bi-function ξ if there exists a constant $0 \le \omega$ such that the following holds:*

$$\langle \mathcal{H}'(u),\ \xi(v,u)\rangle\tilde{+}\omega\|\xi(v,u)\|^2 \succcurlyeq \tilde{0}\ \Rightarrow\ \mathcal{H}(v)\tilde{-}\mathcal{H}(u) \succcurlyeq \tilde{0},\ for\ all\ u,v \in K_\xi. \tag{18}$$

If $\omega = 0$, then from Definition 7 and Definition 8, we obtain the classical definitions of invex-FM and pseudo invex-FM, respectively. If $\xi(v,u) = v - u$, then Definition 7 and Definition 8 reduce to known ones.

Example 3. *We consider the FMs $\mathcal{H} : (0,\ \infty) \to \mathbb{F}_0$ defined by, $\mathcal{H}_\gamma(u) = [\gamma u, (3-2\gamma)u]$, then $\mathcal{H}(u)$ is strongly pseudo invex-FM w.r.t. bifunction $\xi(v,u) = v - u$, with $0 \le \omega = \gamma$, where $u \le v$. We have $\mathcal{H}_*(u,\ \gamma) = \gamma u$ and $\mathcal{H}^*(u,\ \gamma) = (3-2\gamma)u$. Now we compute the following:*

$$\langle \mathcal{H}_*{}'(u,\gamma),\ \xi(v,u)\rangle + \omega\|\xi(v,u)\|^2 = \gamma(v-u) + \omega\|v-u\|^2 \ge 0,$$

for all $u, v \in K_\xi$ and $\gamma \in [0,\ 1]$ with $u \le v$, $0 \le \omega$; which implies the following:

$$\begin{aligned} \mathcal{H}_*(v,\ \gamma) = \gamma v &\ge \gamma u = \mathcal{H}_*(u,\ \gamma), \\ \mathcal{H}_*(v,\ \gamma) &\ge \mathcal{H}_*(u,\ \gamma), \end{aligned}$$

Similarly, it can be easily shown that the following holds:

$$\langle \mathcal{H}_*'(u,\gamma),\ \xi(v,u)\rangle + \omega\|\xi(v,u)\|^2 = (3-2\gamma)(v-u) + \omega\|v-u\|^2 \geq 0,$$

for all $u, v \in K_\xi$ *and* $\gamma \in [0,\ 1]$ *with* $u \leq v, 0 \leq \omega$. *This means that the following holds:*

$$\mathcal{H}^*(v,\ \gamma) = (3-2\gamma)v \geq \gamma u = \mathcal{H}^*(u,\ \gamma),$$

from which, it follows that

$$\mathcal{H}^*(v,\ \gamma) \geq \mathcal{H}^*(u,\ \gamma)$$

Hence, the FM $\mathcal{H}_\gamma(u) = [\gamma u, (3-2\gamma)u]$ *is strongly pseudo invex-FM w.r.t.* $\xi(v,u) = v - u$, *with* $0 \leq \omega$, *where* $u \leq v$. *It can be easily seen that* $\mathcal{H}(u)$ *is not a pseudo invex-FM w.r.t.* ξ.

Theorem 5. *Let* $\mathcal{H} : K_\xi \to \mathbb{F}_0$ *be a G-differentiable and strongly preinvex-FM then* $\mathcal{H}$ *is a strongly invex-FM.*

Proof. Let $\mathcal{H} : K_\xi \to \mathbb{F}_0$ be G-differentiable strongly preinvex-FM. Since $\mathcal{H}$ is strongly preinvex, then for each $u, v \in K_\xi$ and $\tau \in [0,1]$, we have the following:

$$\begin{aligned}\mathcal{H}(u+\tau\xi(v,u)) &\preccurlyeq (1-\tau)\mathcal{H}(u)\tilde{+}\tau\mathcal{H}(v)\tilde{-}\omega\tau(1-\tau)\|\xi(v,u)\|^2,\\ &\preccurlyeq \mathcal{H}(u)\tilde{+}\tau(\mathcal{H}(v)\tilde{-}\mathcal{H}(u))\tilde{-}\omega\tau(1-\tau)\|\xi(v,u)\|^2,\end{aligned}$$

Therefore, for every $\gamma \in [0,1]$, we have the following:

$$\begin{aligned}\mathcal{H}_*(u+\tau\xi(v,u),\gamma) &\leq \mathcal{H}_*(u,\gamma) + \tau(\mathcal{H}_*(v,\gamma) - \mathcal{H}_*(u,\gamma)) - \omega\tau(1-\tau)\|\xi(v,u)\|^2,\\ \mathcal{H}^*(u+\tau\xi(v,u),\gamma) &\leq \mathcal{H}_*(u,\gamma) + \tau(\mathcal{H}^*(v,\gamma) - \mathcal{H}^*(u,\gamma)) - \omega\tau(1-\tau)\|\xi(v,u)\|^2,\end{aligned}$$

which implies that the following:

$$\begin{aligned}\tau(\mathcal{H}_*(v,\gamma) - \mathcal{H}_*(u,\gamma)) &\geq \mathcal{H}_*(u+\tau\xi(v,u),\gamma) - \mathcal{H}_*(u,\gamma) + \omega\tau(1-\tau)\|\xi(v,u)\|^2,\\ \tau(\mathcal{H}^*(v,\gamma) - \mathcal{H}^*(u,\gamma)) &\geq \mathcal{H}^*(u+\tau\xi(v,u),\gamma) - \mathcal{H}^*(u,\gamma) + \omega\tau(1-\tau)\|\xi(v,u)\|^2,\\ \mathcal{H}_*(v,\gamma) - \mathcal{H}_*(u,\gamma) &\geq \tfrac{\mathcal{H}_*(u+\tau\xi(v,u),\gamma) - \mathcal{H}_*(u,\gamma)}{\tau} + \omega(1-\tau)\|\xi(v,u)\|^2,\\ \mathcal{H}^*(v,\gamma) - \mathcal{H}^*(u,\gamma) &\geq \tfrac{\mathcal{H}^*(u+\tau\xi(v,u),\gamma) - \mathcal{H}^*(u,\gamma)}{\tau} + \omega(1-\tau)\|\xi(v,u)\|^2.\end{aligned}$$

Taking the limit in the above inequality as $\tau \to 0$, we have the following:

$$\begin{aligned}\mathcal{H}_*(v,\gamma) - \mathcal{H}_*(u,\gamma) &\geq \langle \mathcal{H}_*'(u,\gamma),\ \xi(v,u)\rangle + \omega\|\xi(v,u)\|^2,\\ \mathcal{H}^*(v,\gamma) - \mathcal{H}^*(u,\gamma) &\geq \langle \mathcal{H}^{*\prime}(u,\gamma),\ \xi(v,u)\rangle + \omega\|\xi(v,u)\|^2,\end{aligned}$$

that is,

$$\mathcal{H}(v)\tilde{-}\mathcal{H}(u) \succcurlyeq \langle \mathcal{H}'(u),\ \xi(v,u)\rangle\tilde{+}\omega\|\xi(v,u)\|^2.$$

As a special case of Theorem 5, when $\omega = 0$, we have the following. □

Corollary 1. *Let* $\mathcal{H} : K_\xi \to \mathbb{F}_0$ *be a G-differentiable preinvex-FM on* K_ξ *[32]. Then,* $\mathcal{H}$ *is an invex-FM.*

It is well known that the differentiable preinvex functions are invex functions, but the converse is not true. However, Mohan and Neogy [13] showed that the preinvex functions and invex functions are equivalent under Condition C. Similarly, the converse of Theorem 5 is not valid; the natural question is how to obtain a strongly preinvex-FM from strongly invex-FM. To prove the converse, we need the following assumption regarding the bi-function ξ, which plays an important role in G-differentiation of the main results.

Condition C.
$$\xi(\nu, u + \tau\xi(\nu, u)) = (1-\tau)\xi(\nu, u),$$
$$\xi(u, u + \tau\xi(\nu, u)) = -\tau\xi(\nu, u).$$

Clearly for $\tau = 0$, we have $\xi(\nu, u) = 0$ if, and only if, $\nu = u$ for all $u,\ \nu \in K_{\xi}$. Additionally, note that from Condition C, we have the following:

$$\xi(u + \tau_2\xi(\nu, u), u + \tau_1\xi(\nu, u)) = (\tau_2 - \tau_1)\xi(\nu, u)$$

For the application of Condition C, see [13–17].

The following Theorem 6 gives the result of the converse of Theorem 5.

Theorem 6. *Let $\mathcal{H} : K_{\xi} \to \mathbb{F}_0$ be a G-differentiable FM on K_{ξ} . Let Condition C holds and $\mathcal{H}(u)$ satisfies the following condition:*

$$\mathcal{H}(u + \tau\xi(\nu, u)) \preccurlyeq \mathcal{H}(\nu), \tag{19}$$

and then, the following are equivalent:
(a) $\mathcal{H}$ is strongly preinvex-FM.

$$\text{(b) } \mathcal{H}(\nu)\tilde{-}\mathcal{H}(u) \succcurlyeq \mathcal{H}'(u),\ \xi(\nu, u)\tilde{+}\omega\|\xi(\nu, u)\|^2, \text{ for all } u,\ \nu \in K_{\xi}, \tag{20}$$

$$\langle\text{(c) } \mathcal{H}'(u),\ \xi(\nu, u)\rangle\left\langle\tilde{+}\mathcal{H}'(\nu),\ \xi(u, \nu)\right\rangle \preccurlyeq \tilde{-}\omega\left\{\|\xi(\nu, u)\|^2 + \|\xi(u, \nu)\|^2\right\} \tag{21}$$

for all $u,\ \nu \in K_{\xi}$.

Proof. (a) implies (b) □

The demonstration is analogous to the demonstration of Theorem 5.

(b) implies (c). Let (b) hold. Then, for every$\gamma \in [0,1]$, we have the following:

$$\begin{aligned}\mathcal{H}_*(\nu, \gamma) - \mathcal{H}_*(u, \gamma) &\geq \langle \mathcal{H}_*{}'(u, \gamma),\ \xi(\nu, u)\rangle + \omega\|\xi(\nu, u)\|^2, \\ \mathcal{H}^*(\nu, \gamma) - \mathcal{H}^*(u, \gamma) &\geq \langle \mathcal{H}^{*\prime}(u, \gamma),\ \xi(\nu, u)\rangle + \omega\|\xi(\nu, u)\|^2,\end{aligned} \tag{22}$$

Then, by replacing ν by u and u by ν in (22), we obtain the following:

$$\begin{aligned}\mathcal{H}_*(u, \gamma) - \mathcal{H}_*(\nu, \gamma) &\geq \langle \mathcal{H}_*{}'(\nu, \gamma),\ \xi(u, \nu)\rangle + \omega\|\xi(u, \nu)\|^2, \\ \mathcal{H}^*(u, \gamma) - \mathcal{H}^*(\nu, \gamma) &\geq \langle \mathcal{H}^{*\prime}(\nu, \gamma),\ \xi(u, \nu)\rangle + \omega\|\xi(u, \nu)\|^2.\end{aligned} \tag{23}$$

Adding (22) and (23), we have the following:

$$\begin{aligned}\langle \mathcal{H}_*{}'(u, \gamma),\ \xi(\nu, u)\rangle + \langle \mathcal{H}_*{}'(\nu, \gamma),\ \xi(u, \nu)\rangle &\leq -\omega\left(\|\xi(\nu, u)\|^2 + \|\xi(u, \nu)\|^2\right), \\ \langle \mathcal{H}^{*\prime}(u, \gamma),\ \xi(\nu, u)\rangle + \langle \mathcal{H}^{*\prime}(\nu, \gamma),\ \xi(u, \nu)\rangle &\leq -\omega\left(\|\xi(\nu, u)\|^2 + \|\xi(u, \nu)\|^2\right),\end{aligned}$$

That is, the following:

$$\langle \mathcal{H}'(u),\ \xi(\nu, u)\rangle\tilde{+}\langle \mathcal{H}'(\nu),\ \xi(u, \nu)\rangle \preccurlyeq \tilde{-}\omega\left\{\|\xi(\nu, u)\|^2 + \|\xi(u, \nu)\|^2\right\}$$

(c) implies (b). Assume that (21) holds. Then, for every $\gamma \in [0,1]$, we have the following:

$$\begin{aligned}\langle \mathcal{H}_*{}'(\nu, \gamma),\ \xi(u, \nu)\rangle &\leq -\langle \mathcal{H}_{*'}(u, \gamma),\ \xi(\nu, u)\rangle - \omega\left(\|\xi(\nu, u)\|^2 + \|\xi(u, \nu)\|^2\right), \\ \langle \mathcal{H}^{*\prime}(\nu,\ \gamma),\ \xi(u, \nu)\rangle &\leq -\langle \mathcal{H}^{*\prime}(u,\ \gamma),\ \xi(\nu, u)\rangle - \omega\left(\|\xi(\nu, u)\|^2 + \|\xi(u, \nu)\|^2\right).\end{aligned} \tag{24}$$

Since, $\nu_\tau = u + \tau\xi(\nu, u) \in K_\xi$ for all $u, \nu \in K_\xi$ and $\tau \in [0, 1]$. Taking $\nu = \nu_\tau$ in (24), we obtain the following:

$$\begin{aligned}\langle \mathcal{H}_*'(u+\tau\xi(\nu,u),\ \gamma),\ \xi(u,u+\tau\xi(\nu,u))\rangle &\le -\langle \mathcal{H}_*'(u,\gamma),\ \xi(u+\tau\xi(\nu,u),u)\rangle \\ &\quad -\omega\Big(\|\xi(u+\tau\xi(\nu,u),u)\|^2 + \|\xi(u,u+\tau\xi(\nu,u))\|^2\Big), \\ \langle \mathcal{H}^{*\prime}(u+\tau\xi(\nu,u),\gamma),\ \xi(u,u+\tau\xi(\nu,u))\rangle &\le -\langle \mathcal{H}^{*\prime}(u,\gamma),\ \xi(u+\tau\xi(\nu,u),u)\rangle \\ &\quad -\omega\Big(\|\xi(u+\tau\xi(\nu,u),u)\|^2 + \|\xi(u,u+\tau\xi(\nu,u))\|^2\Big),\end{aligned}$$

by using Condition C, we have the following:

$$\begin{aligned}\langle \mathcal{H}_*'(u+\tau\xi(\nu,u),\gamma),\tau\xi(\nu,u)\rangle &\ge \langle \mathcal{H}_*'(u,\gamma),\tau\xi(\nu,u)\rangle + 2\omega\tau^2\|\xi(\nu,u)\|^2, \\ \langle \mathcal{H}^{*\prime}(u+\tau\xi(\nu,u),\gamma),\tau\xi(\nu,u)\rangle &\ge \langle \mathcal{H}^{*\prime}(u,\gamma),\tau\xi(\nu,u)\rangle + 2\omega\tau^2\|\xi(\nu,u)\|^2,\end{aligned}$$

$$\begin{aligned}\langle \mathcal{H}_*'(u+\tau\xi(\nu,u),\gamma),\xi(\nu,u)\rangle &\ge \langle \mathcal{H}_*'(u,\gamma),\xi(\nu,u)\rangle + 2\omega\tau\|\xi(\nu,u)\|^2, \\ \langle \mathcal{H}^{*\prime}(u+\tau\xi(\nu,u),\gamma),\xi(\nu,u)\rangle &\ge \langle \mathcal{H}^{*\prime}(u,\gamma),\xi(\nu,u)\rangle + 2\omega\tau\|\xi(\nu,u)\|^2,\end{aligned} \tag{25}$$

Let the following hold:

$$\begin{aligned}H_*(\tau) &= \mathcal{H}_*(u+\tau\xi(\nu,u),\gamma), \\ H^*(\tau) &= \mathcal{H}^*(u+\tau\xi(\nu,u),\gamma).\end{aligned}$$

Taking the derivative w.r.t. τ, we obtain the following:

$$\begin{aligned}H_*'(\tau) &= \mathcal{H}_*'(u+\tau\xi(\nu,u),\gamma).\xi(\nu,u) = \langle \mathcal{H}_*'(u+\tau\xi(\nu,u),\gamma),\xi(\nu,u)\rangle, \\ H^{*\prime}(\tau) &= \mathcal{H}^{*\prime}(u+\tau\xi(\nu,u),\gamma).\xi(\nu,u) = \langle \mathcal{H}^{*\prime}(u+\tau\xi(\nu,u),\gamma),\xi(\nu,u)\rangle,\end{aligned}$$

from which, using (25), we have the following:

$$\begin{aligned}H_*'(\tau) &\ge \langle \mathcal{H}_*'(u,\gamma),\xi(\nu,u)\rangle + 2\omega\tau\|\xi(\nu,u)\|^2, \\ H^{*\prime}(\tau) &\ge \langle \mathcal{H}^{*\prime}(u,\gamma),\xi(\nu,u)\rangle + 2\omega\tau\|\xi(\nu,u)\|^2.\end{aligned} \tag{26}$$

By integrating (26) between 0 to 1, w.r.t. τ, we obtain the following:

$$\begin{aligned}H_*(1) - H_*(0) &\ge \langle \mathcal{H}_*'(u,\gamma),\xi(\nu,u)\rangle + \omega\|\xi(\nu,u)\|^2, \\ H^*(1) - H^*(0) &\ge \langle \mathcal{H}^{*\prime}(u,\gamma),\xi(\nu,u)\rangle + \omega\|\xi(\nu,u)\|^2.\end{aligned}$$

$$\begin{aligned}\mathcal{H}_*(u+\xi(\nu,u),\gamma) - \mathcal{H}_*(u,\gamma) &\ge \langle \mathcal{H}_*'(u,\gamma),\xi(\nu,u)\rangle + \omega\|\xi(\nu,u)\|^2, \\ \mathcal{H}^*(u+\xi(\nu,u),\gamma) - \mathcal{H}^*(u,\gamma) &\ge \langle \mathcal{H}^{*\prime}(u,\gamma),\xi(\nu,u)\rangle + \omega\|\xi(\nu,u)\|^2.\end{aligned}$$

Using (19), we have the following:

$$\begin{aligned}\mathcal{H}_*(\nu,\gamma) - \mathcal{H}_*(u,\gamma) &\ge \langle \mathcal{H}_*'(u,\gamma),\xi(\nu,u)\rangle + \omega\|\xi(\nu,u)\|^2, \\ \mathcal{H}^*(\nu,\gamma) - \mathcal{H}^*(u,\gamma) &\ge \langle \mathcal{H}^{*\prime}(u,\gamma),\xi(\nu,u)\rangle + \omega\|\xi(\nu,u)\|^2,\end{aligned}$$

that is, the following:

$$\mathcal{H}(\nu)\tilde{-}\mathcal{H}(u) \succcurlyeq \langle \mathcal{H}'(u),\tau\xi(\nu,u)\rangle\tilde{+}\omega\|\xi(\nu,u)\|^2, \text{ for all } u,\ \nu \in K_\xi.$$

(b) implies (a). Assume that (20) holds. Since K_ξ, $\nu_\tau = u + \tau\xi(\nu,u) \in K_\xi$ for all $u, \nu \in K_\xi$ and $\tau \in [0, 1]$. Taking $\nu = \nu_\tau$ in (20), we obtain the following:

$$\mathcal{H}(u+\tau\xi(\nu,u))\tilde{-}\mathcal{H}(u) \succcurlyeq \langle \mathcal{H}'(u),\ \xi(u+\tau\xi(\nu,u),u)\rangle\tilde{+}\omega\|\xi(u+\tau\xi(\nu,u),u)\|^2.$$

Therefore, for every $\gamma \in [0,\ 1]$, we have the following:

$$\begin{aligned}\mathcal{H}_*(u+\tau\xi(\nu,u),\gamma) - \mathcal{H}_*(u,\gamma) &\ge \langle \mathcal{H}_*'(u,\gamma),\ \xi(u+\tau\xi(\nu,u),u)\rangle + \omega\|\xi(u+\tau\xi(\nu,u),u)\|^2, \\ \mathcal{H}^*(u+\tau\xi(\nu,u),\ \gamma) - \mathcal{H}^*(u,\gamma) &\ge \langle \mathcal{H}^{*\prime}(u,\gamma),\ \xi(u+\tau\xi(\nu,u),u)\rangle + \omega\|\xi(u+\tau\xi(\nu,u),u)\|^2.\end{aligned}$$

Using Condition C, we have the following:

$$\begin{aligned}\mathcal{H}_*(u+\tau\xi(v,u),\gamma)-\mathcal{H}_*(u,\gamma)\geq(1-\tau)\langle\mathcal{H}_*{}'(u,\gamma),\ \xi(v,u)\rangle+\omega(1-\tau)^2\|\xi(v,u)\|^2,\\ \mathcal{H}^*(u+\tau\xi(v,u),\gamma)-\mathcal{H}^*(u,\gamma)\geq(1-\tau)\langle\mathcal{H}^{*\prime}(u,\gamma),\ \xi(v,u)\rangle+\omega(1-\tau)^2\|\xi(v,u)\|^2.\end{aligned} \tag{27}$$

In a similar way, we have the following:

$$\begin{aligned}\mathcal{H}_*(u,\gamma)-\mathcal{H}_*(u+\tau\xi(v,u),\gamma)\geq-\tau\langle\mathcal{H}_*{}'(u,\gamma),\xi(v,u)\rangle+\omega\tau^2\|\xi(v,u)\|^2,\\ \mathcal{H}^*(u,\gamma)-\mathcal{H}^*(u+\tau\xi(v,u),\gamma)\geq-\tau\langle\mathcal{H}^{*\prime}(u,\gamma),\xi(v,u)\rangle+\omega\tau^2\|\xi(v,u)\|^2.\end{aligned} \tag{28}$$

Multiplying (27) by τ and (28) by $(1-\tau)$, and adding the resultant, we have the following:

$$\begin{aligned}\mathcal{H}_*(u+\tau\xi(v,u),\gamma)\leq(1-\tau)\mathcal{H}_*(u,\gamma)+\tau\mathcal{H}_*(v,\gamma)-\omega\tau(1-\tau)\|\xi(v,u)\|^2,\\ \mathcal{H}^*(u+\tau\xi(v,u),\gamma)\leq(1-\tau)\mathcal{H}^*(u,\gamma)+\tau\mathcal{H}^*(v,\gamma)-\omega\tau(1-\tau)\|\xi(v,u)\|^2,\end{aligned}$$

That is, the following holds:

$$\mathcal{H}(u+\tau\xi(v,u))\preccurlyeq(1-\tau)\mathcal{H}(u)\tilde{+}\tau\mathcal{H}(v)\tilde{-}\omega\tau(1-\tau)\|\xi(v,u)\|^2.$$

Hence, $\mathcal{H}$ is strongly preinvex-FM w.r.t.
Theorems 5 and 6, enable us to define the followings new definitions.

Definition 9. *A G-differentiable FM $\mathcal{H}:K_\xi\to\mathbb{F}_0$ is said to be as follows:*

(i) *Strongly monotone w.r.t. bi-function ξ if, and only if, there exists a constant $0\leq\omega$ such that the following is true:*

$$\langle\mathcal{H}'(u),\ \xi(v,u)\rangle\tilde{+}\langle\mathcal{H}'(v),\ \xi(u,v)\rangle\preccurlyeq\tilde{-}\omega\left\{\|\xi(v,u)\|^2+\|\xi(u,v)\|^2\right\},\textit{for all } u,\ v\in K_\xi$$

(ii) *Strongly pseudo monotone w.r.t. bi-function ξ if, and only if, there exists a constant $0\leq\omega$ such that the following is true:*

$$\langle\mathcal{H}'(u),\ \xi(v,u)\rangle\tilde{+}\omega\|\xi(v,u)\|^2\succcurlyeq\tilde{0}\ \Rightarrow\ \tilde{-}\langle\mathcal{H}'(v),\ \xi(u,v)\rangle\succcurlyeq\tilde{0},\ \textit{for all } u,\ v\in K_\xi.$$

If $\xi(v,u)=-\xi(u,v)$, then Definition 9. reduces to new one.

Example 4. *We consider the FMs $\mathcal{H}:(0,\ \infty)\to\mathbb{F}_0$ defined by the following:*

$$\mathcal{H}(u)(\sigma)=\begin{cases}\frac{\sigma}{2u^2}\ \sigma\in\left[0,\ 2u^2\right]\\ \frac{5u^2-\sigma}{3u^2}\ \sigma\in\left(2u^2,\ 5u^2\right]\\ 0\ \textit{otherwise}.\end{cases}$$

Then, for each $\gamma\in[0,\ 1]$, we have $\mathcal{H}_\gamma(u)=\left[2\gamma u^2,(5-3\gamma)u^2\right]$, where $\mathcal{H}(u)$ is strongly fuzzy pseudomonotone w.r.t. bifunction $\xi(v,u)=u-v$, with $1\leq\omega$, where $v\leq u$. We have $\mathcal{H}_*(u,\ \gamma)=2\gamma u^2$ and $\mathcal{H}^*(u,\ \gamma)=(5-3\gamma)u^2$. Now, we compute the following:

$$\langle\mathcal{H}_*{}'(u,\gamma),\ \xi(v,u)\rangle+\omega\|\xi(v,u)\|^2=4\gamma u(u-v)+\omega u-v^2\geq0,$$

for all $u,v\in K_\xi$ and $\gamma\in[0,\ 1]$ with $v\leq u$, $1\leq\omega$; which implies he following:

$$\begin{aligned}-\langle\mathcal{H}_*{}'(v,\gamma),\ \xi(u,v)=-4\gamma u(v-u)\rangle=4\gamma v(u-v)\geq0,\ \forall\, u,\ v\in K_\xi,\\ -\langle\mathcal{H}^{*\prime}(v,\gamma),\ \xi(u,v)\rangle\geq0.\end{aligned}$$

Similarly, it can be easily shown that the following holds:

$$\langle\mathcal{H}^{*\prime}(u,\gamma),\ \xi(v,u)\rangle+\omega\|\xi(v,u)\|^2=2(5-3\gamma)u(u-v)+\omega\|u-v\|^2\geq0,$$

for all $u, v \in K_\xi$ *and* $\gamma \in [0, 1]$ *with* $v \leq u, 1 \leq \omega$. *This means the following is true:*

$$-\langle \mathcal{H}^{*\prime}(v, \gamma),\ \xi(u, v)\rangle = -2(5-3\gamma)u(v-u) = 2(5-3\gamma)v(u-v) \geq 0,\ \forall\ u,\ v \in K_\xi,$$

From which, it follows that

$$-\langle \mathcal{H}^{*\prime}(v, \gamma),\ \xi(u, v)\rangle \geq 0.$$

Hence, the G-differentiable FM $\mathcal{H}_\gamma(u) = [\gamma u, (5-4\gamma)u]$ *is strongly fuzzy pseudomonotone w.r.t.* $\xi(v, u) = u - v$, *with* $1 \leq \omega$, *where* $v \leq u$. *It can be easily noted that* $\mathcal{H}'(u)$ *is neither fuzzy pseudomonotone nor fuzzy quasimonotone w.r.t.* ξ.

If $\omega = 0$, *then from Theorem 6, we obtain following result.*

Corollary 2. *[36] Let* $\mathcal{H} : K_\xi \to \mathbb{F}_0$ *be a G-differentiable FM on* K_ξ. *Let Condition C holds and* $\mathcal{H}(u)$ *satisfies the following condition:*

$$\mathcal{H}(u + \tau\xi(v, u)) \preccurlyeq \mathcal{H}(v),$$

and then, the following are equivalent:

(a) $\mathcal{H}$ *is invex-FM.*
(b) $\mathcal{H}'$ *is monotone.*

Theorem 7. *Let* $\mathcal{H} : K_\xi \to \mathbb{F}_0$ *be FM on* K_ξ *w.r.t.* ξ *and Condition C hold. Let* $\mathcal{H}(u)$ *is G-differentiable on* K_ξ *with the following conditions:*

(a) $\mathcal{H}(u + \tau\xi(v, u)) \preccurlyeq \mathcal{H}(v)$.
(b) $\mathcal{H}'(u)$ is a strongly fuzzy pseudomonotone.

Then, $\mathcal{H}$ is a strongly pseudo invex-FM.

Proof. Let $\mathcal{H}'$ be strongly pseudomonotone. Then, for all $u,\ v \in K_\xi$, we have the following:

$$\langle \mathcal{H}'(u),\ \xi(v, u)\rangle \widetilde{+} \omega \|\xi(v, u)\|^2 \succcurlyeq \widetilde{0}.$$

Therefore, for every $\gamma \in [0,\ 1]$, we have the following:

$$\begin{aligned} &\langle \mathcal{H}_*{}'(u, \gamma), \xi(v, u)\rangle + \omega\|\xi(v, u)\|^2 \geq 0, \\ &\langle \mathcal{H}^{*\prime}(u, \gamma), \xi(v, u)\rangle + \omega\|\xi(v, u)\|^2 \geq 0, \end{aligned}$$

which implies that the following is true:

$$\begin{aligned} &-\langle \mathcal{H}_*{}'(v, \gamma),\ \xi(u, v)\rangle \geq 0, \\ &-\langle \mathcal{H}^{*\prime}(v, \gamma),\ \xi(u, v)\rangle \geq 0. \end{aligned} \tag{29}$$

Since $v_\tau = u + \tau\xi(v, u) \in K_\xi$ for all $u, v \in K_\xi$ and $\tau \in [0, 1\,]$. Taking $v = v_\tau$ in (29), we obtain the following:

$$\begin{aligned} &-\langle \mathcal{H}_*{}'(u + \tau\xi(v, u), \gamma),\ \xi(u, u + \tau\xi(v, u))\rangle \geq 0, \\ &-\langle \mathcal{H}^{*\prime}(u + \tau\xi(v, u), \gamma),\ \xi(u, u + \tau\xi(v, u))\rangle \geq 0. \end{aligned}$$

By using Condition C, we have the following:

$$\begin{aligned} &\langle \mathcal{H}_*{}'(u + \tau\xi(v, u), \gamma),\ \xi(v, u)\rangle \geq 0, \\ &\langle \mathcal{H}^{*\prime}(u + \tau\xi(v, u), \gamma),\ \xi(v, u)\rangle \geq 0. \end{aligned} \tag{30}$$

Assume the following:

$$H_*(\tau) = \mathcal{H}_*(u + \tau\xi(\nu, u), \gamma),$$
$$H^*(\tau) = \mathcal{H}^*(u + \tau\xi(\nu, u), \gamma),$$

taking G-derivative w.r.t. τ, then using (30), we have the following:

$$\begin{aligned} H_*'(\tau) &= \langle \mathcal{H}_*'(u + \tau\xi(\nu, u), \gamma),\ \xi(\nu, u)\rangle \geq 0, \\ H^{*\prime}(\tau) &= \langle \mathcal{H}^{*\prime}(u + \tau\xi(\nu, u), \gamma),\ \xi(\nu, u)\rangle \geq 0, \end{aligned} \tag{31}$$

Integrating (31) between 0 to 1 w.r.t. τ, we obtain the following:

$$H_*(1) - H_*(0) \geq 0,$$
$$H^*(1) - H^*(0) \geq 0,$$

which implies the following:

$$\mathcal{H}_*(u + \xi(\nu, u), \gamma) - \mathcal{H}_*(u, \gamma) \geq 0,$$
$$\mathcal{H}^*(u + \xi(\nu, u), \gamma) - \mathcal{H}^*(u, \gamma) \geq 0.$$

From condition (i), we have the following:

$$\mathcal{H}_*(\nu, \gamma) - \mathcal{H}_*(u, \gamma) \geq 0,$$
$$\mathcal{H}^*(\nu, \gamma) - \mathcal{H}^*(u, \gamma) \geq 0,$$

that is,

$$\mathcal{H}(\nu)\tilde{-}\mathcal{H}(u) \succcurlyeq \tilde{0},\ \forall\, u, \nu \in K_\xi.$$

Hence, $\mathcal{H}$ is a strongly pseudo invex-FM.

If $\omega = 0$, then Theorem 7 reduces to the following result. □

Corollary 3. *Let $\mathcal{H} : K_\xi \to \mathbb{F}_0$ be a FM on K_ξ w.r.t. ξ and Condition C hold [36]. Let $\mathcal{H}(u)$ be G-differentiable on K_ξ with the following conditions:*

(a) $\mathcal{H}(u + \tau\xi(\nu, u)) \preccurlyeq \mathcal{H}(\nu)$.

(b) $\mathcal{H}'(.)$ *is fuzzy pseudomonotone.*

Then, $\mathcal{H}$ is a pseudo invex-FM.

The fuzzy optimality requirement for G-differentiable strongly preinvex-FMs, which is the fundamental impetus for our findings, is now discussed.

4. Fuzzy Mixed Variational-like and Integral Inequalities

The variational inequality problem has a close relationship with the optimization problem, which is a well-known fact in mathematical programming. Similarly, the fuzzy variational inequality problem and the fuzzy optimization problem have a strong link.

Consider the following unconstrained fuzzy optimization problem:

$$\min_{u \in K_\xi} \mathcal{H}(u),$$

where K_ξ is a subset of $\mathbb{R}$, $\mathcal{H} : K_\xi \to \mathbb{F}_0$ and is a FM.

A feasible point is defined, as $u \in K_\xi$ is called an optimal solution, a global optimal solution, or simply a solution to the fuzzy optimization problem if $u \in K_\xi$ and no $\nu \in K_\xi$, $\mathcal{H}(u) \preccurlyeq \mathcal{H}(\nu)$.

The fuzzy optimality criterion for G-differentiable preinvex-FMs is discussed in the following theorems, and this is the fundamental rationale for the results.

Theorem 8. *Let $\mathcal{H}$ be a G-differentiable strongly preinvex-FM modulus $0 \leq \omega$. If $u \in K_{\xi}$ is the minimum of the FM $\mathcal{H}$, then the following holds:*

$$\mathcal{H}(v)\widetilde{-}\mathcal{H}(u) \succcurlyeq \omega\|\xi(v,u)\|^2, \text{ for all } u,v \in K_{\xi}. \tag{32}$$

Proof: Let $u \in K$ be a minimum of $\mathcal{H}$. Then

$\mathcal{H}(u) \preccurlyeq \mathcal{H}(v)$, for all $v \in K_{\xi}$.

Therefore, for every $\gamma \in [0,\ 1]$, we have the following:

$$\begin{aligned} \mathcal{H}_*(u,\gamma) \leq \mathcal{H}_*(v,\gamma), \\ \mathcal{H}^*(u,\gamma) \leq \mathcal{H}^*(v,\gamma). \end{aligned} \tag{33}$$

For all $u,\ v \in K_{\xi}$, $\tau \in [0,\ 1]$, we have the following:

$$v_\tau = u + \tau\xi(v,u) \in K_{\xi}$$

Taking $v = v_\tau$ in (33), and dividing by "τ", we obtain the following:

$$\begin{aligned} 0 \leq \frac{\mathcal{H}_*(u+\tau\xi(v,u),\gamma) - \mathcal{H}_*(u,\gamma)}{\tau}, \\ 0 \leq \frac{\mathcal{H}^*(u+\tau\xi(v,u),\gamma) - \mathcal{H}^*(u,\gamma)}{\tau}. \end{aligned}$$

Taking limit in the above inequality as $\tau \to 0$, we obtain the following:

$$\begin{aligned} 0 \leq \langle \mathcal{H}_*{}'(u,\gamma),\ \xi(v,u)\rangle, \\ 0 \leq \langle \mathcal{H}^{*\prime}(u,\gamma),\ \xi(v,u)\rangle. \end{aligned} \tag{34}$$

Since $\mathcal{H} : K_{\xi} \to \mathbb{F}_0$ is a G-differentiable strongly preinvex-FM, we have the following:

$$\begin{aligned} \mathcal{H}_*(u+\tau\xi(v,u),\gamma) \leq (1-\tau)\mathcal{H}_*(u,\gamma) + \tau\mathcal{H}_*(v,\gamma) - \omega\tau(1-\tau)\|\xi(v,u)\|^2, \\ \mathcal{H}^*(u+\tau\xi(v,u),\gamma) \leq (1-\tau)\mathcal{H}^*(u,\gamma) + \tau\mathcal{H}^*(v,\gamma) - \omega\tau(1-\tau)\|\xi(v,u)\|^2, \end{aligned}$$

$$\begin{aligned} \mathcal{H}_*(v,\gamma) - \mathcal{H}_*(u,\gamma) \geq \frac{\mathcal{H}_*(u+\tau\xi(v,u),\gamma) - \mathcal{H}_*(u,\gamma)}{\tau} + \omega(1-\tau)\|\xi(v,u)\|^2, \\ \mathcal{H}^*(v,\gamma) - \mathcal{H}^*(u,\gamma) \geq \frac{\mathcal{H}^*(u+\tau\xi(v,u),\gamma) - \mathcal{H}^*(u,\gamma)}{\tau} + \omega(1-\tau)\|\xi(v,u)\|^2, \end{aligned}$$

Again, taking the limit in the above inequality as $\tau \to 0$, we obtain the following:

$$\begin{aligned} \mathcal{H}_*(v,\gamma) - \mathcal{H}_*(u,\gamma) \geq \langle \mathcal{H}_*{}'(u,\gamma),\ \xi(v,u)\rangle + \omega\|\xi(v,u)\|^2, \\ \mathcal{H}^*(v,\gamma) - \mathcal{H}^*(u,\gamma) \geq \langle \mathcal{H}^{*\prime}(u,\gamma),\ \xi(v,u)\rangle + \omega\|\xi(v,u)\|^2, \end{aligned}$$

from which, using (34), we have the following:

$$\begin{aligned} \mathcal{H}_*(v,\gamma) - \mathcal{H}_*(u,\gamma) \geq \omega\|\xi(v,u)\|^2 \geq 0, \\ \mathcal{H}^*(v,\gamma) - \mathcal{H}^*(u,\gamma) \geq \omega\|\xi(v,u)\|^2 \geq 0, \end{aligned}$$

that is,

$$\mathcal{H}(v)\widetilde{-}\mathcal{H}(u) \succcurlyeq \widetilde{0}.$$

Hence, the result follows. □

Theorem 9. *Let $\mathcal{H}$ be a G-differentiable strongly preinvex-FM modulus $0 \leq \omega$, and*

$$\langle \mathcal{H}'(u),\ \xi(v,u)\rangle \widetilde{+} \omega\|\xi(v,u)\|^2 \succcurlyeq \widetilde{0}, \text{for all } u,v \in K_{\xi}, \tag{35}$$

then $u \in K_{\xi}$ is the minimum of the FM $\mathcal{H}$.

Proof. Let $\mathcal{H} : K_{\xi} \to \mathbb{F}_0$ be a G-differentiable strongly preinvex-FM and $u \in K_{\xi}$ satisfies (35). Then, by Theorem 5, we have the following:

$$\mathcal{H}(v)\tilde{-}\mathcal{H}(u) \succcurlyeq \langle \mathcal{H}'(u),\ \xi(v,u)\rangle \tilde{+} \omega\|\xi(v,u)\|^2,$$

Therefore, for every $\gamma \in [0,\ 1]$, we have the following:

$$\mathcal{H}_*(v,\gamma) - \mathcal{H}_*(u,\gamma) \geq \langle \mathcal{H}_*{}'(u,\gamma),\ \xi(v,u)\rangle + \omega\|\xi(v,u)\|^2,$$
$$\mathcal{H}^*(v,\gamma) - \mathcal{H}^*(u,\gamma) \geq \langle \mathcal{H}^{*\prime}(u,\gamma),\ \xi(v,u)\rangle + \omega\|\xi(v,u)\|^2,$$

from which, using (35), we have the following:

$$\mathcal{H}_*(v,\gamma) - \mathcal{H}_*(u,\gamma) \geq 0,$$
$$\mathcal{H}^*(v,\gamma) - \mathcal{H}^*(u,\gamma) \geq 0,$$

that is,

$$\mathcal{H}(u) \preccurlyeq \mathcal{H}(v).$$

□

If $\omega = 0$, then Theorem 9 reduces to the following result:

Corollary 4. *Let $\mathcal{H}$ be a G-differentiable preinvex-FM w.r.t. ξ [32]. Then, $u \in K_{\xi}$ is the minimum of $\mathcal{H}$ if, and only if, $u \in K_{\xi}$ satisfies the following:*

$$\langle \mathcal{H}'(u),\ \xi(v,u)\rangle \succcurlyeq \tilde{0}, for\ all\ u,v \in K_{\xi}.$$

Remark 2. *The inequality of the type (35) is called a strongly variational-like inequality. It is very important to note that the optimality condition of preinvex-FMs cannot be obtained with the help of (35). So, this idea inspires us to introduce a more general form of a fuzzy variational-like inequality of which (35) is a special case. To be more unambiguous, for given FM Ψ, bi function $\xi(.,.)$ and a $0 \leq \omega$, consider the problem of finding $u \in K_{\xi}$, such that the following holds:*

$$\langle \Psi(u),\ \xi(v,u)\rangle \tilde{+} \omega\|\xi(v,u)\|^2 \succcurlyeq \tilde{0},\ \forall\, v \in K_{\xi}. \tag{36}$$

This inequality is called a strongly fuzzy variational-like inequality.
We look at the functional $I(v)$, which is defined as follows:

$$I(v) = \mathcal{H}(v) \tilde{+} \mathcal{J}(v),\ \forall\, v \in \mathbb{R}, \tag{37}$$

where $\mathcal{H}$ is a G-differentiable preinvex-FM and $\mathcal{J}$ is a strongly preinvex-FM, which is non-G-differentiable.

The following theorem shows that the functional $I(v)$ minimum can be distinguished by a class of variational-like inequalities.

Theorem 10. *Let $\mathcal{H} : K_{\xi} \to \mathbb{F}_0$ be a G-differentiable preinvex-FM and $\mathcal{J} : K_{\xi} \to \mathbb{F}_0$ be a non-G-differentiable strongly preinvex-FM. Then, the functional $I(v)$ has minimum $u \in K_{\xi}$, if and only if $u \in K_{\xi}$ satisfies the following:*

$$\langle \mathcal{H}'(u),\ \xi(v,u)\rangle \tilde{+} \mathcal{J}(v) \tilde{-} \mathcal{J}(u) \tilde{+} \omega\|\xi(v,u)\|^2 \succcurlyeq \tilde{0},\ \forall\, v \in K_{\xi}. \tag{38}$$

Proof: Let $u \in K_{\xi}$ be the smallest value of I.

Therefore, for every $\gamma \in [0, 1]$, we have the following:

$$\begin{aligned} I_*(u,\gamma) &\leq I_*(v,\gamma), \\ I^*(u,\gamma) &\leq I^*(v,\gamma). \end{aligned} \tag{39}$$

Since $v_\tau = u + \tau\xi(v,u)$, for all $u,\ v \in K_\xi$ and $\tau \in [0, 1]$. Replacing v by v_τ in (39), we obtain the following:

$$\begin{aligned} I_*(u,\gamma) &\leq I_*(u+\tau\xi(v,u),\gamma), \\ I^*(u,\gamma) &\leq I^*(u+\tau\xi(v,u),\gamma). \end{aligned}$$

which implies that, using (37), the following holds:

$$\begin{aligned} \mathcal{H}_*(u,\gamma)+\mathcal{J}_*(u,\gamma) &\leq \mathcal{H}_*(u+\tau\xi(v,u),\gamma)+\mathcal{J}_*(u+\tau\xi(v,u),\gamma), \\ \mathcal{H}^*(u,\gamma)+\mathcal{J}^*(u,\gamma) &\leq \mathcal{H}^*(u+\tau\xi(v,u),\gamma)+\mathcal{J}^*(u+\tau\xi(v,u),\gamma). \end{aligned}$$

Since $\mathcal{J}$ is strongly preinvex-FM, then the following holds:

$$\begin{aligned} \mathcal{H}_*(u,\gamma)+\mathcal{J}_*(u,\gamma) \leq{}& \mathcal{H}_*(u+\tau\xi(v,u),\gamma)+(1-\tau)\mathcal{J}_*(u,\gamma)+\tau\mathcal{J}_*(v,\gamma) \\ &+\omega\tau(1-\tau)\|\xi(v,u)\|^2, \\ \mathcal{H}^*(u,\gamma)+\mathcal{J}^*(u,\gamma) \leq{}& \mathcal{H}^*(u+\tau\xi(v,u),\gamma)+(1-\tau)\mathcal{J}^*(u,\gamma)+\tau\mathcal{J}^*(v,\gamma) \\ &+\omega\tau(1-\tau)\|\xi(v,u)\|^2, \end{aligned}$$

that is

$$\begin{aligned} 0 &\leq \mathcal{H}_*(u+\tau\xi(v,u),\gamma)-\mathcal{H}_*(u,\gamma)+\tau(\mathcal{J}_*(v,\gamma)-\mathcal{J}_*(u,\gamma))+\omega\tau(1-\tau)\|\xi(v,u)\|^2, \\ 0 &\leq \mathcal{H}^*(u+\tau\xi(v,u),\gamma)-\mathcal{H}^*(u,\gamma)+\tau(\mathcal{J}^*(v,\gamma)-\mathcal{J}^*(u,\gamma))+\omega\tau(1-\tau)\|\xi(v,u)\|^2, \end{aligned}$$

Now dividing by "τ" and taking $\lim\limits_{\tau\to 0}$, we have the following:

$$\begin{aligned} 0 &\leq \lim_{\tau\to 0}\left\{\tfrac{\mathcal{H}_*(u+\tau\xi(v,u),\gamma)-\mathcal{H}_*(u,\gamma)}{\tau}+\mathcal{J}_*(v,\gamma)-\mathcal{J}_*(u,\gamma)+\omega(1-\tau)\|\xi(v,u)\|^2\right\}, \\ 0 &\leq \lim_{\tau\to 0}\left\{\tfrac{\mathcal{H}^*(u+\tau\xi(v,u),\gamma)-\mathcal{H}^*(u,\gamma)}{\tau}+\mathcal{J}^*(v,\gamma)-\mathcal{J}^*(u,\gamma)+\omega(1-\tau)\|\xi(v,u)\|^2\right\}, \end{aligned}$$

then

$$\begin{aligned} 0 &\leq \langle \mathcal{H}_*{}'(u,\gamma),\ \xi(v,u)\rangle+\mathcal{J}_*(v,\gamma)-\mathcal{J}_*(u,\gamma)+\omega\|\xi(v,u)\|^2, \\ 0 &\leq \langle \mathcal{H}^{*\prime}(u,\gamma),\ \xi(v,u)\rangle+\mathcal{J}^*(v,\gamma)-\mathcal{J}^*(u,\gamma)+\omega\|\xi(v,u)\|^2, \end{aligned}$$

that is,

$$\tilde{0} \preccurlyeq \langle \mathcal{H}'(u),\ \xi(v,u)\rangle \tilde{+} \mathcal{J}(v) \tilde{-} \mathcal{J}(u) \tilde{+} \omega\|\xi(v,u)\|^2.$$

Conversely, let (38) be satisfied to prove that $u \in K_\xi$ is a minimum of I. Assume that for all $v \in K_\xi$, we have $I(u)\tilde{-}I(v) = \mathcal{H}(u)\tilde{+}\mathcal{J}(u)\tilde{-}\mathcal{H}(v)\tilde{-}\mathcal{J}(v), = \mathcal{H}(u)\tilde{-}\mathcal{H}(v)\tilde{+}\mathcal{J}(u)\tilde{-}\mathcal{J}(v)$,
Therefore, for every $\gamma \in [0, 1]$, we have the following:

$$\begin{aligned} I_*(u,\gamma)-I_*(v,\gamma) &= \mathcal{H}_*(u,\gamma)-\mathcal{H}_*(v,\gamma)+\mathcal{J}_*(u,\gamma)-\mathcal{J}_*(v,\gamma), \\ I^*(u,\gamma)-I^*(v,\gamma) &= \mathcal{H}^*(u,\gamma)-\mathcal{H}^*(v,\gamma)+\mathcal{J}^*(u,\gamma)-\mathcal{J}^*(v,\gamma). \end{aligned}$$

By Corollary 1, we have the following:

$$\begin{aligned} I_*(u,\gamma)-I_*(v,\gamma) &\leq -[\langle \mathcal{H}_*{}'(u,\gamma),\ \xi(v,u)\rangle+\mathcal{J}_*(v,\gamma)-\mathcal{J}_*(u,\gamma)], \\ I^*(u,\gamma)-I^*(v,\gamma) &\leq -[\langle \mathcal{H}^{*\prime}(u,\gamma),\ \xi(v,u)\rangle+\mathcal{J}^*(v,\gamma)-\mathcal{J}^*(u,\gamma)], \end{aligned}$$

from which, using (38), we have the following:

$$\begin{aligned} I_*(u,\gamma)-I_*(v,\gamma) &\leq -\omega\|\xi(v,u)\|^2 \leq 0, \\ I^*(u,,\gamma)-I^*(v,\gamma) &\leq -\omega\|\xi(v,u)\|^2 \leq 0, \end{aligned}$$

that is, $I(u)\tilde{-}I(v) \preccurlyeq \tilde{0}$, hence, $I(u) \preccurlyeq I(v)$. □

Note that (38) are called strongly fuzzy mixed variational-like inequalities. This result shows that the minimum of fuzzy functional $I(\nu)$ can be characterized by a strongly fuzzy mixed variational-like inequality. It is very important to observe that the optimality conditions of preinvex-FMs and strongly preinvex-FMs cannot be obtained with the help of (38). This idea encourages us to introduce a more general type of fuzzy variational-like inequality of which (38) is a particular case. In order to be more precise, for given FMs Ψ, ϖ, bi function $\xi(.,.)$ and a $0 \leq \omega$, consider the problem of finding $u \in K_\xi$, such that the following holds:

$$\langle \Psi(u),\ \xi(\nu,u)\rangle \widetilde{+} \varpi(\nu) \widetilde{-} \varpi(u) \widetilde{+} \omega \|\xi(\nu,u)\|^2 \succcurlyeq \widetilde{0}, \forall\, \nu \in K_\xi. \tag{40}$$

This inequality is called a strongly fuzzy mixed variational-like inequality.

Now, we look at a few specific types of strongly fuzzy mixed variational-like inequalities:

If $\xi(\nu,u) = \nu - u$, then (40) is called a strongly fuzzy mixed variational inequality such as the following:

$$\langle \Psi(u), \nu - u\rangle \widetilde{+} \varpi(\nu) \widetilde{-} \varpi(u) \widetilde{+} \omega \|\nu - u\|^2 \succcurlyeq \widetilde{0},\ \forall\, \nu \in K_\xi.$$

If $\omega = 0$, then (40) is called fuzzy mixed variational-like inequality such as the following:

$$\langle \Psi(u),\ \xi(\nu,u)\rangle \widetilde{+} \varpi(\nu) \widetilde{-} \varpi(u) \succcurlyeq \widetilde{0},\ \forall\, \nu \in K_\xi.$$

If $\xi(\nu,u) = \nu - u$ and $\omega = 0$, then (40) is called a fuzzy mixed variational inequality such as the following:

$$\langle \Psi(u),\ \nu - u\rangle \widetilde{+} \varpi(\nu) \widetilde{-} \varpi(u) \succcurlyeq \widetilde{0},\ \forall\, \nu \in K_\xi.$$

Similarly, we can obtain a fuzzy variational inequality and fuzzy variational-like inequality in [32] as special cases of (40). In a similar way, some special cases of strongly fuzzy variational-like inequality (36) can also be discussed.

Remark 3. *The inequalities (36) and (40) show that the variational-like inequalities arise naturally in connection with the minimization of the G-differentiable preinvex-FMs, subject to certain constraints.*

The Theorem 11 provides the Hermite –Hadamard inequality for strongly preinvex-FM. This inequality provides a lower and an upper estimation for the average of strongly preinvex-FM defined on a compact interval.

Theorem 11. *Let $\mathcal{H} : [u,\ u + \xi(\nu,\ u)] \to \mathbb{F}_0$ be a strongly preinvex-FM with $\mathcal{H}(z) \succcurlyeq \widetilde{0}$. If $\mathcal{H}$ is fuzzy integrable and $\xi(.,\ .)$ satisfies Condition C, then the following holds:*

$$\mathcal{H}\left(\frac{2u + \xi(\nu,\ u)}{2}\right) \widetilde{+} \frac{\omega}{12}\|\xi(\nu,u)\|^2 \preccurlyeq \frac{1}{\xi(\nu,\ u)}\ (FR) \int_u^{u+\xi(\nu,\ u)} \mathcal{H}(z)dz \preccurlyeq \frac{\mathcal{H}(u) \widetilde{+} \mathcal{H}(\nu)}{2} \widetilde{-} \frac{\omega}{6}\|\xi(\nu,u)\|^2. \tag{41}$$

If $\mathcal{H}$ is preconcave FM then, inequality (41) reduces to the following inequality:

$$\mathcal{H}\left(\frac{2u + \xi(\nu,\ u)}{2}\right) \widetilde{+} \frac{\omega}{12}\|\xi(\nu,u)\|^2 \succcurlyeq \frac{1}{\xi(\nu,\ u)}\ (FR) \int_u^{u+\xi(\nu,\ u)} \mathcal{H}(z)dz \succcurlyeq \frac{\mathcal{H}(u) \widetilde{+} \mathcal{H}(\nu)}{2} \widetilde{-} \frac{\omega}{6}\|\xi(\nu,u)\|^2.$$

Proof. Let $\mathcal{H} : [u,\ u + \xi(\nu,\ u)] \to \mathbb{F}_0$ be a strongly preinvex-FM. Then, by hypothesis, we have the following:

$$\begin{aligned} 2\mathcal{H}\left(\tfrac{2u+\xi(\nu,\ u)}{2}\right) & \\ &\preccurlyeq \mathcal{H}(u + (1-\tau)\xi(\nu,\ u)) \\ &\widetilde{+}\ \mathcal{H}(u + \tau\xi(\nu,\ u)) \widetilde{-} \tfrac{\omega}{2}(1-2\tau)^2\|\xi(\nu,u)\|^2. \end{aligned}$$

Therefore, for every $\gamma \in (0,\ 1]$, we have the following:

$$2\mathcal{H}_*\left(\tfrac{2u+\xi(v,\ u)\|}{2},\ \gamma\right) \leq \mathcal{H}_*(u+(1-\tau)\xi(v,\ u),\ \gamma) + \mathcal{H}_*(u+\tau\xi(v,\ u),\ \gamma) - \tfrac{\omega}{2}(1-2\tau)^2\|\xi(v,u)\|^2$$
$$2\mathcal{H}^*\left(\tfrac{2u+\xi(v,\ u)\|}{2},\ \gamma\right) \leq \mathcal{H}^*(u+(1-\tau)\xi(v,\ u),\gamma) + \mathcal{H}^*(u+\tau\xi(v,\ u),\ \gamma) - \tfrac{\omega}{2}(1-2\tau)^2\|\xi(v,u)\|^2.$$

Then

$$2\int_0^1 \mathcal{H}_*\left(\tfrac{2u+\xi(v,\ u)}{2},\ \gamma\right)d\tau \leq \int_0^1 \mathcal{H}_*(u+(1-\tau)\xi(v,\ u),\ \gamma)d\tau + \int_0^1 \mathcal{H}_*(u+\tau\xi(v,\ u),\gamma)d\tau - \tfrac{\omega}{6}\|\xi(v,u)\|^2,$$
$$2\int_0^1 \mathcal{H}^*\left(\tfrac{2u+\xi(v,\ u)}{2},\gamma\right)d\tau \leq \int_0^1 \mathcal{H}^*(u+(1-\tau)\xi(v,\ u),\gamma)d\tau + \int_0^1 \mathcal{H}^*(u+\tau\xi(v,\ u),\ \gamma)d\tau - \tfrac{\omega}{2}\|\xi(v,u)\|^2.$$

It follows that

$$\mathcal{H}_*\left(\tfrac{2u+\xi(v,\ u)\|}{2},\ \gamma\right) + \tfrac{\omega}{12}\|\xi(v,u)\|^2 \leq \tfrac{1}{\xi(v,\ u)} \int_u^{u+\xi(v,\ u)} \mathcal{H}_*(z,\ \gamma)dz,$$
$$\mathcal{H}^*\left(\tfrac{2u+\xi(v,\ u)\|}{2},\ \gamma\right) + \tfrac{\omega}{12}\|\xi(v,u)\|^2 \leq \tfrac{1}{\xi(v,\ u)} \int_u^{u+\xi(v,\ u)} \mathcal{H}^*(z,\ \gamma)dz.$$

That is

$$\left[\mathcal{H}_*\left(\tfrac{2u+\xi(v,\ u)}{2},\ \gamma\right),\ \mathcal{H}^*\left(\tfrac{2u+\xi(v,\ u)}{2},\ \gamma\right)\right] + \tfrac{\omega}{12}\|\xi(v,u)\|^2 \leq_I \tfrac{1}{\xi(v,\ u)}\left[\int_u^{u+\xi(v,\ u)} \mathcal{H}_*(z,\ \gamma)dz,\ \int_u^{u+\xi(v,\ u)} \mathcal{H}^*(z,\ \gamma)dz\right].$$

Thus,

$$\mathcal{H}\left(\frac{2u+\xi(v,\ u)}{2}\right) + \frac{\omega}{12}\|\xi(v,u)\|^2 \preccurlyeq \frac{1}{\xi(v,\ u)}\ (FR)\int_u^{u+\xi(v,\ u)} \mathcal{H}(z)dz. \tag{42}$$

In a similar way as above, we have the following:

$$\frac{1}{\xi(v,\ u)}\ (FR)\int_u^{u+\xi(v,\ u)} \mathcal{H}(z)dz \preccurlyeq \frac{\mathcal{H}(u)\ \tilde{+}\ \mathcal{H}(v)}{2} - \frac{\omega}{6}\|\xi(v,u)\|^2. \tag{43}$$

Combining (42) and (43), we have the following:

$$\mathcal{H}\left(\frac{2u+\xi(v,\ u)}{2}\right) \tilde{+} \frac{\omega}{12}\|\xi(v,u)\|^2 \preccurlyeq \frac{1}{\xi(v,\ u)}\ (FR)\int_u^{u+\xi(v,\ u)} \mathcal{H}(z)dz \preccurlyeq \frac{\mathcal{H}(u)\ \tilde{+}\ \mathcal{H}(v)}{2} \tilde{-} \frac{\omega}{6}\|\xi(v,u)\|^2.$$

This completes the proof. □

Remark 4. *If $\omega = 0$, then Theorem 11 reduces to the result for preinvex convex-FM as follows:*

$$\mathcal{H}\left(\frac{2u+\xi(v,\ u)}{2}\right) \preccurlyeq \frac{1}{\xi(v,\ u)}\ (FR)\int_u^{u+\xi(v,\ u)} \mathcal{H}(z)dz \preccurlyeq \frac{\mathcal{H}(u)\ \tilde{+}\ \mathcal{H}(v)}{2}.$$

If $\xi(v,u) = v - u$, then Theorem 11 reduces to the result for strongly convex-FM as follows:

$$\mathcal{H}\left(\frac{u+v}{2}\right) \tilde{+} \frac{\omega}{12} v - u^2 \preccurlyeq \frac{1}{v-u}\ (FR)\int_u^v \mathcal{H}(z)dz \preccurlyeq \frac{\mathcal{H}(u)\ \tilde{+}\ \mathcal{H}(v)}{2} \tilde{-} \frac{\omega}{6}\|v-u\|^2.$$

If $\xi(\nu, u) = \nu - u$ and $\omega = 0$, then Theorem 11 reduces to the result for convex-FM in [55] *as follows:*

$$\mathcal{H}\left(\frac{u+\nu}{2}\right) \preccurlyeq \frac{1}{\nu - u} (FR) \int_u^\nu \mathcal{H}(z)dz \preccurlyeq \frac{\mathcal{H}(u) \widetilde{+} \mathcal{H}(\nu)}{2}. \tag{44}$$

If $\mathcal{H}_*(u, \gamma) = \mathcal{H}^*(\nu, \gamma)$ with $\omega = 0$ and $\gamma = 1$, then Theorem 11 reduces to the result for preinvex function as follows (see [36]):

$$\mathcal{H}\left(\frac{2u + \xi(\nu, u)}{2}\right) \le \frac{1}{\xi(\nu, u)} (R) \int_u^{u+\xi(\nu, u)} \mathcal{H}(z)dz \le \frac{\mathcal{H}(u) + \mathcal{H}(\nu)}{2}. \tag{45}$$

If $\mathcal{H}_*(u, \gamma) = \mathcal{H}^*(\nu, \gamma)$ with $\xi(\nu, u) = \nu - u$, $\omega = 0$ and $\gamma = 1$, then Theorem 11 reduces to the result for convex function as follows (see [42,43]):

$$\mathcal{H}\left(\frac{u+\nu}{2}\right) \le \frac{1}{\nu - u} (R) \int_u^\nu \mathcal{H}(z)dz \le \frac{\mathcal{H}(u) + \mathcal{H}(\nu)}{2}. \tag{46}$$

Example 5. *We consider the fuzzy-IVF $\mathcal{H} : [u, u + \xi(\nu, u)] = [0, \xi(2, 0)] \to \mathbb{F}_0$ defined by the following:*

$$\mathcal{H}(z)(\sigma) = \begin{cases} \frac{\sigma}{2z^2}, \ \sigma \in [0, 2z^2], \\ \frac{4z^2 - \sigma}{2z^2}, \ \sigma \in (2z^2, 4z^2], \\ 0, \ otherwise, \end{cases}$$

Then, for each $\gamma \in [0, 1]$, we have $\mathcal{H}_\gamma(z) = [2\gamma z^2, (4 - 2\gamma)z^2]$. Since for each $\gamma \in [0, 1]$, $\mathcal{H}_*(z, \gamma) = 2\gamma z^2$, $\mathcal{H}^*(z, \gamma) = (4 - 2\gamma)z^2$ are preinvex functions w.r.t. $\xi(\nu, u) = \nu - u$ and $\omega = \frac{2}{3}\gamma$. Hence $\mathcal{H}(z)$ is preinvex fuzzy-IVF w.r.t. $\xi(\nu, u) = \nu - u$. We now compute the following:

$$\mathcal{H}_*\left(\frac{2u+\xi(\nu, u)}{2}, \gamma\right) + \frac{\omega}{12}\|\xi(\nu, u)\|^2 = \mathcal{H}_*(1, \gamma) = \frac{8\gamma}{3},$$
$$\frac{1}{\xi(\nu, u)} \int_u^{u+\xi(\nu, u)} \mathcal{H}_*(z, \gamma)dz = \frac{1}{2} \int_0^2 2\gamma z^2 dz = \frac{8\gamma}{3},$$
$$\frac{\mathcal{H}_*(u, \gamma) + \mathcal{H}_*(\nu, \gamma)}{2} - \frac{\omega}{6}\|\xi(\nu, u)\|^2 = \frac{32\gamma}{9},$$

for all $\gamma \in [0, 1]$. That means the following holds:

$$\frac{8\gamma}{3} \le \frac{8\gamma}{3} \le \frac{32\gamma}{9}.$$

Similarly, it can be easily shown that the following holds:

$$\mathcal{H}^*\left(\frac{2u + \xi(\nu, u)}{2}, \gamma\right) \le \frac{1}{\xi(\nu, u)} \int_u^{u+\xi(\nu, u)} \mathcal{H}^*(z, \gamma)dz \le \frac{\mathcal{H}^*(u, \gamma) + \mathcal{H}^*(\nu, \gamma)}{2}.$$

for all $\gamma \in [0, 1]$, such that we have the following:

$$\mathcal{H}^*\left(\frac{2u+\xi(\nu, u)}{2}, \gamma\right) + \frac{\omega}{12}\xi(\nu, u)^2 = \mathcal{H}_*(1, \gamma) = \frac{36-16\gamma}{9},$$
$$\frac{1}{\xi(\nu, u)} \int_u^{u+\xi(\nu, u)} \mathcal{H}^*(z, \gamma)dz = \frac{1}{2} \int_0^2 (4 - 2\gamma)z^2 dz = \frac{8(2-\gamma)}{3},$$
$$\frac{\mathcal{H}^*(u, \gamma) + \mathcal{H}^*(\nu, \gamma)}{2} - \frac{\omega}{6}\|\xi(\nu, u)\|^2 = \frac{72-22\gamma}{9}.$$

From which, it follows that

$$\frac{36 - 16\gamma}{9} \le \frac{8(2 - \gamma)}{3} \le \frac{72 - 22\gamma}{9},$$

that is

$$\left[\frac{8\gamma}{3}, \frac{36-16\gamma}{9}\right] \leq_I \left[\frac{8\gamma}{3}, \frac{8(2-\gamma)}{3}\right] \leq_I \left[\frac{32\gamma}{9}, \frac{72-22\gamma}{9}\right], \text{ for all } \gamma \in [0, 1],$$

and hence, the Theorem 11 is verified.

5. Conclusions

In this study, we introduced and studied a new class of preinvex-FMs called strongly preinvex-FMs. Using Condition C, we obtained the equivalence relation between strongly preinvex- and strongly invex-FMs. To characterize the optimality condition of the sum of preinvex-FMs and strongly preinvex-FMs, we introduced the strong fuzzy mixed variational-like inequality. Moreover, we established a strong relationship between strongly preinvex-FM and the Hermite–Hadamard inequality. There is much room for further study to explore this concept in fuzzy convex and non-convex theory, such as the existence of a unique solution of strong fuzzy mixed variational-like inequalities and some iterative algorithms, which can also obtained under some mild conditions. From last two sections, we can conclude that these classes of FMs will play an important and significant role in fuzzy optimization and their related areas.

Author Contributions: Conceptualization, M.B.K.; validation, H.M.S. and P.O.M.; formal analysis, H.M.S. and P.O.M.; investigation, M.B.K. and P.O.M.; resources, M.B.K. and J.L.G.G.; writing—original draft, M.B.K. and J.L.G.G.; writing—review and editing, M.B.K. and P.O.M.; visualization, M.B.K. and P.O.M.; supervision, M.B.K. and P.O.M.; project administration, J.L.G.G. and H.M.S. and P.O.M. All authors contributed equally to the writing of this paper. All authors have read and agreed to the published version of the manuscript.

Funding: This paper has been partially supported by Ministerio de Ciencia, Innovaci ón y Universidades, grant number PGC2018-097198-B-I00, and by Fundaci ón Séneca of Región de Murcia, grant number 20783/PI/18.

Institutional Review Board Statement: Not applicable.

Informed Consent Statement: Not applicable.

Data Availability Statement: Not applicable.

Conflicts of Interest: The authors declare no conflict of interest.

References

1. Polyak, B.T. Existence theorems and convergence of minimizing sequences in extremum problems with restrictions. *Sov. Math. Dokl.* **1966**, *7*, 2–75.
2. Karamardian, S. The nonlinear complementarity problem with applications, Part 2. *J. Optim. Theory Appl.* **1969**, *4*, 167–181. [CrossRef]
3. Qu, G.; Li, N. On the exponentially stability of primal-dual gradeint dynamics. *IEEE Control Syst. Lett.* **2019**, *3*, 43–48. [CrossRef]
4. Nikodem, K.; Pales, Z.S. Characterizations of inner product spaces by strongly convex functions. *Banach J. Math. Anal.* **2011**, *1*, 83–87. [CrossRef]
5. Adamek, M. On a problem connected with strongly convex functions. *Math. Inequal. Appl.* **2016**, *19*, 1287–1293. [CrossRef]
6. Angulo, H.; Gimenez, J.; Moeos, A.M. On strongly h-convex functions. *Ann. Funct. Anal.* **2011**, *2*, 85–91. [CrossRef]
7. Awan, M.U.; Noor, M.A.; Noor, K.I. Hermite-Hadamard inequalities for exponentially convex functions. *Appl. Math. Inf. Sci.* **2018**, *12*, 405–409. [CrossRef]
8. Awan, M.U.; Noor, M.A.; Set, E. On strongly (p, h)-convex functions. *TWMS J. Pure Appl. Math.* **2019**, *10*, 145–153.
9. Azcar, A.; Gimnez, J.; Nikodem, K.; Sánchez, J.L. On strongly midconvex functions. *Opusc. Math.* **2011**, *31*, 15–26. [CrossRef]
10. Jovanovic, M.V. A note on strongly convex and strongly quasi convex functions. *Math. Notes* **1966**, *60*, 584–585. [CrossRef]
11. Hanson, M.A. On sufficiency of the Kuhn-Tucker conditions. *J. Math. Anal. Appl.* **1980**, *80*, 545–550. [CrossRef]
12. Ben-Isreal, A.; Mond, B. What is invexity? *Anziam J.* **1986**, *28*, 1–9. [CrossRef]
13. Mohan, M.S.; Neogy, S.K. On invex sets and preinvex functions. *J. Math. Anal. Appl.* **1995**, *189*, 901–908. [CrossRef]
14. Noor, M.A.; Noor, K.I. On strongly generalized preinvex functions. *J. Inequal. Pure Appl. Math.* **2005**, *6*, 1–20.
15. Noor, M.A.; Noor, K.I. Some characterization of strongly preinvex functions. *J. Math. Anal. Appl.* **2006**, *316*, 697–706. [CrossRef]

16. Noor, M.A.; Noor, K.I. Generalized preinvex functions and their properties. *Int. J. Stoch. Anal.* **2006**, *2006*, 12736. [CrossRef]
17. Khan, M.B.; Noor, M.A.; Mohammed, P.O.; Guirao, J.L.; Noor, K.I. Some Integral Inequalities for Generalized Convex Fuzzy-Interval-Valued Functions via Fuzzy Riemann Integrals. *Int. J. Comput. Intell. Syst.* **2021**, *14*, 1–15. [CrossRef]
18. Noor, M.A.; Noor, K.I.; Iftikhar, S. Integral inequaliies for differentiable harmonic preinvex functions (survey). *J. Inequal. Pure Appl. Math.* **2016**, *7*, 3–19.
19. Weir, T.; Mond, B. Preinvex functions in multiobjective optimization. *J. Math. Anal. Appl.* **1986**, *136*, 29–38. [CrossRef]
20. Zadeh, L.A. Fuzzy sets. *Inf. Control* **1965**, *8*, 338–353. [CrossRef]
21. Liu, Y.M. Some properties of convex fuzzy sets. *J. Math. Anal. Appl.* **1985**, *111*, 119–129. [CrossRef]
22. Lowen, R. Convex fuzzy sets. *Fuzzy Sets Syst.* **1980**, *3*, 291–310. [CrossRef]
23. Ammar, E.; Metz, J. On fuzzy convexity and parametric fuzzy optimization. *Fuzzy Sets Syst.* **1992**, *49*, 135–141. [CrossRef]
24. Ammar, E.E. Some properties of convex fuzzy sets and convex fuzzy cones. *Fuzzy Sets Syst.* **1999**, *106*, 381–386. [CrossRef]
25. Dubois, D.; Prade, H. Operations on fuzzy numbers. *Int. J. Syst. Sci.* **1978**, *9*, 613–626. [CrossRef]
26. Goetschel, R., Jr.; Voxman, W. Topological properties of fuzzy numbers. *Fuzzy Sets Syst.* **1983**, *10*, 87–99. [CrossRef]
27. Nanda, S.; Kar, K. Convex fuzzy mappings. *Fuzzy Sets Syst.* **1992**, *48*, 129–132. [CrossRef]
28. Syau, Y.R. On convex and concave fuzzy mappings. *Fuzzy Sets Syst.* **1999**, *103*, 163–168. [CrossRef]
29. Furukawa, N. Convexity and local Lipschitz continuity of fuzzy-valued mappings. *Fuzzy Sets Syst.* **1998**, *93*, 113–119. [CrossRef]
30. Yan, H.; Xu, J. A class of convex fuzzy mappings. *Fuzzy Sets Syst.* **2002**, *129*, 47–56. [CrossRef]
31. Goetschel, R., Jr.; Voxman, W. Elementary fuzzy calculus. *Fuzzy Sets Syst.* **1986**, *18*, 31–43. [CrossRef]
32. Noor, M.A. Fuzzy preinvex functions. *Fuzzy Sets Syst.* **1994**, *64*, 95–104. [CrossRef]
33. Syau, Y.R. (ϕ1, ϕ2)-convex fuzzy mappings. *Fuzzy Sets Syst.* **2003**, *138*, 617–625. [CrossRef]
34. Syau, Y.R.; Lee, E.S. Fuzzy Weirstrass theorem and convex fuzzy mappings. *Comput. Math. Appl.* **2006**, *51*, 1741–1750. [CrossRef]
35. Bede, B.; Gal, S.G. Generalizations of the differentiability of fuzzy-number-valued functions with applications to fuzzy differential equations. *Fuzzy Sets Syst.* **2005**, *151*, 581–599. [CrossRef]
36. Ruiz-Garzón, G.; Osuna-Gómez, R.; Rufián-Lizana, A. Generalized invex monotonicity. *Eur. J. Oper. Res.* **2003**, *144*, 501–512. [CrossRef]
37. Stefanini, L.; Bede, B. Generalized Hukuhara differentiability of interval-valued functions and interval differential equations. *Nonlinear Anal. Theory Methods Appl.* **2009**, *71*, 1311–1328. [CrossRef]
38. Syau, Y.R. Preinvex fuzzy mappings. *Comp. Math. Appl.* **1999**, *37*, 31–39. [CrossRef]
39. Noor, M.A. Variational-like inequalities. *Optimization* **1994**, *30*, 323–330. [CrossRef]
40. Noor, M.A.; Noor, K.I.; Awan, M.U. Some quantum integral inequalities via preinvex functions. *Appl. Math. Comput.* **2015**, *269*, 242–251. [CrossRef]
41. Rashid, S.; Khalid, A.; Rahman, G.; Nisar, K.S.; Chu, Y.M. On new modifications governed by quantum Hahn's integral operator pertaining to fractional calculus. *J. Funct. Spaces* **2020**, *2020*, 1–12. [CrossRef]
42. Hadamard, J. Étude sur les propriétés des fonctions entières et en particulier d'une fonction considérée par Riemann. *J. Math. Pures Appl.* **1893**, *7*, 171–215.
43. Hermite, C. Sur deux limites d'une intégrale définie. *Mathesis* **1883**, *3*, 82–97.
44. Iscan, I. A new generalization of some integral inequalities for (α, m)-convex functions. *Math. Sci.* **2013**, *7*, 1–8. [CrossRef]
45. Iscan, I. Hermite–Hadamard type inequalities for harmonically convex functions. *Hacet. J. Math. Stat.* **2014**, *43*, 935–942. [CrossRef]
46. Iscan, I. Hermite–Hadamard type inequalities for p-convex functions. *Int. J. Anal. Appl.* **2016**, *11*, 137–145.
47. Pachpatte, B.G. On some inequalities for convex functions. *RGMIA Res. Rep. Coll.* **2003**, *6*, 1–9.
48. Kaleva, O. Fuzzy differential equations. *Fuzzy Sets Syst.* **1987**, *24*, 301–317. [CrossRef]
49. Kulish, U.; Miranker, W. *Computer Arithmetic in Theory and Practice*; Academic Press: New York, NY, USA, 2014.
50. Osuna-G'omez, R.; Jim'enez-Gamero, M.D.; Chalco-Cano, Y.; Rojas-Medar, M.A. Hadamard and Jensen Inequalities for s−Convex Fuzzy Processes. In *Soft Methodology and Random Information Systems (Advances in Soft Computing)*; Springer: Berlin/Heidelberg, Germany, 2004.
51. Costa, T.M. Jensen's inequality type integral for fuzzy-interval-valued functions. *Fuzzy Sets Syst.* **2017**, *327*, 31–47. [CrossRef]
52. Costa, T.M.; Román-Flores, H.; Chalco-Cano, Y. Opial-type inequalities for interval-valued functions. *Fuzzy Sets Syst.* **2019**, *358*, 48–63. [CrossRef]
53. Moore, R.E. *Interval Analysis*; Prentice Hall: Englewood Cliffs, NJ, USA, 1966.
54. Costa, T.M.; Roman-Flores, H. Some integral inequalities for fuzzy-interval-valued functions. *Inf. Sci.* **2017**, *420*, 110–125. [CrossRef]
55. Khan, M.B.; Noor, M.A.; Noor, K.I.; Chu, Y.-M. New Hermite-Hadamard Type Inequalities for (h1, h2)-Convex Fuzzy-Interval-Valued Functions. *Adv. Differ. Equat.* **2021**, *2021*, 6–20.
56. Liu, P.; Khan, M.B.; Noor, M.A.; Noor, K.I. New Hermite–Hadamard and Jensen inequalities for log-s-convex fuzzy-interval-valued functions in the second sense. *Complex Intell. Syst.* **2021**, *2021*, 1–15.
57. Khan, M.B.; Noor, M.A.; Abdullah, L.; Noor, K.I. New Hermite-Hadamard and Jensen Inequalities for Log-h-Convex Fuzzy-Interval-Valued Functions. *Int. J. Comput. Intell. Syst.* **2021**, *14*, 155. [CrossRef]

58. Khan, M.B.; Noor, M.A.; Abdullah, L.; Chu, Y.M. Some New Classes of Preinvex Fuzzy-Interval-Valued Functions and Inequalities. *Int. J. Comput. Intell. Syst.* **2021**, *14*, 1403–1418. [CrossRef]
59. Khan, M.B.; Mohammed, P.O.; Noor, M.A.; Hamed, Y.S. New Hermite–Hadamard inequalities in fuzzy-interval fractional calculus and related inequalities. *Symmetry* **2021**, *13*, 673. [CrossRef]
60. Khan, M.B.; Mohammed, P.O.; Noor, M.A.; Abuahalnaja, K. Fuzzy Integral Inequalities on Coordinates of Convex Fuzzy Interval-Valued Functions. *Math. Biosci. Eng.* **2021**, *18*, 6552–6580. [CrossRef]
61. Khan, M.B.; Mohammed, P.O.; Noor, M.A.; Hameed, Y.; Noor, K.I. New Fuzzy-Interval Inequalities in Fuzzy-Interval Fractional Calculus by Means of Fuzzy Order Relation. *AIMS Math.* **2021**, *6*, 10964–10988. [CrossRef]
62. Khan, M.B.; Mohammed, P.O.; Noor, M.A.; Baleanu, D.; Guirao, J.L.G. Some New Fractional Estimates of Inequalities for LR-p-Convex Interval-Valued Functions by Means of Pseudo Order Relation. *Axioms* **2021**, *10*, 175. [CrossRef]
63. Srivastava, H.M.; El-Deeb, S.M. Fuzzy differential subordinations based upon the Mittag-Leffler type Borel distribution. *Symmetry* **2021**, *13*, 1023. [CrossRef]
64. Touchent, K.; Hammouch, Z.; Mekkaoui, T. A modified invariant subspace method for solving partial differential equations with non-singular kernel fractional derivatives. *Appl. Math. Nonlinear Sci.* **2020**, *5*, 35–48. [CrossRef]
65. Vanli, A.; Ünal, I.; Özdemir, D. Normal complex contact metric manifolds admitting a semi symmetric metric connection. *Appl. Math. Nonlinear Sci.* **2020**, *5*, 49–66. [CrossRef]
66. Sharifi, M.; Raesi, B. Vortex Theory for Two Dimensional Boussinesq Equations. *Appl. Math. Nonlinear Sci.* **2020**, *5*, 67–84. [CrossRef]
67. Rajesh Kanna, M.; Pradeep Kumar, R.; Nandappa, S.; Cangul, I. On Solutions of Fractional order Telegraph Partial Differential Equation by Crank-Nicholson Finite Difference Method. *Appl. Math. Nonlinear Sci.* **2020**, *5*, 85–98. [CrossRef]
68. Harisha, R.P.; Ranjini, P.S.; Lokesha, V.; Kumar, S. Degree Sequence of Graph Operator for some Standard Graphs. *Appl. Math. Nonlinear Sci.* **2020**, *5*, 99–108. [CrossRef]

symmetry

MDPI

Article

A Quadruple Definite Integral Expressed in Terms of the Lerch Function

Robert Reynolds * and Allan Stauffer

Department of Mathematics and Statistics, York University, 4700 Keele Street, Toronto, ON M3J 1P3, Canada; stauffer@yorku.ca
* Correspondence: milver@my.yorku.ca

Abstract: A quadruple integral involving the logarithmic, exponential and polynomial functions is derived in terms of the Lerch function. Special cases of this integral are evaluated in terms of special functions and fundamental constants. Almost all Lerch functions have an asymmetrical zero-distribution. The majority of the results in this work are new.

Keywords: Lerch function; quadruple integral; contour integral; logarithmic function

Citation: Reynolds, R.; Stauffer, A. A Quadruple Definite Integral Expressed in Terms of the Lerch Function. *Symmetry* **2021**, *13*, 1638. https://doi.org/10.3390/sym13091638

Academic Editor: Emilio Acerbi

Received: 4 August 2021
Accepted: 2 September 2021
Published: 6 September 2021

1. Significance Statement

Quadruple definite integrals are widely used in a vast number of areas spanning mathematics and physics, from integrating over a four-dimensional volume, integrating over a Lagrangian density in field theory and four-dimensional Fourier transforms of a function of spacetime (x, y, z, t).

Some interesting areas where these integrals are used are in asymptotic expansion [1], calculating the mean distance between two independent points within a circle [2], providing a classical derivation of the Compton effect [3], the radiation impedance computations of a square piston in a rigid infinite baffle [4], the acoustic radiation impedance of a rectangular panel [5], the statistical basis for the theory of stellar scintillation [6], modelling in three dimensions of a guiding center plasma within the purview of gyroelastic magnetohydrodynamics [7], and the formulation of an axisymmetric potential problem for a plane circular electrode [8].

After perusing the current literature, the authors found many applications of quadruple integrals. In some cases these integrals were separable and in some cases asymptotic expansions were used to attain a solution. To the best of our knowledge the authors were unable to find quadruple definite integrals involving the logarithmic, exponential and polynomial functions derived in terms of a closed form solution.

In this present work we provide a formal derivation for a quadruple integral not present in the current literature. This integral features a kernel with the product of the logarithmic, exponential and polynomial functions. The log term mixes the variables so that the integral is not separable except for special values of k.

In this work our goal is to expand upon the current literature of definite quadruple integrals by providing a formal derivation in terms of the Lerch function.

2. Introduction

In this paper we derive the quadruple definite integral given by

$$\int_0^\infty \int_0^\infty \int_0^\infty \int_0^\infty (t+z)^{-m}(x+y)^{m-1}e^{-p(x+z)-q(t+y)} \log^k\left(\frac{a(x+y)}{t+z}\right)dxdydzdt \quad (1)$$

where the parameters k, a, p, q and m are general complex numbers. This definite integral will be used to derive special cases in terms of special functions and fundamental constants.

The derivations follow the method used by us in [9]. This method involves using a form of the generalized Cauchy's integral formula given by

$$\frac{y^k}{\Gamma(k+1)} = \frac{1}{2\pi i}\int_C \frac{e^{wy}}{w^{k+1}}dw. \tag{2}$$

where C is in general an open contour in the complex plane where the bilinear concomitant has the same value at the end points of the contour. We then multiply both sides by a function of x, y, z and t, then take a definite quadruple integral of both sides. This yields a definite integral in terms of a contour integral. Then we multiply both sides of Equation (2) by another function of x, y, z and t and take the infinite sums of both sides such that the contour integral of both equations are the same.

3. Definite Integral of the Contour Integral

We use the method in [9]. The variable of integration in the contour integral is $\alpha = w + m$. The cut and contour are in the second quadrant of the complex α-plane. The cut approaches the origin from the interior of the second quadrant and the contour goes round the origin with zero radius and is on opposite sides of the cut. Using a generalization of Cauchy's integral formula we form the quadruple integral by replacing y by $\log\left(\frac{a(x+y)}{t+z}\right)$ and multiplying by $(t+z)^{-m}(x+y)^{m-1}e^{-p(x+z)-q(t+y)}$ then taking the definite integral with respect to $x \in [0,\infty), y \in [0,\infty), z \in [0,\infty)$ and $t \in [0,\infty)$ to obtain

$$\begin{aligned}
&\frac{1}{\Gamma(k+1)}\int_0^\infty\int_0^\infty\int_0^\infty\int_0^\infty (t+z)^{-m}(x+y)^{m-1}e^{-p(x+z)-q(t+y)}\log^k\left(\frac{a(x+y)}{t+z}\right)dxdydzdt \\
&= \frac{1}{2\pi i}\int_0^\infty\int_0^\infty\int_0^\infty\int_0^\infty \left(\int_C a^w w^{-k-1}(t+z)^{-m-w}(x+y)^{m+w-1}e^{-p(x+z)-q(t+y)}dw\right)dxdydzdt \\
&= \frac{1}{2\pi i}\int_C\left(\int_0^\infty\int_0^\infty\int_0^\infty\int_0^\infty a^w w^{-k-1}(t+z)^{-m-w}(x+y)^{m+w-1}e^{-p(x+z)-q(t+y)}dxdydzdt\right)dw \\
&= \frac{1}{2\pi i}\int_C \frac{\pi a^w w^{-k-1}\csc(\pi(m+w))\left(p^{-m-w+1}-q^{-m-w+1}\right)\left(p^{m+w}-q^{m+w}\right)}{pq(p-q)^2}dw
\end{aligned} \tag{3}$$

from Equation (3.1.3.7) in [10] where $0 < Re(w+m)$ and using the reflection Formula (8.334.3) in [11] for the Gamma function. We are able to switch the order of integration over α, x, y, z and t using Fubini's theorem since the integrand is of bounded measure over the space $\mathbb{C} \times [0,\infty) \times [0,\infty) \times [0,\infty) \times [0,\infty)$.

4. The Lerch Function and Infinite Sum of the Contour Integral

In this section we use Equation (2) to derive the contour integral representations for the Lerch function.

4.1. The Lerch Function

The Lerch function has a series representation given by

$$\Phi(z,s,v) = \sum_{n=0}^{\infty}(v+n)^{-s}z^n \tag{4}$$

where $|z| < 1, v \neq 0, -1, \ldots$ and is continued analytically by its integral representation given by

$$\Phi(z,s,v) = \frac{1}{\Gamma(s)}\int_0^\infty \frac{t^{s-1}e^{-vt}}{1-ze^{-t}}dt = \frac{1}{\Gamma(s)}\int_0^\infty \frac{t^{s-1}e^{-(v-1)t}}{e^t - z}dt \tag{5}$$

where $Re(v) > 0$, and either $|z| \leq 1, z \neq 1, Re(s) > 0$, or $z = 1, Re(s) > 1$.

4.2. Derivation of the First Contour Integral

In this section we will derive the contour integral given by

$$\frac{1}{2\pi i}\int_C \frac{\pi a^w w^{-k-1}\csc(\pi(m+w))}{p(p-q)^2}dw \tag{6}$$

Using Equation (2) and replacing y by $\log(a) + i\pi(2y+1)$ then multiplying both sides by $-\frac{2i\pi e^{i\pi m(2y+1)}}{p(p-q)^2}$ taking the infinite sum over $y \in [0,\infty)$ and simplifying in terms of the Lerch function we obtain

$$\begin{aligned} &-\frac{(2i\pi)^{k+1}e^{i\pi m}\Phi\left(e^{2im\pi},-k,\frac{\pi-i\log(a)}{2\pi}\right)}{\Gamma(k+1)p(p-q)^2} \\ &= -\frac{1}{2\pi i}\sum_{y=0}^{\infty}\int_C \frac{2i\pi a^w w^{-k-1}e^{i\pi(2y+1)(m+w)}}{p(p-q)^2}dw \\ &= -\frac{1}{2\pi i}\int_C \sum_{y=0}^{\infty} \frac{2i\pi a^w w^{-k-1}e^{i\pi(2y+1)(m+w)}}{p(p-q)^2}dw \\ &= \frac{1}{2\pi i}\int_C \frac{\pi a^w w^{-k-1}\csc(\pi(m+w))}{p(p-q)^2}dw \end{aligned} \tag{7}$$

from Equation (1.232.3) in [11] and $Im(w+m) > 0$ in order for the sum to converge.

4.3. Derivation of the Second Contour Integral

In this section we will derive the contour integral given by

$$\frac{1}{2\pi i}\int_C \frac{\pi a^w w^{-k-1}\csc(\pi(m+w))}{q(p-q)^2}dw \tag{8}$$

Using Equation (2) and replacing y with $\log(a) + i\pi(2y+1)$ then multiplying both sides by $-\frac{2i\pi e^{i\pi m(2y+1)}}{q(p-q)^2}$ taking the infinite sum over $y \in [0,\infty)$ and simplifying in terms of the Lerch function we obtain

$$\begin{aligned} &-\frac{(2i\pi)^{k+1}e^{i\pi m}\Phi\left(e^{2im\pi},-k,\frac{\pi-i\log(a)}{2\pi}\right)}{\Gamma(k+1)q(p-q)^2} \\ &= -\frac{1}{2\pi i}\sum_{y=0}^{\infty}\int_C \frac{2i\pi a^w w^{-k-1}e^{i\pi(2y+1)(m+w)}}{q(p-q)^2}dw \\ &= -\frac{1}{2\pi i}\int_C \sum_{y=0}^{\infty} \frac{2i\pi a^w w^{-k-1}e^{i\pi(2y+1)(m+w)}}{q(p-q)^2}dw \\ &= \frac{1}{2\pi i}\int_C \frac{\pi a^w w^{-k-1}\csc(\pi(m+w))}{q(p-q)^2}dw \end{aligned} \tag{9}$$

from Equation (1.232.3) in [11] and $Im(w+m) > 0$ in order for the sum to converge.

4.4. Derivation of the Third Contour Integral

In this section we will derive the contour integral given by

$$-\frac{1}{2\pi i}\int_C \frac{\pi a^w w^{-k-1}p^{m+w-1}q^{-m-w}\csc(\pi(m+w))}{(p-q)^2}dw \tag{10}$$

Using Equation (2) and replacing y with $\log(a) + \log(p) - \log(q) + i\pi(2y+1)$ then multiplying both sides by $\frac{2i\pi p^{m-1}q^{-m}e^{i\pi m(2y+1)}}{(p-q)^2}$ taking the infinite sum over $y \in [0,\infty)$ and simplifying in terms of the Lerch function we obtain

$$\begin{aligned} &\frac{(2i\pi)^{k+1}e^{i\pi m}p^{m-1}q^{-m}\Phi\left(e^{2im\pi},-k,\frac{-i\log(a)-i\log(p)+i\log(q)+\pi}{2\pi}\right)}{\Gamma(k+1)(p-q)^2} \\ &= \frac{1}{2\pi i}\sum_{y=0}^{\infty}\int_C \frac{2i\pi w^{-k-1}p^{m-1}q^{-m}\exp(w(\log(a)+\log(p)-\log(q))+i\pi(2y+1)(m+w))}{(p-q)^2}dw \\ &= \frac{1}{2\pi i}\int_C \sum_{y=0}^{\infty} \frac{2i\pi w^{-k-1}p^{m-1}q^{-m}\exp(w(\log(a)+\log(p)-\log(q))+i\pi(2y+1)(m+w))}{(p-q)^2}dw \\ &= -\frac{1}{2\pi i}\int_C \frac{\pi a^w w^{-k-1}p^{m+w-1}q^{-m-w}\csc(\pi(m+w))}{(p-q)^2}dw \end{aligned} \tag{11}$$

from Equation (1.232.3) in [11] and $Im(w+m) > 0$ in order for the sum to converge.

4.5. Derivation of the Fourth Contour Integral

In this section we will derive the contour integral given by

$$-\frac{1}{2\pi i}\int_C \frac{\pi a^w w^{-k-1} p^{-m-w} q^{m+w-1}\csc(\pi(m+w))}{(p-q)^2}dw \tag{12}$$

Using Equation (2) and replacing y by $\log(a)-\log(p)+\log(q)+i\pi(2y+1)$ then multiplying both sides by $\frac{2i\pi p^{-m}q^{m-1}e^{i\pi m(2y+1)}}{(p-q)^2}$ taking the infinite sum over $y \in [0,\infty)$ and simplifying in terms of the Lerch function we obtain

$$\begin{aligned}
&\frac{(2i\pi)^{k+1}e^{i\pi m}p^{-m}q^{m-1}\Phi\left(e^{2im\pi},-k,\frac{-i\log(a)+i\log(p)-i\log(q)+\pi}{2\pi}\right)}{\Gamma(k+1)(p-q)^2} \\
&= \frac{1}{2\pi i}\sum_{y=0}^{\infty}\int_C \frac{2i\pi w^{-k-1}p^{-m}q^{m-1}\exp(w(\log(a)-\log(p)+\log(q))+i\pi(2y+1)(m+w))}{(p-q)^2}dw \\
&= \frac{1}{2\pi i}\int_C\sum_{y=0}^{\infty} \frac{2i\pi w^{-k-1}p^{-m}q^{m-1}\exp(w(\log(a)-\log(p)+\log(q))+i\pi(2y+1)(m+w))}{(p-q)^2}dw \\
&= -\frac{1}{2\pi i}\int_C \frac{\pi a^w w^{-k-1}p^{-m-w}q^{m+w-1}\csc(\pi(m+w))}{(p-q)^2}dw
\end{aligned} \tag{13}$$

from Equation (1.232.3) in [11] and $Im(w+m)>0$ in order for the sum to converge.

5. Definite Integral in Terms of the Lerch Function

Theorem 1. *For all $k,a,p,q \in \mathbb{C}, -1 < Re(m) < 1$,*

$$\begin{aligned}
&\int_0^\infty\int_0^\infty\int_0^\infty\int_0^\infty (t+z)^{-m}(x+y)^{m-1}e^{-p(x+z)-q(t+y)}\log^k\left(\frac{a(x+y)}{t+z}\right)dxdydzdt \\
&= \frac{1}{(p-q)^2}(2i\pi)^{k+1}e^{i\pi m}p^{-m-1}q^{-m-1}\left(-p^m q^m(p+q)\Phi\left(e^{2im\pi},-k,\frac{\pi-i\log(a)}{2\pi}\right)+\right. \\
&\qquad qp^{2m}\Phi\left(e^{2im\pi},-k,\frac{-i\log(a)-i\log(p)+i\log(q)+\pi}{2\pi}\right)+ \\
&\qquad \left. pq^{2m}\Phi\left(e^{2im\pi},-k,\frac{-i\log(a)+i\log(p)-i\log(q)+\pi}{2\pi}\right)\right)
\end{aligned} \tag{14}$$

Proof. Since the right-hand side of Equation (3) is equal to the addition of the right-hand side of Equations (7), (9), (11) and (13) we can equate the left-hand sides and simplify the gamma function to obtain the stated result. □

Corollary 1.

$$\begin{aligned}
&\int_0^\infty\int_0^\infty\int_0^\infty\int_0^\infty \frac{e^{-p(x+z)-q(t+y)}\log^k\left(\frac{a(x+y)}{t+z}\right)}{\sqrt{t+z}\sqrt{x+y}}dxdydzdt \\
&= \frac{1}{pq(p-q)^2}i^k 2^{2k+1}\pi^{k+1}\left((p+q)\zeta\left(-k,\frac{\pi-i\log(a)}{4\pi}\right)\right. \\
&\qquad -(p+q)\zeta\left(-k,\frac{3}{4}-\frac{i\log(a)}{4\pi}\right) \\
&\qquad +\sqrt{p}\sqrt{q}\left(-\zeta\left(-k,\frac{-i\log(a)+i\log(p)-i\log(q)+\pi}{4\pi}\right)\right. \\
&\qquad +\zeta\left(-k,\frac{-i\log(a)+i\log(p)-i\log(q)+3\pi}{4\pi}\right) \\
&\qquad -\zeta\left(-k,\frac{-i\log(a)-i\log(p)+i\log(q)+\pi}{4\pi}\right) \\
&\qquad \left.\left.+\zeta\left(-k,\frac{-i\log(a)-i\log(p)+i\log(q)+3\pi}{4\pi}\right)\right)\right)
\end{aligned} \tag{15}$$

Proof. Use Equation (14) and set $m = 1/2$ and simplify in terms of the Hurwitz zeta function $\zeta(s,v)$ using entry (4) below the Table in [12]. □

Corollary 2.

$$\begin{aligned}
&\int_0^\infty\int_0^\infty\int_0^\infty\int_0^\infty (t+z)^{-m}(x+y)^{m-1}e^{-p(x+z)-q(t+y)}dxdydzdt \\
&= -\frac{\pi p^{-m-1}q^{-m-1}\csc(\pi m)(p^m-q^m)(qp^m-pq^m)}{(p-q)^2}
\end{aligned} \tag{16}$$

Proof. Use Equation (14) and set $k=0$ and simplify using entry (2) below Table in [12]. □

Corollary 3.

$$\begin{aligned}&\int_0^\infty\int_0^\infty\int_0^\infty\int_0^\infty (t+z)^{-m}(x+y)^{m-1}\log\left(\frac{x+y}{t+z}\right)e^{-p(x+z)-q(t+y)}dxdydzdt\\&=\frac{1}{(-1+e^{2i\pi m})^2(p-q)^2}4\pi e^{2i\pi m}p^{-m-1}q^{-m-1}(-\pi\cos(\pi m)(p^m-q^m)(qp^m\\&\quad-pq^m)-\sin(\pi m)(pq^{2m}-qp^{2m})(\log(p)-\log(q)))\end{aligned} \tag{17}$$

Proof. Use Equation (14) and set $k=1$ and simplify using entry (3) in the Table below (64:12:7) [12]. □

Corollary 4.

$$\begin{aligned}&\int_0^\infty\int_0^\infty\int_0^\infty\int_0^\infty \frac{(t-x-y+z)\log^k\left(\frac{x+y}{t+z}\right)e^{-p(x+z)-q(t+y)}}{\sqrt{t+z}(x+y)^{3/2}}dxdydzdt\\&=\frac{1}{p^{3/2}q^{3/2}(p-q)^2}i^k2^{2k+1}\pi^{k+1}(p+q)(-2\sqrt{p}\sqrt{q}\zeta(-k,\tfrac{1}{4})\\&+2\sqrt{p}\sqrt{q}\zeta(-k,\tfrac{3}{4})+p\zeta(-k,\tfrac{i\log(p)-i\log(q)+\pi}{4\pi})\\&-p\zeta(-k,\tfrac{i\log(p)-i\log(q)+3\pi}{4\pi})+q\zeta(-k,\tfrac{-i\log(p)+i\log(q)+\pi}{4\pi})\\&-q\zeta(-k,\tfrac{-i\log(p)+i\log(q)+3\pi}{4\pi}))\end{aligned} \tag{18}$$

Proof. Use Equation (14) and form a second equation by replacing $m\to n$ and take their difference. Using the resulting equation set $m=1/2, n=-1/2, a=1$ and simplify in terms of the Hurwitz zeta function $\zeta(s,v)$ using entry (4) in the Table below (64:12:7) in [12]. □

Lemma 1.

$$\begin{aligned}&\int_0^\infty\int_0^\infty\int_0^\infty\int_0^\infty \frac{e^{-\frac{t}{2}-x-\frac{y}{2}-z}(t-x-y+z)}{\sqrt{t+z}(x+y)^{3/2}\log\left(\frac{x+y}{t+z}\right)}dxdydzdt\\&=3i\left(4\pi+\sqrt{2}\left(-H_{-\frac{1}{4}-\frac{i\log(2)}{4\pi}}+H_{-\frac{3}{4}-\frac{i\log(2)}{4\pi}}+2H_{-\frac{3}{4}+\frac{i\log(2)}{4\pi}}-2H_{-\frac{1}{4}+\frac{i\log(2)}{4\pi}}\right)\right)\end{aligned} \tag{19}$$

Proof. Use Equation (18) apply l'Hopital's rule to the right-hand side as $k\to-1$ and set $p=1, q=1/2$ and simplify in terms of the Harmonic number function H_n using Equations (44:1:1) and (64:4:1) in [12]. □

Lemma 2.

$$\int_0^\infty\int_0^\infty\int_0^\infty\int_0^\infty \frac{e^{-2t-x-2y-z}\left(\log^2\left(\frac{x+y}{t+z}\right)-\pi^2\right)}{\sqrt{t+z}\sqrt{x+y}\left(\log^2\left(\frac{x+y}{t+z}\right)+\pi^2\right)^2}dxdydzdt=\pi\left(\frac{47}{16}-\frac{\sqrt{2}}{\log^2(2)}\right) \tag{20}$$

and

Lemma 3.

$$\int_0^\infty\int_0^\infty\int_0^\infty\int_0^\infty \frac{2\pi e^{-2t-x-2y-z}\log\left(\frac{x+y}{t+z}\right)}{\sqrt{t+z}\sqrt{x+y}\left(\log^2\left(\frac{x+y}{t+z}\right)+\pi^2\right)^2}dxdydzdt=0 \tag{21}$$

Proof. Use Equation (15) and set $k=-2, a=-1, p=1, q=2$ and simplify by rationalizing the denominator and comparing real and imaginary parts and using Equation (9.521.1) in [11]. Note the integrand in Equation (21) is highly oscillatory. □

Corollary 5.

$$\int_0^\infty\int_0^\infty\int_0^\infty\int_0^\infty \frac{e^{-p(x+z)-q(t+y)}}{\sqrt{t+z}\sqrt{x+y}\log^2\left(-\frac{x+y}{t+z}\right)}dxdydzdt=\frac{\pi\left(\frac{p\left(-p^2+26pq+23q^2\right)}{(p-q)^2}-\frac{48\sqrt{p}\sqrt{q}}{(\log(p)-\log(q))^2}-q\right)}{24pq(p-q)^2} \tag{22}$$

Proof. Use Equation (15) and set $k = -2, a = -1$ and simplify using Equation (9.521.1) in [11]. □

Corollary 6.

$$\int_0^\infty\int_0^\infty\int_0^\infty\int_0^\infty \frac{e^{-t-2x-y-2z}(t-x-y+z)}{\sqrt{t+z}(x+y)^{3/2}\log\left(\frac{x+y}{t+z}\right)}dxdydzdt$$
$$= \frac{3}{8}i\left(4\pi+\sqrt{2}\left(-H_{-\frac{1}{4}-\frac{i\log(2)}{4\pi}}+H_{-\frac{3}{4}-\frac{i\log(2)}{4\pi}}+2H_{-\frac{3}{4}+\frac{i\log(2)}{4\pi}}-2H_{-\frac{1}{4}+\frac{i\log(2)}{4\pi}}\right)\right) \tag{23}$$

Proof. Use Equation (14) and form a second equation by replacing $m \to n$ and take their difference. Next, using the resulting equation set $k = -1$, $a = 1$, $m = 1/2$, $n = -1/2$, $p = 2$, $q = 1$ and simplify using Equations (44:1:1) and (64:4:1) in [12]. □

Corollary 7.

$$\int_0^\infty\int_0^\infty\int_0^\infty\int_0^\infty \frac{e^{-2(t+y)-x-z}\log^k\left(-\frac{x+y}{t+z}\right)}{\sqrt{t+z}\sqrt{x+y}}dxdydzdt$$
$$= i^k 2^{k-\frac{1}{2}}\pi^{k+1}(3\sqrt{2}\zeta(-k)+2^{k+1}(-\zeta(-k,\tfrac{1}{2}-\tfrac{i\log(2)}{4\pi})+\zeta(-k,1-\tfrac{i\log(2)}{4\pi})$$
$$-\zeta(-k,\tfrac{1}{2}+\tfrac{i\log(2)}{4\pi})+\zeta(-k,1+\tfrac{i\log(2)}{4\pi})-3\sqrt{2}\zeta(-k))) \tag{24}$$

Proof. Use Equation (14) and set $m = 1/2$ and simplify in terms of the Hurwitz zeta function $\zeta(s,v)$ using entry (4) in the Table below (64:12:7) in [12]. Next set $a = -1, p = 1, q = 2$ and simplify. □

Example 1.

$$\int_0^\infty\int_0^\infty\int_0^\infty\int_0^\infty \frac{e^{-2(t+y)-x-z}\log(\log(-\frac{x+y}{t+z}))}{\sqrt{t+z}\sqrt{x+y}}dxdydzdt$$
$$= \pi(\log(8)+\sqrt{2}(-\log\Gamma(-\tfrac{i\log(2)}{4\pi})-\log\Gamma(\tfrac{i\log(2)}{4\pi})+\log\Gamma(-\tfrac{1}{2}-\tfrac{i\log(2)}{4\pi})$$
$$+\log\Gamma(-\tfrac{1}{2}+\tfrac{i\log(2)}{4\pi})+\log(\pi+\tfrac{\log^2(2)}{4\pi})-2\log(\log(2))))$$
$$+\tfrac{1}{4}i(3-2\sqrt{2})\pi^2 \tag{25}$$

Proof. Use Equation (24) take the first partial derivative with respect to k and set $k = 0$ and simplify using Equation (64:10:2) in [12]. Note the integrand is highly oscillatory. □

Lemma 4.

$$\int_0^\infty\int_0^\infty\int_0^\infty\int_0^\infty \frac{e^{-t-2x-y-2z}}{\sqrt{t+z}\sqrt{x+y}\left(\log^2\left(\frac{x+y}{t+z}\right)+\pi^2\right)}dxdydzdt$$
$$= \frac{\log(64)+\sqrt{2}\left(-H_{-\frac{i\log(2)}{4\pi}}-H_{\frac{i\log(2)}{4\pi}}+H_{-\frac{1}{2}-\frac{i\log(2)}{4\pi}}+H_{-\frac{1}{2}+\frac{i\log(2)}{4\pi}}\right)}{4\pi} \tag{26}$$

and

Lemma 5.

$$\int_0^\infty\int_0^\infty\int_0^\infty\int_0^\infty \frac{e^{-t-2x-y-2z}\log\left(\frac{x+y}{t+z}\right)}{\sqrt{t+z}\sqrt{x+y}\left(\log^2\left(\frac{x+y}{t+z}\right)+\pi^2\right)}dxdydzdt = 0 \tag{27}$$

Proof. Use Equation (15) apply l'Hopitals' rule as $k \to -1$ and set $a = -1, p = 1, q = 2$ and simplify by rationalizing the denominator and comparing real and imaginary parts and using Equation (9.521.1) in [11]. □

Lemma 6.

$$
\begin{aligned}
&\int_0^\infty \int_0^\infty \int_0^\infty \int_0^\infty \frac{\left(\log^2\left(\frac{x+y}{t+z}\right)-\pi^2\right)e^{-p(x+z)-q(t+y)}}{\sqrt{t+z}\sqrt{x+y}\left(\log^2\left(\frac{x+y}{t+z}\right)+\pi^2\right)^2}dxdydzdt \\
&= \frac{\pi\left(-\frac{p^2}{q}-\frac{(p-q)^2}{p}-\frac{48(p-q)^2}{\sqrt{p}\sqrt{q}(\log(p)-\log(q))^2}+26p+23q\right)}{24(p-q)^4}
\end{aligned} \tag{28}
$$

and

Lemma 7.

$$
\int_0^\infty \int_0^\infty \int_0^\infty \int_0^\infty \frac{\pi \log\left(\frac{x+y}{t+z}\right)e^{-p(x+z)-q(t+y)}}{\sqrt{t+z}\sqrt{x+y}\left(\log^2\left(\frac{x+y}{t+z}\right)+\pi^2\right)^2}dxdydzdt = 0 \tag{29}
$$

Proof. Use Equation (15) set $k = -2, a = -1$ and simplify using Equation (64:3:5) in [12]. □

Lemma 8.

$$
\begin{aligned}
&\int_0^\infty \int_0^\infty \int_0^\infty \int_0^\infty \frac{e^{-2(t+y)-3(x+z)}\left(\log^2\left(\frac{x+y}{t+z}\right)-\pi^2\right)}{\sqrt{t+z}\sqrt{x+y}\left(\log^2\left(\frac{x+y}{t+z}\right)+\pi^2\right)^2}dxdydzdt \\
&= \frac{1}{24}\pi\left(\frac{715}{6}-\frac{8\sqrt{6}}{\log^2\left(\frac{3}{2}\right)}\right)
\end{aligned} \tag{30}
$$

Proof. Use Equation (28) set $p = 3, q = 2$ and simplify. Note the integrand is highly oscillatory. □

Lemma 9.

$$
\int_0^\infty \int_0^\infty \int_0^\infty \int_0^\infty \frac{e^{-\frac{t}{2}-x-\frac{y}{2}-z}\left(\log^2\left(\frac{x+y}{t+z}\right)-\pi^2\right)}{\sqrt{t+z}\sqrt{x+y}\left(\log^2\left(\frac{x+y}{t+z}\right)+\pi^2\right)^2}dxdydzdt = \pi\left(\frac{47}{2}-\frac{8\sqrt{2}}{\log^2(2)}\right) \tag{31}
$$

Proof. Use Equation (28) set $p = 1, q = 1/2$ and simplify. Note the integrand is highly oscillatory. □

6. Discussion

In the current work, the authors use their contour integration method to derive a quadruple integral based on the Lerch function that does not exist in the current literature. The formulae derived in this work use our method [9], which can be used to derive other quadruple integrals. The authors will use their method in future work to generate more multiple definite integrals. Wolfram Mathematica was used to verify numerical values of the parameters in the integral formulae.

Author Contributions: Conceptualization, R.R.; methodology, R.R.; draft preparation, R.R.; funding acquisition, A.S.; supervision, A.S. Both authors have read and agreed to the published version of the manuscript.

Funding: This research is supported by NSERC Canada under Grant 504070.

Data Availability Statement: Not applicable.

Institutional Review Board Statement: Not applicable.

Informed Consent Statement: Not applicable.

Conflicts of Interest: The authors declare no conflict of interest.

References

1. McClure, J.P.; Wong, R. Asymptotic expansion of a quadruple integral involving a Bessel function. *J. Comput. Appl. Math.* **1990**, *33*, 199–215. [CrossRef]
2. Watson, G.N. A Quadruple Integral. *Math. Gaz.* **1959**, *43*, 280–283. [CrossRef]
3. Raman, C.V. A classical derivation of the Compton effect. *Indian J. Phys.* **1928**, *3*, 357–369.
4. Lee, J.; Seo, I. Radiation impedance computations of a square piston in a rigid infinite baffle. *J. Sound Vib.* **1996**, *198*, 299–312. [CrossRef]
5. Davy, J.L.; Larner, D.J.; Wareing, R.R.; Pearse, J.R. The acoustic radiation impedance of a rectangular panel. *Build. Environ.* **2015**, *92*, 743–755. [CrossRef]
6. Chandrasekhar, S. A Statistical Basis for the Theory of Stellar Scintillation. *Mon. Not. R. Astron. Soc.* **1952**, *112*, 475–483. [CrossRef]
7. Kerbel, G.D. *Gyroelastic Fluids*; Lawrence Livermore Lab.: Livermore, CA, USA, 1981. [CrossRef]
8. Temme, N.M; de Bruin, R. *Quadruple Integral Equations for the Charged Disc and Coplanar Annulus*; Toegepaste Wiskunde; Stichting Mathematisch Centrum: Amsterdam, The Netherlands, 1981.
9. Reynolds, R.; Stauffer, A. A Method for Evaluating Definite Integrals in Terms of Special Functions with Examples. *Int. Math. Forum* **2020**, *15*, 235–244. [CrossRef]
10. Prudnikov, A.P.; Brychkov, Y.A.; Marichev, O.I. *Integrals and Series, More Special Functions*; USSR Academy of Sciences: Moscow, Russia, 1990; Volume 1.
11. Gradshteyn, I.S.; Ryzhik, I.M. *Tables of Integrals, Series and Products*, 6th ed.; Academic Press: Cambridge, MA, USA, 2000.
12. Oldham, K.B.; Myland, J.C.; Spanier, J. *An Atlas of Functions: With Equator, the Atlas Function Calculator*, 2nd ed.; Springer: New York, NY, USA, 2009.

symmetry

MDPI

Article

On New Generalized Dunkel Type Integral Inequalities with Applications

Dong-Sheng Wang [1], Huan-Nan Shi [2], Chun-Ru Fu [3] and Wei-Shih Du [4,*]

1 Basic Courses Department, Beijing Polytechnic, Beijing 100176, China; wangdongshen@bpi.edu.cn
2 Department of Electronic Information, Teacher's College, Beijing Union University, Beijing 100011, China; sfthuannan@buu.edu.cn
3 Institute of Mathematics and Physics, Beijing Union University, Beijing 100101, China; bytchunru@buu.edu.cn
4 Department of Mathematics, National Kaohsiung Normal University, Kaohsiung 82444, Taiwan
* Correspondence: wsdu@mail.nknu.edu.tw

Abstract: In this paper, by applying majorization theory, we study the Schur convexity of functions related to Dunkel integral inequality. We establish some new generalized Dunkel type integral inequalities and their applications to inequality theory.

Keywords: Dunkel type integral inequality; Schur-convexity; majorization theory; arithmetic mean-geometric mean (AM-GM) inequality

MSC: 26A51; 26E60; 26D15; 26D05; 34K38; 39B62

Citation: Wang, D.-S.; Shi, H.-N.; Fu, C.-R.; Du, W.-S. On New Generalized Dunkel Type Integral Inequalities with Applications. *Symmetry* **2021**, *13*, 1576. https://doi.org/10.3390/sym13091576

Academic Editor: Paolo Emilio Ricci

Received: 6 August 2021
Accepted: 23 August 2021
Published: 27 August 2021

Publisher's Note: MDPI stays neutral with regard to jurisdictional claims in published maps and institutional affiliations.

1. Introduction and Preliminaries

Over the last half a century, rapid developments in inequality theory and its applications have contributed greatly to many branches of mathematics such as linear and nonlinear analysis, differential equations, finance, statistics, physics, fractional calculus, and so on; for more details, one can refer to [1–4] and the references therein.

The original Dunkel integral inequality can be stated as follows.

Theorem 1 (original Dunkel integral inequality; see [1–3,5,6]). *Let $f(x)$ be a continuous real-valued function on $[a,b]$ which is not identically zero and satisfies $0 \le f(x) \le M$ for all $x \in [a,b]$. Then*

$$0 < \left(\int_a^b f(x)dx\right)^2 - \left(\int_a^b f(x)\cos x dx\right)^2 - \left(\int_a^b f(x)\sin x dx\right)^2 \le \frac{1}{12}M^2(b-a)^2. \quad (1)$$

There are many ways to prove Dunkel integral inequality (see [1–3,5,6] and references therein). Some interesting proofs of Dunkel integral inequality are the probabilistic method (see, e.g., [1]), re-integral method (see [2,3]), and so on.

In fact, if $f(x)$ is a nonnegative continuous real-valued function on $[a,b]$ (here, f is allowed to be a zero function), then from (1) one deduces the following fascinating concise inequality:

$$\left(\int_a^b f(x)\cos x dx\right)^2 + \left(\int_a^b f(x)\sin x dx\right)^2 \le \left(\int_a^b f(x)dx\right)^2. \quad (2)$$

In 1923, Professor Issai Schur first systematically studied the functions preserving the ordering of majorization. In Schur's honor, such functions are named to have "Schur-convexity". During the previous more than four decades, majorization theory and Schur-convexity have been applied widely in many areas of mathematics including integral inequality, stochastic matrices, rearrangement theory, analytic inequalities, information theory, quantum correlations, quantum cryptography, combinatorial optimization, and other related fields (see, e.g., [7–12]).

Let us recall some basic definitions and notation that will be needed in this paper.

Definition 1 (see [4,8]). *Let Ω be a nonempty subset of $\mathbb{R}^n$.*

(*i*) *Let $x = (x_1, \ldots, x_n)$ and $y = (y_1, \ldots, y_n) \in \mathbb{R}^n$. x is said to be majorized by y (in symbols $x \prec y$) if $\sum_{i=1}^{k} x_{[i]} \leq \sum_{i=1}^{k} y_{[i]}$ for $k = 1, 2, \ldots, n-1$ and $\sum_{i=1}^{n} x_i = \sum_{i=1}^{n} y_i$, where $x_{[1]} \geq \cdots \geq x_{[n]}$ and $y_{[1]} \geq \cdots \geq y_{[n]}$ are rearrangements of x and y in a descending order;*

(*ii*) *Ω is called convex if $\alpha x + \beta y \in \Omega$ for any $x, y \in \Omega$ and $\alpha, \beta \geq 0$ with $\alpha + \beta = 1$;*

(*iii*) *Ω is called symmetric if $x \in \Omega$ implies $Px \in \Omega$ for every $n \times n$ permutation matrix P;*

(*iv*) *A function $\varphi : \Omega \to \mathbb{R}$ is called symmetric if for every permutation matrix P, $\varphi(Px) = \varphi(x)$ for all $x \in \Omega$;*

(*v*) *A function $\varphi : \Omega \to \mathbb{R}$ is said to be Schur convex on Ω if $x \prec y$ on Ω implies $\varphi(x) \leq \varphi(y)$. φ is said to be Schur concave on Ω if and only if $-\varphi$ is Schur convex.*

The paper is divided into five sections. In Sections 2 and 3, by applying majorization theory, we present some new generalized Dunkel type integral inequalities and new Dunkel (p)-type integral inequalities for $p \geq 2$. As applications of our new results, some new integral inequalities are established in Section 4. Finally, some summary and conclusions are given in Section 5.

2. Some Generalizations of Dunkel Integral Inequality

The following two known results are important for proving our new theorem.

Lemma 1 (see [4]). *Let $a \leq b$. Let $u(t) := ta + (1-t)b$ and $v(t) := tb + (1-t)a$ for $\frac{1}{2} \leq t_1 \leq t_2 \leq 1$ or $0 \leq t_1 \leq t_2 \leq \frac{1}{2}$. Then*

$$\left(\frac{a+b}{2}, \frac{a+b}{2}\right) \prec (u(t_2), v(t_2)) \prec (u(t_1), v(t_1)) \prec (a, b).$$

Lemma 2 (see [4,7]). *Let $\Omega \subset \mathbb{R}^n$ be a nonempty convex set and has a nonempty interior set Ω°. Let $\varphi : \Omega \to \mathbb{R}$ be continuous on Ω and differentiable in Ω°. Then, φ is a Schur convex (resp. Schur concave) function, if and only if it is symmetric on Ω and*

$$(x_1 - x_2)\left(\frac{\partial \varphi}{\partial x_1} - \frac{\partial \varphi}{\partial x_2}\right) \geq 0 \quad (resp. \leq 0)$$

holds for any $x = (x_1, \cdots, x_n) \in \Omega^\circ$.

Remark 1. *It is worth noticing that Lemma 2 is equivalent to the following:*
φ is a Schur convex (resp. Schur concave) function, if and only if it is symmetric on Ω and

$$\frac{\partial \varphi}{\partial x_i} \geq \frac{\partial \varphi}{\partial x_{i+1}} \quad (resp. \leq 0), i = 1, 2, \ldots, n-1.$$

for all $x \in D \cap \Omega$, where $D = \{x : x_1 \geq \cdots \geq x_n\}$.

With the help of Lemmas 1 and 2, we can establish the following crucial result.

Theorem 2. *Let I be an interval of $\mathbb{R}$. Assume that $f(x)$ and $g(x)$ are two nonnegative continuous real-valued functions on I, and $\kappa(x)$ and $\lambda(x)$ are two continuous real-valued functions on I. Define $L : I \times I \to \mathbb{R}$ by*

$$\begin{aligned} L(a,b) = & \int_a^b f(x)dx \int_a^b g(x)dx \\ & - \int_a^b f(x)\kappa(x)dx \int_a^b g(x)\kappa(x)dx - \int_a^b f(x)\lambda(x)dx \int_a^b g(x)\lambda(x)dx \end{aligned}$$

for any $(a,b) \in I \times I$. Then the following holds:

(i) If $\kappa(x) \cdot \kappa(b) + \lambda(x) \cdot \lambda(b) \leq 1$ and $\kappa(x) \cdot \kappa(a) + \lambda(x) \cdot \lambda(a) \leq 1$ for $x, a, b \in I$, then L is Schur convex on $I \times I$;

(ii) If $\kappa(x) \cdot \kappa(b) + \lambda(x) \cdot \lambda(b) \geq 1$ and $\kappa(x) \cdot \kappa(a) + \lambda(x) \cdot \lambda(a) \geq 1$ for $x, a, b \in I$, then L is Schur concave on $I \times I$.

Proof. Obviously, $L(a,b)$ is a symmetric operator for $a, b \in I$. So, without loss of generality, we may assume that $b \geq a$. Since

$$\begin{aligned} \frac{\partial L}{\partial b} = & f(b) \int_a^b g(x)dx + g(b) \int_a^b f(x)dx \\ & - f(b)\kappa(b) \int_a^b g(x)\kappa(x)dx - g(b)\kappa(b) \int_a^b f(x)\kappa(x)dx \\ & - f(b)\lambda(b) \int_a^b g(x)\lambda(x)dx - g(b)\lambda(b) \int_a^b f(x)\lambda(x)dx \end{aligned}$$

and

$$\begin{aligned} \frac{\partial L}{\partial a} = & - f(a) \int_a^b g(x)dx - g(a) \int_a^b f(x)dx \\ & + f(a)\kappa(a) \int_a^b g(x)\kappa(x)dx + g(a)\kappa(a) \int_a^b f(x)\kappa(x)dx \\ & + f(a)\lambda(a) \int_a^b g(x)\lambda(x)dx + g(a)\lambda(a) \int_a^b f(x)\lambda(x)dx, \end{aligned}$$

we have

$$\begin{aligned} \Delta := & (b-a)\left(\frac{\partial L}{\partial b} - \frac{\partial L}{\partial a}\right) \\ = & (f(a) + f(b)) \int_a^b g(x)dx + (g(a) + g(b)) \int_a^b f(x)dx \\ & - (f(b)\kappa(b) + f(a)\kappa(a)) \int_a^b g(x)\kappa(x)dx - (f(b)\lambda(b) + f(a)\lambda(a)) \int_a^b g(x)\lambda(x)dx \\ & - (g(b)\kappa(b) + g(a)\kappa(a)) \int_a^b f(x)\kappa(x)dx - (g(b)\lambda(b) + g(a)\lambda(a)) \int_a^b f(x)\lambda(x)dx \\ = & f(b) \int_a^b g(x)(1 - \kappa(b)\kappa(x) - \lambda(b)\lambda(x))dx + f(a) \int_a^b g(x)(1 - \kappa(a)\kappa(x) \\ & - \lambda(a)\lambda(x))dx + g(b) \int_a^b f(x)(1 - \kappa(b)\kappa(x) - \lambda(b)\lambda(x))dx \\ & + g(a) \int_a^b f(x)(1 - \kappa(a)\kappa(x) - \lambda(a)\lambda(x))dx. \end{aligned}$$

(i) When $\kappa(x) \cdot \kappa(b) + \lambda(x) \cdot \lambda(b) \leq 1$ and $\kappa(x) \cdot \kappa(a) + \lambda(x) \cdot \lambda(a) \leq 1$, we have $\Delta \geq 0$. By Lemma 2, L is Schur convex on $I \times I$.

(ii) When $\kappa(x) \cdot \kappa(b) + \lambda(x) \cdot \lambda(b) \geq 1$ and $\kappa(x) \cdot \kappa(a) + \lambda(x) \cdot \lambda(a) \geq 1$, we have $\Delta \leq 0$. By Lemma 2, L is Schur concave on $I \times I$.

The proof is completed. □

We now present the following generalized Dunkel type integral inequality which is one of the main results of this paper.

Theorem 3. *Let I be an interval of $\mathbb{R}$. Assume that $f(x)$ and $g(x)$ are two nonnegative continuous real-valued functions on I, and $\kappa(x)$ and $\lambda(x)$ are two continuous real-valued functions on I. Let $u(t) := ta + (1-t)b$ and $v(t) := tb + (1-t)a$, for $\frac{1}{2} \leq t \leq 1$. Then the following holds:*

(i) If $\kappa(x) \cdot \kappa(b) + \lambda(x) \cdot \lambda(b) \leq 1$ and $\kappa(x) \cdot \kappa(a) + \lambda(x) \cdot \lambda(a) \leq 1$ for $x, a, b \in I$, then

$$\begin{aligned}
&\int_a^b f(x)\kappa(x)dx \int_a^b g(x)\kappa(x)dx + \int_a^b f(x)\lambda(x)dx \int_a^b g(x)\lambda(x)dx \\
\leq &\int_a^b f(x)dx \int_a^b g(x)dx - \int_{u(t)}^{v(t)} f(x)dx \int_{u(t)}^{v(t)} g(x)dx \\
&+ \int_{u(t)}^{v(t)} f(x)\kappa(x)dx \int_{u(t)}^{v(t)} g(x)\kappa(x)dx + \int_{u(t)}^{v(t)} f(x)\lambda(x)dx \int_{u(t)}^{v(t)} g(x)\lambda(x)dx \\
\leq &\int_a^b f(x)dx \int_a^b g(x)dx;
\end{aligned}$$

(ii) If $\kappa(x) \cdot \kappa(b) + \lambda(x) \cdot \lambda(b) \geq 1$ and $\kappa(x) \cdot \kappa(a) + \lambda(x) \cdot \lambda(a) \geq 1$ for $x, a, b \in I$, then

$$\begin{aligned}
&\int_a^b f(x)\kappa(x)dx \int_a^b g(x)\kappa(x)dx + \int_a^b f(x)\lambda(x)dx \int_a^b g(x)\lambda(x)dx \\
\geq &\int_a^b f(x)dx \int_a^b g(x)dx - \int_{u(t)}^{v(t)} f(x)dx \int_{u(t)}^{v(t)} g(x)dx \\
&+ \int_{u(t)}^{v(t)} f(x)\kappa(x)dx \int_{u(t)}^{v(t)} g(x)\kappa(x)dx + \int_{u(t)}^{v(t)} f(x)\lambda(x)dx \int_{u(t)}^{v(t)} g(x)\lambda(x)dx \\
\geq &\int_a^b f(x)dx \int_a^b g(x)dx.
\end{aligned}$$

Proof. We only show case (i) and a similar argument could be made for the case (ii). Define $L : I \times I \to \mathbb{R}$ by

$$\begin{aligned}
L(a,b) = &\int_a^b f(x)dx \int_a^b g(x)dx \\
&- \int_a^b f(x)\kappa(x)dx \int_a^b g(x)\kappa(x)dx - \int_a^b f(x)\lambda(x)dx \int_a^b g(x)\lambda(x)dx
\end{aligned}$$

for any $(a,b) \in I \times I$. If $\kappa(x) \cdot \kappa(b) + \lambda(x) \cdot \lambda(b) \leq 1$ and $\kappa(x) \cdot \kappa(a) + \lambda(x) \cdot \lambda(a) \leq 1$ for $x, a, b \in I$, by applying Theorem 2 (i), we show that L is Schur convex on $I \times I$. On the other hand, by using Lemma 1, we get

$$L(a,b) \geq L(u(t), v(t)) \geq L\left(\frac{a+b}{2}, \frac{a+b}{2}\right) = 0.$$

Hence, we obtain

$$\begin{aligned}
&\int_a^b f(x)dx \int_a^b g(x)dx - \int_a^b f(x)\kappa(x)dx \int_a^b g(x)\kappa(x)dx \\
&- \int_a^b f(x)\lambda(x)dx \int_a^b g(x)\lambda(x)dx \\
&\geq \int_{u(t)}^{v(t)} f(x)dx \int_{u(t)}^{v(t)} g(x)dx - \int_{u(t)}^{v(t)} f(x)\kappa(x)dx \int_{u(t)}^{v(t)} g(x)\kappa(x)dx \\
&- \int_{u(t)}^{v(t)} f(x)\lambda(x)dx \int_{u(t)}^{v(t)} g(x)\lambda(x)dx \\
&\geq 0,
\end{aligned}$$

which implies

$$\begin{aligned}
&\int_a^b f(x)\kappa(x)dx \int_a^b g(x)\kappa(x)dx + \int_a^b f(x)\lambda(x)dx \int_a^b g(x)\lambda(x)dx \\
&\leq \int_a^b f(x)dx \int_a^b g(x)dx - \int_{u(t)}^{v(t)} f(x)dx \int_{u(t)}^{v(t)} g(x)dx \\
&+ \int_{u(t)}^{v(t)} f(x)\kappa(x)dx \int_{u(t)}^{v(t)} g(x)\kappa(x)dx + \int_{u(t)}^{v(t)} f(x)\lambda(x)dx \int_{u(t)}^{v(t)} g(x)\lambda(x)dx \\
&\leq \int_a^b f(x)dx \int_a^b g(x)dx.
\end{aligned}$$

The proof is completed. □

As a direct consequence of Theorem 3, we can obtain the following generalized Dunkel integral inequality.

Theorem 4 (Generalized Dunkel integral inequality). *Let $f(x)$ and $g(x)$ be two nonnegative continuous real-valued functions on $[a,b]$ and m be any real number. Then*

$$\begin{aligned}
&\left(\int_a^b f(x)\cos mx dx\right)\left(\int_a^b g(x)\cos mx dx\right) + \left(\int_a^b f(x)\sin mx dx\right)\left(\int_a^b g(x)\sin mx dx\right) \\
&\leq \left(\int_a^b f(x)dx\right)\left(\int_a^b g(x)dx\right).
\end{aligned} \tag{3}$$

Proof. In theorem 3, we take $I = [a,b]$, $\kappa(x) = \cos mx$, and $\lambda(x) = \sin mx$ for $x \in I$. Thus, $\kappa(x)$ and $\lambda(x)$ are two continuous real-valued functions on I. Clearly, we have

$$\begin{aligned}
\kappa(x)\cdot\kappa(b) + \lambda(x)\cdot\lambda(b) &= \cos mx \cdot \cos mb + \sin mx \cdot \sin mb \\
&= \cos m(b-x) \leq 1
\end{aligned}$$

and

$$\begin{aligned}
\kappa(x)\cdot\kappa(a) + \lambda(x)\cdot\lambda(a) &= \cos mx \cdot \cos ma + \sin mx \cdot \sin ma \\
&= \cos m(a-x) \leq 1.
\end{aligned}$$

Thus, all the assumptions of Theorem 3 (i) are satisfied. Therefore the desired conclusion follows immediately from Theorem 3. □

The following generalized Dunkel integral inequality is an immediate consequence of Theorem 4.

Corollary 1 (Generalized Dunkel integral inequality). *Let $f(x)$ be a continuous nonnegative real-valued function on $[a,b]$ and m be any real number. Then*

$$\left(\int_a^b f(x)\cos mxdx\right)^2 + \left(\int_a^b f(x)\sin mxdx\right)^2 \leq \left(\int_a^b f(x)dx\right)^2. \tag{4}$$

Remark 2. *It is worth noticing that inequality (3) in Theorem 4 and inequality (4) in Corollary 1 are real generalizations of inequality (2).*

3. A New Dunkel (p)-Type Integral Inequality for $p \geq 2$

In this section, we will present a new Dunkel (p)-type integral inequality for $p \geq 2$. In order to prove our results, we need the following important auxiliary lemma.

Lemma 3. *Let $k \in \mathbb{N} \cup \{0\}$. Denote $I_k := \left[2k\pi, 2k\pi + \frac{\pi}{2}\right]$. Assume that $f(x)$ is a nonnegative continuous real-valued function on I_k. Define $M : I_k \times I_k \to \mathbb{R}$ by*

$$\begin{aligned} M(a,b) =& \left[(b-a)\int_a^b f(x)\sin xdx\right]^p \\ &+ \left[(b-a)\int_a^b f(x)\cos xdx\right]^p - \left[(b-a)\int_a^b f(x)dx\right]^p \end{aligned}$$

for $(a,b) \in I_k \times I_k$. If $p \geq 2$, then M is Schur concave on $I_k \times I_k$.

Proof. It is obvious that $M(a,b)$ is symmetric for a, b. Hence, without loss of generality, we may assume that $b \geq a$. By Corollary 1, we have

$$\begin{aligned} \frac{\partial M}{\partial b} =& p\left[(b-a)\int_a^b f(x)\sin xdx\right]^{p-1}\left[\int_a^b f(x)\sin xdx + (b-a)f(b)\sin b\right] \\ &+ p\left[(b-a)\int_a^b f(x)\cos xdx\right]^{p-1}\left[\int_a^b f(x)\cos xdx + (b-a)f(b)\cos b\right] \\ &- p\left[(b-a)\int_a^b f(x)dx\right]^{p-1}\left[\int_a^b f(x)dx + (b-a)f(b)\right] \\ =& p(b-a)^{p-1}\left[\int_a^b f(x)\sin xdx\right]^{p-2} \\ &\times \left[\left(\int_a^b f(x)\sin xdx\right)^2 + (b-a)f(b)\sin b\int_a^b f(x)\sin xdx\right] \\ &+ p(b-a)^{p-1}\left[\int_a^b f(x)\cos xdx\right]^{p-2} \\ &\times \left[\left(\int_a^b f(x)\cos xdx\right)^2 + (b-a)f(b)\cos b\int_a^b f(x)\cos xdx\right] \\ &- p(b-a)^{p-1}\left[\int_a^b f(x)dx\right]^{p-2}\left[\left(\int_a^b f(x)dx\right)^2 + (b-a)f(b)\int_a^b f(x)dx\right] \end{aligned}$$

$$
\begin{aligned}
&\leq p(b-a)^{p-1}\left(\int_a^b f(x)dx\right)^{p-2} \\
&\times \left[\left(\int_a^b f(x)\sin x dx\right)^2 + \left(\int_a^b f(x)\cos x dx\right)^2 - \left(\int_a^b f(x)dx\right)^2\right. \\
&\left. +(b-a)f(b)\int_a^b f(x)(\sin b\cdot\sin x + \cos b\cdot\cos x - 1)dx\right] \\
&= p(b-a)^{p-1}\left(\int_a^b f(x)dx\right)^{p-2} \\
&\times \left[\left(\int_a^b f(x)\sin x dx\right)^2 + \left(\int_a^b f(x)\cos x dx\right)^2 - \left(\int_a^b f(x)dx\right)^2\right. \\
&\left. +(b-a)f(b)\int_a^b f(x)(\cos(b-x)) - 1)dx\right] \\
&\leq 0
\end{aligned}
$$

and

$$
\begin{aligned}
\frac{\partial M}{\partial a} &= p\left[(b-a)\int_a^b f(x)\sin x dx\right]^{p-1}\left[-\int_a^b f(x)\sin x dx - (b-a)f(a)\sin a\right] \\
&+ p\left[(b-a)\int_a^b f(x)\cos x dx\right]^{p-1}\left[-\int_a^b f(x)\cos x dx - (b-a)f(a)\cos a\right] \\
&- p\left[(b-a)\int_a^b f(x)dx\right]^{p-1}\left[-\int_a^b f(x)dx - (b-a)f(a)\right] \\
&= -p(b-a)^{p-1}\left[\int_a^b f(x)\sin x dx\right]^{p-2} \\
&\times \left[\left(\int_a^b f(x)\sin x dx\right)^2 + (b-a)f(a)\sin a\int_a^b f(x)\sin x dx\right] \\
&- p(b-a)^{p-1}\left[\int_a^b f(x)\cos x dx\right]^{p-2} \\
&\times \left[\left(\int_a^b f(x)\cos x dx\right)^2 + (b-a)f(a)\cos a\int_a^b f(x)\cos x dx\right] \\
&+ p(b-a)^{p-1}\left[\int_a^b f(x)dx\right]^{p-2}\left[\left(\int_a^b f(x)dx\right)^2 + (b-a)f(a)\int_a^b f(x)dx\right] \\
&\geq -p(b-a)^{p-1}\left(\int_a^b f(x)dx\right)^{p-2} \\
&\times \left[\left(\int_a^b f(x)\sin x dx\right)^2 + \left(\int_a^b f(x)\cos x dx\right)^2 - \left(\int_a^b f(x)dx\right)^2\right. \\
&\left. +(b-a)f(a)\int_a^b f(x)(\sin a\cdot\sin x + \cos a\cdot\cos x - 1)dx\right]
\end{aligned}
$$

$$
\begin{aligned}
&= -p(b-a)^{p-1}\left(\int_a^b f(x)dx\right)^{p-2} \\
&\times \left[\left(\int_a^b f(x)\sin x dx\right)^2 + \left(\int_a^b f(x)\cos x dx\right)^2 - \left(\int_a^b f(x)dx\right)^2\right. \\
&\left. +(b-a)f(a)\int_a^b f(x)(\cos(a-x)) - 1)dx\right] \\
&\geq 0,
\end{aligned}
$$

which deduce

$$\Delta' := (b-a)\left(\frac{\partial M}{\partial b} - \frac{\partial M}{\partial a}\right) \leq 0.$$

By Lemma 2, M is Schur concave on $I_k \times I_k$. The proof is completed. □

The following result is a new Dunkel (p)-type integral inequality for $p \geq 2$.

Theorem 5. *Let $k \in \mathbb{N} \cup \{0\}$. Denote $I_k := \left[2k\pi, 2k\pi + \frac{\pi}{2}\right]$. Assume that $f(x)$ is a nonnegative continuous real-valued function on I_k. If $p \geq 2$ and $[a,b] \subseteq I_k$, then*

$$\left(\int_a^b f(x)\cos x dx\right)^p + \left(\int_a^b f(x)\sin x dx\right)^p \leq \left(\int_a^b f(x)dx\right)^p.$$

Proof. Define $M : I_k \times I_k \to \mathbb{R}$ by

$$
\begin{aligned}
M(a,b) =& \left[(b-a)\int_a^b f(x)\sin x dx\right]^p \\
&+ \left[(b-a)\int_a^b f(x)\cos x dx\right]^p - \left[(b-a)\int_a^b f(x)dx\right]^p
\end{aligned}
$$

for $(a,b) \in I_k \times I_k$. By Lemmas 1 and 3, we obtain

$$M(a,b) \leq M\left(\frac{a+b}{2}, \frac{a+b}{2}\right) = 0,$$

which means that

$$\left(\int_a^b f(x)\cos x dx\right)^p + \left(\int_a^b f(x)\sin x dx\right)^p \leq \left(\int_a^b f(x)dx\right)^p.$$

□

The following result is immediate from Theorem 5.

Corollary 2. *Let $n \in \mathbb{N}$. Let $k_1, k_2, \cdots, k_n \in \mathbb{N} \cup \{0\}$. Assume that $f_i(x)$ is a nonnegative continuous real-valued function on $\left[2k_i\pi, 2k_i\pi + \frac{\pi}{2}\right]$ and $[a_i, b_i] \subseteq \left[2k_i\pi, 2k_i\pi + \frac{\pi}{2}\right]$ for any $1 \leq i \leq n$. If $p \geq 2$, then*

$$\sum_{i=1}^n \left(\int_{a_i}^{b_i} f_i(x)\cos x dx\right)^p + \sum_{i=1}^n \left(\int_{a_i}^{b_i} f_i(x)\sin x dx\right)^p \leq \sum_{i=1}^n \left(\int_{a_i}^{b_i} f_i(x)dx\right)^p.$$

4. Some New Integral Inequalities

In this section, we will provide some new integral inequalities by applying our main results.

Lemma 4 (Bessel inequality; see [1]). *Let $f(x)$ be a continuous or a piecewise continuous nonnegative function on $[0,2\pi]$. The Fourier series of $f(x)$ is*

$$\frac{a_0}{2}+\sum_{m=1}^{\infty}(a_m\cos mx+b_m\sin mx),$$

where $a_0=\frac{1}{\pi}\int_0^{2\pi}f(x)dx$, $a_m=\frac{1}{\pi}\int_0^{2\pi}f(x)\cos mxdx$, and $b_m=\frac{1}{\pi}\int_0^{2\pi}f(x)\sin mxdx$, for $m\in\mathbb{N}$. Then

$$\frac{a_0^2}{2}+\sum_{m=1}^{n}(a_m^2+b_m^2)\le\frac{1}{\pi}\int_0^{2\pi}f^2(x)dx.$$

Lemma 5 (see [1]). *Let $f(x)$ be a nonnegative integrable concave function on $[a,b]$. If $p\ge 1$, then*

$$\int_a^b f^p(x)dx\le\frac{2^p}{(b-a)^{p-1}(p+1)}\left(\int_a^b f(x)dx\right)^p.$$

Theorem 6. *Let $f(x)$ be a nonnegative continuous concave function on $[0,2\pi]$. Then*

$$\left(\int_0^{2\pi}f(x)\sin xdx\right)^2+\left(\int_0^{2\pi}f(x)\cos xdx\right)^2\le\frac{4}{9\pi^2}\left(\int_0^{2\pi}f(x)dx\right)^2.$$

Proof. Using the notations in Lemma 4 and applying Theorem 4, we get

$$\sum_{m=1}^{n}(a_m^2+b_m^2)\le na_0^2. \tag{5}$$

By combining (5) with Bessel inequality (see Lemma 4), we obtain

$$\sum_{m=1}^{n}(a_m^2+b_m^2)\le 2n\left(\frac{1}{\pi}\int_0^{2\pi}f^2(x)dx-\sum_{m=1}^{n}(a_m^2+b_m^2)\right)$$

which implies

$$\sum_{m=1}^{n}(a_m^2+b_m^2)\le\frac{2n}{(2n+1)\pi}\int_0^{2\pi}f^2(x)dx.$$

Let $n=1$. By applying Lemma 5, we obtain

$$\begin{aligned}\left(\int_0^{2\pi}f(x)\sin xdx\right)^2+\left(\int_0^{2\pi}f(x)\cos xdx\right)^2&\le\frac{2}{3\pi}\int_0^{2\pi}f^2(x)dx\\&\le\frac{4}{9\pi^2}\left(\int_0^{2\pi}f(x)dx\right)^2.\end{aligned}$$

The proof is completed. □

Theorem 7. *Let $f(x)$ be a nonnegative continuous function on $[a,b]$. If $0<a\le x\le b<\frac{\pi}{2}$, then*

$$\begin{aligned}&\left(\int_a^b f(x)\tan xdx\right)^2+\left(\int_a^b f(x)\cot xdx\right)^2\\&\ge 2\left(\int_a^b f(x)dx\right)^2+\left(\int_{u(t)}^{v(t)}f(x)\tan xdx\right)^2+\left(\int_{u(t)}^{v(t)}f(x)\cot xdx\right)^2-2\left(\int_{u(t)}^{v(t)}f(x)dx\right)^2\\&\ge 2\left(\int_a^b f(x)dx\right)^2,\end{aligned}$$

where $u(t) = ta + (1-t)b$ *and* $v(t) = tb + (1-t)a$, *for* $\frac{1}{2} \le t \le 1$.

Proof. Let $\kappa(x) = \frac{\sqrt{2}}{2}\tan x$ and $\lambda(x) = \frac{\sqrt{2}}{2}\cot x$ for $x \in [a,b]$. By the arithmetic mean-geometric mean (AM-GM) inequality, we have

$$\begin{aligned}\kappa(x)\cdot\kappa(b) + \lambda(x)\cdot\lambda(b) =& \frac{1}{2}\tan x \cdot \tan b + \frac{1}{2}\cot x \cdot \cot b \\ \ge& (\tan x \cdot \tan b \cdot \cot x \cdot \cot b)^{\frac{1}{2}} = 1.\end{aligned}$$

In the same way, we also have $\kappa(x)\cdot\kappa(a) + \lambda(x)\cdot\lambda(a) \ge 1$. By Theorem 2 (ii), we obtain

$$\begin{aligned}&\left(\int_a^b f(x)\frac{\sqrt{2}\tan x}{2}dx\right)^2 + \left(\int_a^b f(x)\frac{\sqrt{2}\cot x}{2}dx\right)^2 \\ \ge& \left(\int_a^b f(x)dx\right)^2 + \left(\int_{u(t)}^{v(t)} f(x)\frac{\sqrt{2}\tan x}{2}dx\right)^2 \\ &+ \left(\int_{u(t)}^{v(t)} f(x)\frac{\sqrt{2}\cot x}{2}dx\right)^2 - \left(\int_{u(t)}^{v(t)} f(x)dx\right)^2 \\ \ge& \left(\int_a^b f(x)dx\right)^2,\end{aligned}$$

which deduces

$$\begin{aligned}&\left(\int_a^b f(x)\tan x dx\right)^2 + \left(\int_a^b f(x)\cot x dx\right)^2 \\ \ge& 2\left(\int_a^b f(x)dx\right)^2 + \left(\int_{u(t)}^{v(t)} f(x)\tan x dx\right)^2 \\ &+ \left(\int_{u(t)}^{v(t)} f(x)\cot x dx\right)^2 - 2\left(\int_{u(t)}^{v(t)} f(x)dx\right)^2 \\ \ge& 2\left(\int_a^b f(x)dx\right)^2.\end{aligned}$$

The proof is completed. □

Theorem 8. *Let* $0 \le a < b \le 1$ *and* $f(x)$ *be a nonnegative continuous function on* $[a,b]$. *If* $\beta \ge \frac{1}{2}$, *then*

$$\left(\int_a^b x^\beta f(x)dx\right)^2 + \left(\int_a^b (1-x)^\beta f(x)dx\right)^2 \le \left(\int_a^b f(x)dx\right)^2. \tag{6}$$

Proof. Let $\kappa(x) = x^\beta$ and $\lambda(x) = (1-x)^\beta$ for $x \in [a,b]$. For any $x \in [a,b]$, since $\beta \ge \frac{1}{2}$, we have

$$\begin{aligned}\kappa(x)\cdot\kappa(b) + \lambda(x)\cdot\lambda(b) &= x^\beta \cdot b^\beta + (1-x)^\beta \cdot (1-b)^\beta \\ &\le x^{\frac{1}{2}} \cdot b^{\frac{1}{2}} + (1-x)^{\frac{1}{2}} \cdot (1-b)^{\frac{1}{2}} \\ &\le \frac{x+b+1-x+1-b}{2} \\ &= 1.\end{aligned}$$

In the same way, we can also show that $\kappa(x)\cdot\kappa(a) + \lambda(x)\cdot\lambda(a) \le 1$ for $x \in [a,b]$. Therefore, the desired inequality (6) follows immediately from Theorem 2 (i). □

Theorem 9. *Let $p \geq 1$ and $0 \leq a < b \leq 1$. Assume that $f(x)$ is a nonnegative continuous function on $[a,b]$. If $f(x)$ is decreasing and $xf(x)$ is increasing, or $f(x)$ is increasing and $(1-x)f(x)$ is decreasing, then*

$$2m^p\left(\int_a^b f(x)dx\right)^2 \leq \left(\int_a^b x^p f(x)dx\right)^2 + \left(\int_a^b (1-x)^p f(x)dx\right)^2 \leq \left(\int_a^b f(x)dx\right)^2, \quad (7)$$

where $m := \min\{a(1-a), b(1-b)\}$.

Proof. From Theorem 8, we know that the right side of the desired inequality (7) holds. Next, we verify that the left side of desired inequality (7) also holds. By the AM-GM inequality, we have

$$\left(\int_a^b x^p f(x)dx\right)^2 + \left(\int_a^b (1-x)^p f(x)dx\right)^2 \geq 2\int_a^b x^p f(x)dx \int_a^b (1-x)^p f(x)dx.$$

Let $\varphi(x) = x^p f(x)$ and $\mu(x) = (1-x)^p f(x)$ for $x \in [a,b]$. Thus, we get

$$\varphi'(x) = x^{p-1}(pf(x) + xf'(x))$$

and

$$\mu'(x) = (1-x)^{p-1}[-pf(x) + (1-x)f'(x)].$$

Since $p \geq 1$, if $f(x)$ is decreasing and $xf(x)$ is increasing, we obtain $\varphi'(x) \geq 0$ and $\mu'(x) \leq 0$. Similarly, if $f(x)$ is increasing and $(1-x)f(x)$ is decreasing, we also have $\varphi'(x) \geq 0$ and $\mu'(x) \leq 0$. By the Chebyshev inequality, we have

$$\begin{aligned}\left(\int_a^b x^p f(x)dx\right)^2 + \left(\int_a^b (1-x)^p f(x)dx\right)^2 &\geq 2\int_a^b x^p f(x)dx \int_a^b (1-x)^p f(x)dx \\ &\geq 2(b-a)\int_a^b [x(1-x)]^p f(x)dx.\end{aligned}$$

Since $h(x) = x(1-x)$ is concave, h attains its minimum value $h(a)$ or $h(b)$. Due to $m = \min\{a(1-a), b(1-b)\}$, we obtain

$$\begin{aligned}&\left(\int_a^b x^p f(x)dx\right)^2 + \left(\int_a^b (1-x)^p f(x)dx\right)^2 \\ \geq& 2(b-a)\int_a^b [x(1-x)]^p f(x)dx \\ \geq& 2(b-a)m^p \int_a^b f^2(x)dx \\ \geq& 2m^p\left(\int_a^b f(x)dx\right)^2.\end{aligned}$$

The proof is completed. □

5. Conclusions

In this paper, we establish the following two important main results for the generalized Dunkel type integral inequality:

- (Generalized Dunkel integral inequality; see Theorem 4.)

Let $f(x)$ and $g(x)$ be two nonnegative continuous real-valued functions on $[a,b]$ and m be any real number. Then

$$\left(\int_a^b f(x)\cos mxdx\right)\left(\int_a^b g(x)\cos mxdx\right)+\left(\int_a^b f(x)\sin mxdx\right)\left(\int_a^b g(x)\sin mxdx\right)$$
$$\leq\left(\int_a^b f(x)dx\right)\left(\int_a^b g(x)dx\right).$$

- (Dunkel (p)-type integral inequality for $p\geq 2$; see Theorem 5.)

Let $k\in\mathbb{N}\cup\{0\}$. Denote $I_k:=\left[2k\pi,2k\pi+\frac{\pi}{2}\right]$. Assume that $f(x)$ is a nonnegative continuous real-valued function on I_k. If $p\geq 2$ and $[a,b]\subseteq I_k$, then

$$\left(\int_a^b f(x)\cos xdx\right)^p+\left(\int_a^b f(x)\sin xdx\right)^p\leq\left(\int_a^b f(x)dx\right)^p.$$

As applications of our new results, some new integral inequalities are presented in Section 4.

Author Contributions: Writing original draft, D.-S.W., H.-N.S., C.-R.F. and W.-S.D. All authors have read and agreed to the published version of the manuscript.

Funding: The fourth author is partially supported by Grant No. MOST 110-2115-M-017-001 of the Ministry of Science and Technology of the Republic of China.

Institutional Review Board Statement: Not applicable.

Informed Consent Statement: Not applicable.

Data Availability Statement: Not applicable.

Acknowledgments: The authors wish to express their hearty thanks to the anonymous referees for their valuable suggestions and comments.

Conflicts of Interest: The authors declare no conflict of interest.

References

1. Kuang, J.C. *Applied Inequalities*, 4th ed.; Shandong Press of Science and Technology: Jinan, China, 2010. (In Chinese)
2. Wang, W.L. *Approaches to Prove Inequalities*; Press of Harbin Industrial University: Harbin, China, 2011. (In Chinese)
3. Mitrinović, D.S.; Vasić, P.M. *Analytic Inequalities*; Springer: Berlin/Heidelberg, Germany, 1970.
4. Shi, H.-N. *Schur Convex Functions and Inequalities*; Press of Harbin Industrial University: Harbin, China, 2017. (In Chinese)
5. Dunkel, O.; Bennett, A.A. Problem 3104. *Am. Math. Mon.* **1925**, *32*, 319–321. [CrossRef]
6. Wang, W.L.; Zhang, Q. Dunkel type inequalities involving hyperbolic functions. *Proc. Jangjeon Math. Soc.* **2018**, *21*, 205–219.
7. Wang, B.Y. *Foundations of Majorization Inequalities*; Press of Beijing Normal Univ: Beijing, China, 1990. (In Chinese)
8. Marshall, A.M.; Olkin, I. *Inequalities: Theory of Majorization and Its Application*; Academies Press: New York, NY, USA, 1979.
9. Marshall, A.W.; Olkin, I.; Arnold, B.C. *Inequalities: Theory of Majorization and Its Application*, 2nd ed.; Springer: New York, NY, USA, 2011.
10. Shi, H.-N.; Du, W.-S. Schur-convexity for a class of completely symmetric function dual. *Adv. Theory Nonlinear Anal. Appl.* **2019**, *3*, 74–89.
11. Shi, H.-N.; Du, W.-S. Schur-power convexity of a completely symmetric function dual. *Symmetry* **2019**, *11*, 897. [CrossRef]
12. He, L. Two new mappings associated with inequalities of Hadamard-type for convex function. *J. Inequal. Pure Appl. Math.* **2009**, *10*, 81.

symmetry

MDPI

Article

A Subclass of Multivalent Janowski Type q-Starlike Functions and Its Consequences

Qiuxia Hu [1], Hari M. Srivastava [2,3,4,5], Bakhtiar Ahmad [6], Nazar Khan [7,*], Muhammad Ghaffar Khan [8], Wali Khan Mashwani [8] and Bilal Khan [9]

1 Department of Mathematics, Luoyang Normal University, Luoyang 471934, China; huqiuxia306@163.com
2 Department of Mathematics and Statistics, University of Victoria, Victoria, BC V8W 3R4, Canada; harimsri@math.uvic.ca
3 Department of Medical Research, China Medical University Hospital, China Medical University, Taichung 40402, Taiwan
4 Department of Mathematics and Informatics, Azerbaijan University, 71 Jeyhun Hajibeyli Street, Baku AZ1007, Azerbaijan
5 Section of Mathematics, International Telematic University Uninettuno, I-00186 Rome, Italy
6 Department of Mathematics, Government Degree College Mardan, Marden 23200, Pakistan; pirbakhtiarbacha@gmail.com
7 Department of Mathematics, Abbottabad University of Science and Technology, Abbottabad 22010, Pakistan
8 Institute of Numerical Sciences, Kohat University of Science and Technology, Kohat 26000, Pakistan; ghaffarkhan020@gmail.com (M.G.K.); walikhan@kust.edu.pk (W.K.M.)
9 School of Mathematical Sciences and Shanghai Key Laboratory of PMMP, East China Normal University, 500 Dongchuan Road, Shanghai 200241, China; bilalmaths789@gmail.com
* Correspondence: nazarmaths@aust.edu.pk

Abstract: In this article, by utilizing the theory of quantum (or q-) calculus, we define a new subclass of analytic and multivalent (or p-valent) functions class $\mathcal{A}_p$, where class $\mathcal{A}_p$ is invariant (or symmetric) under rotations. The well-known class of Janowski functions are used with the help of the principle of subordination between analytic functions in order to define this subclass of analytic and p-valent functions. This function class generalizes various other subclasses of analytic functions, not only in classical Geometric Function Theory setting, but also some q-analogue of analytic multivalent function classes. We study and investigate some interesting properties such as sufficiency criteria, coefficient bounds, distortion problem, growth theorem, radii of starlikeness and convexity for this newly-defined class. Other properties such as those involving convex combination are also discussed for these functions. In the concluding part of the article, we have finally given the well-demonstrated fact that the results presented in this article can be obtained for the $(\mathfrak{p}, q)$-variations, by making some straightforward simplification and will be an inconsequential exercise simply because the additional parameter $\mathfrak{p}$ is obviously unnecessary.

Keywords: analytic functions; multivalent (or p-valent) functions; differential subordination; q-derivative (or q-difference) operator

MSC: Primary 30C45; 30C50; 30C80; Secondary 11B65; 47B38

Citation: Hu, Q.; Srivastava, H.M.; Ahmad, B.; Khan, N.; Khan, M.G.; Mashwani, W.K.; Khan, B. A Subclass of Multivalent Janowski Type q-Starlike Functions and Its Consequences. *Symmetry* **2021**, *13*, 1275. https://doi.org/10.3390/sym13071275

Academic Editor: Rahmat Ellahi

Received: 19 May 2021
Accepted: 13 July 2021
Published: 16 July 2021

1. Introduction, Definitions and Motivation

The calculus without the notion of limits, which is known as the quantum (or q-) calculus, has influenced many scientific fields due to its important applications. The generalizations of the derivative and integral operators in q-calculus, which are known as the q-derivative and q-integral operators, were introduced and studied by Jackson [1,2].

Recently, Anastassiu [3] and Aral [4] generalized some complex-valued operators, which are known as the q-Picard and q-Gauss–Weierstrass singular integral operators. Geometric Function Theory is no exception in this regard and many authors have already made a substantial contribution to the field of Complex Analysis. Ismail et al. (see [5]) presented the q-deformation of the familiar class $\mathcal{S}^*$ of starlike functions. However, in the context of Geometric Function Theory in 1989, the usage of the q-difference (or the q-derivative) operator D_q was systematically given by Srivastava [6]. Furthermore, the survey-cum-expository review article by Srivastava [7] is potentially useful for those who are interested in Geometric Function Theory. In this review article, many various applications of the the fractional q-calculus, in Geometric Function Theory were systematically highlighted. Moreover, the triviality of the so-called $(\mathfrak{p}, q)$-calculus involving an obviously redundant and inconsequential additional parameter $\mathfrak{p}$ was revealed and exposed (see, for details, [7], p. 340).

Based on the aforementioned works [5,7], a number of researches got inspiration to gave and their finding to Geometric Function Theory of Complex Analysis. For example, Srivastava and Bansal [8] used the q-derivatives and gave close-to-convexity for certain Mittag-Leffer type functions. Kanas and Răducanu [9] defined the q-analogue of the Ruscheweyh derivative operator and they discussed its various important properties. The applications of this q-derivative operator were further studied by Mahmood and Sokół [10]. More recently, Srivastava et al. [11,12] first defined certain subclasses of q-starlike functions and then studied their various properties including for example some coefficient inequalities, inclusion properties, and a number of sufficient conditions. Moreover, the subclasses of q-starlike functions associated with the Janwoski or some other functions have been studied by the many authors (see, for example, [13–20]). For some more recent investigations based upon the q-calculus, we may refer the interested reader to the works in [21–38]. Our present research is a continuation of some of these earlier developments. It is fairly general in its nature as it not only generalizes many known classes, but also gives a different direction to the study of such classes.

In this article, we are essentially motivated by the recently published paper of Khan et al. in *Symmetry* (see [27]) and some other related works on this subject, which we have mentioned above. We first introduce a new subclass of analytic and multivalent (or p-valent) functions by using the concept of the q-calculus in association with the Janowski functions. We then study some of its geometric properties such as sufficiency criteria, coefficient bounds, radii problems, distortion theorem and growth theorem, and so on. Before stating and proving our main results, we give a brief discussion on the basics of this area which will be beneficial in understanding the work to follow.

Let $\mathcal{A}_p$ be the class of analytic and multivalent (or p-valent) functions $f(z)$ in the open unit disk

$$\mathbb{D} = \{z : z \in \mathbb{C} \quad \text{and} \quad |z| < 1\},$$

with the series representation given by

$$f(z) = z^p + \sum_{n=1}^{\infty} a_{n+p} z^{n+p} \qquad (z \in \mathbb{D};\ p \in \mathbb{N} := \{1, 2, 3, \cdots\}). \tag{1}$$

We note for $p = 1$ that

$$\mathcal{A}(1) = \mathcal{A},$$

where $\mathcal{A}$ is the familiar class of normalized analytic functions in $\mathbb{D}$ and the class $\mathcal{A}$ is invariant (or symmetric) under rotations.

For analytic functions f and g in open unit disk $\mathbb{D}$, the function f is said to subordinate to the function g and written as

$$f \prec g \quad \text{or} \quad f(z) \prec g(z),$$

if there exists a Schwarz function w, which is analytic in $\mathbb{D}$ with

$$w(0) = 0 \qquad \text{and} \qquad |w(z)| < 1,$$

such that

$$f(z) = g(w(z)).$$

Furthermore, if the function g is univalent in $\mathbb{D}$, then it follows that

$$f(z) \prec g(z) \quad (z \in \mathbb{D}) \implies f(0) = g(0) \text{ and } f(\mathbb{D}) \subset g(\mathbb{D}).$$

Definition 1. *Let the* $\Upsilon(z)$ *is analytic in* $\mathbb{D}$ *with* $\Upsilon(0) = 1$ *then* $\Upsilon(z)$ *said to be in the class* $\mathcal{P}[A, B]$, *if*

$$\Upsilon(z) \prec \frac{1 + Az}{1 + Bz} \qquad (1 \leqq B < A \leqq 1).$$

Equivalently, we can write

$$\left| \frac{\Upsilon(z) - 1}{A - B\Upsilon(z)} \right| < 1.$$

The class $\mathcal{P}[A, B]$ *was introduced by Janowski [39].*

Definition 2. *Let* $q \in (0, 1)$ *and define the q-number* $[\lambda]_q$ *by*

$$[\lambda]_q = \begin{cases} \dfrac{1 - q^\lambda}{1 - q} & (\lambda \in \mathbb{C}) \\ \sum\limits_{k=0}^{n-1} q^k = 1 + q + q^2 + \cdots + q^{n-1} & (\lambda = n \in \mathbb{N}). \end{cases}$$

The q-derivative operator D_q, also known as the q-difference operator, for a function f is defined by

$$D_q f(z) = \frac{f(z) - f(qz)}{z(1 - q)} \qquad (z \neq 0), \tag{2}$$

where $0 < q < 1$. One can easily see for $n \in \mathbb{N}$ and $z \in \mathbb{D}$ that

$$D_q \left\{ \sum_{n=1}^{\infty} a_n z^n \right\} = \sum_{n=1}^{\infty} [n]_q a_n z^{n-1}, \tag{3}$$

where

$$[n]_q = \frac{1 - q^n}{1 - q} = 1 + \sum_{l=1}^{n-1} q^l \quad \text{and} \quad [0]_q = 0.$$

Motivated by the above-cited works in [39–44], we now define a new subclass $\mathcal{S}_q(p, \alpha, A, B)$ of $\mathcal{A}_p$ as follows.

Definition 3. *A function* $f \in \mathcal{A}_p$ *is said to be in the class* $\mathcal{S}_q(p, \alpha, A, B)$, *if it satisfies the following subordination condition:*

$$\frac{1}{1 - \alpha} \left(\frac{zD_q f(z)}{[p]_q f(z)} - \alpha \frac{z^2 D_q^2 f(z)}{[p]_q [p-1]_q f(z)} \right) \prec \frac{1 + Az}{1 + Bz} \tag{4}$$

or, equivalently,

$$\left| \frac{\frac{zD_q f(z)}{[p]_q f(z)} - \alpha \frac{z^2 D_q^2 f(z)}{[p]_q [p-1]_q f(z)} - (1 - \alpha)}{(1 - \alpha)A - B\left(\frac{zD_q f(z)}{[p]_q f(z)} - \alpha \frac{z^2 D_q^2 f(z)}{[p]_q [p-1]_q f(z)} \right)} \right| < 1, \tag{5}$$

where $-1 \leqq B < A \leqq 1$, $\alpha \geqq 0$ *and* $q \in (0, 1)$.

Remark 1. *First of all, it is easily seen that*

$$\lim_{q \to 1-} \mathcal{S}_q(1, 0, A, B) = \mathcal{S}^*[A, B],$$

where $\mathcal{S}^*[A, B]$ *is the function class introduced and studied by Janowski [39]. Secondly, we have*

$$\mathcal{S}_q(1, 0, A, B) = \mathcal{S}_q^*[A, B],$$

where $\mathcal{S}_q^*[A, B]$ *is the function class introduced and studied by Srivastava et al. [19]. Thirdly, we have*

$$\lim_{q \to 1-} \mathcal{S}_q(1, 0, 1, -1) = \mathcal{S}^*,$$

where $\mathcal{S}^*$ *is the well-known class of starlike functions.*

For proving our main results we will need the following lemma due to Rogosinski [45].

Lemma 1. *(see [45]) Let the function* $h(z)$ *be given by*

$$h(z) = 1 + \sum_{n=1}^{\infty} d_n z^n$$

and let another function $k(z)$ *be given by*

$$k(z) = 1 + \sum_{n=1}^{\infty} k_n z^n.$$

Suppose also that

$$h(z) \prec k(z) \qquad (z \in \mathbb{D}).$$

If $k(z)$ *is univalent in* $\mathbb{D}$ *and* $k(\mathbb{D})$ *is convex, then*

$$|d_n| \leqq |k_1| \qquad (n \geqq 1).$$

2. The Main Results and Their Consequences

This section is devoted to our main results. Throughout our discussion, we assume that

$$-1 \leqq B < A \leqq 1 \qquad \text{and} \qquad q \in (0, 1)$$

and that

$$\lambda_1 = [p-1]_q \qquad \text{and} \qquad \lambda_2 = [p+n]_q. \tag{6}$$

Theorem 1. *Let* $f \in \mathcal{A}_p$ *be of the form (1). Then the function* $f \in \mathcal{S}_q(p, \alpha, A, B)$*, if and only if the following inequality holds true:*

$$\sum_{n=1}^{\infty} \left((1+B)\left(\lambda_1\lambda_2 + \alpha\lambda_2[p+n-1]_q\right) + (1-\alpha)(1+A)[p]_q\lambda_1 \right) |a_{n+p}| \\ \leqq (1-\alpha)(A-B)[p]_q\lambda_1, \tag{7}$$

where λ_1 *and* λ_2 *are given in (6).*

Proof. Let us suppose that the inequality in (7) holds true. Then, in order to show that $f \in \mathcal{S}_q(p, \alpha, A, B)$, we only need to prove the inequality (5). For this purpose, we consider

$$\left| \frac{\frac{zD_qf(z)}{[p]_qf(z)} - \alpha\frac{z^2D_q^2f(z)}{[p]_q\lambda_1f(z)} - (1-\alpha)}{(1-\alpha)A - B\left(\frac{zD_qf(z)}{[p]_qf(z)} - \alpha\frac{z^2D_q^2f(z)}{[p]_q\lambda_1f(z)}\right)} \right| = \left| \frac{\lambda_1 zD_qf(z) - \alpha z^2D_q^2f(z) - (1-\alpha)[p]_q\lambda_1f(z)}{A(1-\alpha)[p]_q\lambda_1f(z) - B\left(\lambda_1 zD_qf(z) - \alpha z^2D_q^2f(z)\right)} \right|.$$

Now, with the help of (1)–(3), and after some simplification, the above equation can be written as follows:

$$\begin{aligned}
&\left| \frac{\sum_{n=1}^{\infty}\left(\lambda_1\lambda_2 - \alpha\lambda_2[p+n-1]_q - (1-\alpha)[p]_q\lambda_1\right)a_{n+p}z^{n+p}}{(A-B)(1-\alpha)[p]_q\lambda_1z^p + \sum_{n=1}^{\infty}\Lambda_q(\lambda_1,\lambda_2)a_{n+p}z^{n+p}} \right| \\
&= \left| \frac{\sum_{n=1}^{\infty}\left(\lambda_1\lambda_2 - \alpha\lambda_2[p+n-1]_q - (1-\alpha)[p]_q\lambda_1\right)a_{n+p}z^{n}}{(A-B)(1-\alpha)[p]_q\lambda_1 + \sum_{n=1}^{\infty}\Lambda_q(\lambda_1,\lambda_2)a_{n+p}z^{n}} \right| \\
&\leqq \frac{\sum_{n=1}^{\infty}\left(\lambda_1\lambda_2 + \alpha\lambda_2[p+n-1]_q + (1-\alpha)[p]_q\lambda_1\right)|a_{n+p}|}{(A-B)(1-\alpha)[p]_q\lambda_1 - \sum_{n=1}^{\infty}\Lambda_q(\lambda_1,\lambda_2)|a_{n+p}|} \\
&\leqq 1,
\end{aligned}$$

where

$$\Lambda_q(\lambda_1,\lambda_2) = A(1-\alpha)[p]_q\lambda_1 - B\lambda_2\left(\lambda_1 - B\alpha[p+n-1]_q\right).$$

This last inequality can be rewritten as follows:

$$\begin{aligned}
&\left| \frac{\sum_{n=1}^{\infty}\left(\lambda_1\lambda_2 - \alpha\lambda_2[p+n-1]_q - (1-\alpha)[p]_q\lambda_1\right)a_{n+p}z^{n+p}}{(A-B)(1-\alpha)[p]_q\lambda_1z^p + \sum_{n=1}^{\infty}\Lambda_q(\lambda_1,\lambda_2)a_{n+p}z^{n+p}} \right| \\
&= \left| \frac{\sum_{n=1}^{\infty}\left(\lambda_1\lambda_2 - \alpha\lambda_2[p+n-1]_q - (1-\alpha)[p]_q\lambda_1\right)a_{n+p}z^{n}}{(A-B)(1-\alpha)[p]_q\lambda_1 + \sum_{n=1}^{\infty}\Lambda_q(\lambda_1,\lambda_2)a_{n+p}z^{n}} \right| \\
&\leqq \frac{\sum_{n=1}^{\infty}\left(\lambda_1\lambda_2 + \alpha\lambda_2[p+n-1]_q + (1-\alpha)[p]_q\lambda_1\right)|a_{n+p}|}{(A-B)(1-\alpha)[p]_q\lambda_1 - \sum_{n=1}^{\infty}\Lambda_q(\lambda_1,\lambda_2)|a_{n+p}|} \\
&< 1,
\end{aligned}$$

where we have used the inequality (7). This completes the direct part of the result asserted by Theorem 1.

Conversely, let $f \in \mathcal{S}_q(p, \alpha, A, B)$ be given by (1). Then, from (5), we find for $z \in \mathbb{D}$ that

$$\left| \frac{\frac{zD_qf(z)}{[p]_qf(z)} - \alpha\frac{z^2D_q^2f(z)}{[p]_q[p-1]_qf(z)} - (1-\alpha)}{(1-\alpha)A - B\left(\frac{zD_qf(z)}{[p]_qf(z)} - \alpha\frac{z^2D_q^2f(z)}{[p]_q[p-1]_qf(z)}\right)} \right| = \left| \frac{\sum_{n=1}^{\infty}\left(\lambda_1\lambda_2 - \alpha\lambda_2[p+n-1]_q - (1-\alpha)[p]_q\lambda_1\right)a_{n+p}z^{n}}{(A-B)(1-\alpha)[p]_q\lambda_1 + \sum_{n=1}^{\infty}\Lambda_q(\lambda_1,\lambda_2)a_{n+p}z^{n}} \right|.$$

Since

$$\Re(z) \leqq |z|,$$

therefore, we have

$$\Re\left\{\frac{\sum_{n=1}^{\infty}\left(\lambda_1\lambda_2-\alpha\lambda_2[p+n-1]_q-(1-\alpha)[p]_q\lambda_1\right)a_{n+p}z^n}{(A-B)(1-\alpha)[p]_q\lambda_1+\sum_{n=1}^{\infty}\Lambda_q(\lambda_1,\lambda_2)a_{n+p}z^n}\right\}$$
$$<1. \tag{8}$$

We now choose values of z on the real axis in the complex z-plane, so that

$$\frac{1}{1-\alpha}\left(\frac{zD_qf(z)}{[p]_qf(z)}-\alpha\frac{z^2D_q^2f(z)}{[p]_q[p-1]_qf(z)}\right)$$

is real. Upon clearing the denominator in (8) and letting $z\to 1-$ through real values, we obtain (7). This completes the proof of Theorem 1. □

Theorem 2. *If the function $f(z)$, given by (1), belongs to the class $\mathcal{S}_q(p,\alpha,A,B)$, then*

$$|a_{p+1}|\leqq\frac{[p]_q\lambda_1(1-\alpha)(A-B)}{l(1)},$$

$$|a_{p+2}|\leqq\frac{[p]_q\lambda_1(1-\alpha)(A-B)}{l(2)}+\frac{\left([p]_q\lambda_1(1-\alpha)(A-B)\right)^2}{l(1)l(2)},$$

$$|a_{p+3}|\leqq\frac{[p]_q\lambda_1(1-\alpha)(A-B)}{l(3)}+\frac{\left([p]_q\lambda_1(1-\alpha)(A-B)\right)^2}{l(3)l(1)}+\frac{\left([p]_q\lambda_1(1-\alpha)(A-B)\right)^2}{l(3)l(2)}+\frac{\left([p]_q\lambda_1(1-\alpha)(A-B)\right)^3}{l(3)l(2)l(1)},$$

where

$$\lambda_1=[p-1]_q$$

and

$$l(n)=\lambda_1[p+n]_q-\alpha[p+n-1]_q[p+n]_q-(1-\alpha)[p]_q\lambda_1.$$

Proof. Let $f\in\mathcal{S}_q(p,\alpha,A,B)$. Then

$$\frac{1}{1-\alpha}\left(\frac{zD_qf(z)}{[p]_qf(z)}-\alpha\frac{z^2D_q^2f(z)}{[p]_q\lambda_1f(z)}\right)=h(z), \tag{9}$$

where

$$h(z)\prec\frac{1+Az}{1+Bz}=1+(A-B)z+\cdots,$$

is of the form given by

$$h(z)=1+\sum_{n=1}^{\infty}d_nz^n.$$

Thus, by the Rogosinski Lemma, we get

$$|d_n|\leqq(A-B). \tag{10}$$

Now, using the series expansions of $h(z)$ and $f(z)$ in (9), together with simplification and comparsion of the coefficients of like powers of z, we get

$$\begin{aligned}&\frac{\left(\lambda_1-\alpha[p+n-1]_q\right)\lambda_2}{1-\alpha}a_{p+n}\\&\quad=[p]_q\lambda_1\left(a_{p+n}+a_{p+n-1}d_1+a_{p+n-2}d_2+\ldots+a_{p+1}d_{n-1}+d_n\right),\end{aligned}$$

which can be written as follows:

$$\begin{aligned}&\frac{\left(\lambda_1-\alpha[p+n-1]_q\right)\lambda_2-(1-\alpha)[p]_q\lambda_1}{1-\alpha}a_{p+n}\\&\quad=[p]_q\lambda_1\left(a_{p+n-1}d_1+a_{p+n-2}d_2+\ldots+a_{p+1}d_{n-1}+d_n\right).\end{aligned}$$

Next, by first using the triangle inequality for the modulus and then applying (10), we obtain

$$\begin{aligned}&\frac{\left(\lambda_1-\alpha[p+n-1]_q\right)\lambda_2-(1-\alpha)[p]_q\lambda_1}{1-\alpha}|a_{p+n}|\\&\quad\leqq[p]_q\lambda_1(A-B)\sum_{i=0}^{n-1}|a_{p+i}|,\end{aligned}$$

which implies that

$$|a_{p+n}|\leqq\frac{(1-\alpha)[p]_q\lambda_1(A-B)}{\left(\lambda_1-\alpha[p+n-1]_q\right)\lambda_2-(1-\alpha)[p]_q\lambda_1}\sum_{i=0}^{n-1}|a_{p+i}|. \tag{11}$$

If we now put $n=1,2$ and 3 in (11) and use the fact that $a_p=1$, we get the required result asserted by Theorem 2. □

Theorem 3. *Let the function $f(z)$, given by (1), belong to the class $\mathcal{S}_q(p,\alpha,A,B)$. Then, for $|z|=r$,*

$$[p]_q r^{p-1}-\tau_1 r^p\leqq|D_qf(z)|\leqq[p]_q r^{p-1}+\tau_1 r^p,$$

where

$$\tau_1=\frac{(1-\alpha)(A-B)[p+1]_q[p]_q\lambda_1}{(1+B)\left(\lambda_1+\alpha[p]_q\right)[p+1]_q+(1-\alpha)(1+A)[p]_q\lambda_1}$$

and

$$\lambda_1=[p-1]_q.$$

Proof. From (1) we can write

$$D_qf(z)=[p]_q z^{p-1}+\sum_{n=1}^{\infty}\lambda_2 a_{n+p}\,z^{n+p-1},$$

so that, by applying the triangle inequality, we have

$$|D_qf(z)|\leqq[p]_q|z|^{p-1}+\sum_{n=1}^{\infty}\lambda_2|a_{n+p}|\,|z|^{n+p-1}$$

Since $|z|=r<1$ so that $r^{n+p-1}\leqq r^p$, and hence we have

$$D_qf(z)\leqq[p]_q r^{p-1}+r^p\sum_{n=1}^{\infty}\lambda_2|a_{n+p}|. \tag{12}$$

Similarly, we get

$$D_q f(z) \geqq [p]_q r^{p-1} - r^p \sum_{n=1}^{\infty} \lambda_2 |a_{n+p}|. \tag{13}$$

We know from (7) that

$$\sum_{n=1}^{\infty} \left[(1+B)(\lambda_1 + \alpha[p+n-1]_q) + \frac{(1-\alpha)(1+A)\lambda_1 [p]_q}{\lambda_2} \right] \lambda_2 |a_{n+p}| \\ \leqq (1-\alpha)(A-B)[p]_q \lambda_1.$$

We also know that

$$\left[(1+B)\left(\lambda_1 + \alpha[p]_q\right) + \frac{(1-\alpha)(1+A)[p]_q \lambda_1}{[p+1]_q} \right] \sum_{n=1}^{\infty} \lambda_2 |a_{n+p}| \\ \leqq \sum_{n=1}^{\infty} \left[(1+B)\left(\lambda_1 + \alpha[p+n-1]_q\right) + \frac{(1-\alpha)(1+A)[p]_q \lambda_1}{\lambda_2} \right] \lambda_2 |a_{n+p}|.$$

Hence, by the transitive property, we get

$$\left[(1+B)\left(\lambda_1 + \alpha[p]_q\right) + \frac{(1-\alpha)(1+A)[p]_q \lambda_1}{[p+1]_q} \right] \sum_{n=1}^{\infty} \lambda_2 |a_{n+p}| \\ \leqq (1-\alpha)(A-B)[p]_q \lambda_1,$$

which implies that

$$\sum_{n=1}^{\infty} \lambda_2 |a_{n+p}| \leqq \frac{(1-\alpha)(A-B)[p]_q \lambda_1 [p+1]_q}{(1+B)[p+1]_q\left(\lambda_1 + \alpha[p]_q\right) + (1-\alpha)(1+A)[p]_q \lambda_1}.$$

Now, by using the above inequility in (12) and (13), we obtain the required result asserted by Theorem 3. □

Theorem 4. *Let the function $f(z)$, given by (1), belong to the class $\mathcal{S}_q(p, \alpha, A, B)$. Then, for $|z| = r$,*

$$r^p(1-\tau_2) \leqq |f(z)| \leqq r^p(1+\tau_2),$$

where

$$\tau_2 = \frac{(1-\alpha)(A-B)[p]_q \lambda_1}{[p+1]_q(1+B)\left(\lambda_1 + \alpha[p]_q\right) + (1-\alpha)(1+A)[p]_q \lambda_1}.$$

Proof. By applying the triangle inequality in (1), and using the fact that $|z| = r$, we have

$$|f(z)| \leqq r^p + \sum_{n=1}^{\infty} |a_{n+p}|\, r^{n+p}$$

Since $|z| = r < 1$ so that $r^{n+p} < r^p$, therefore, the above relation becomes

$$|f(z)| \leqq r^p + r^p \sum_{n=1}^{\infty} |a_{n+p}|. \tag{14}$$

Similarly, we get

$$|f(z)| \geqq r^p - r^p \sum_{n=1}^{\infty} |a_{n+p}|. \tag{15}$$

We know from (7) that

$$\sum_{n=1}^{\infty} \left[\lambda_2(1+B)\left(\lambda_1 + \alpha[p+n-1]_q\right) + (1-\alpha)(1+A)[p]_q\lambda_1\right] |a_{n+p}| \leqq (1-\alpha)(A-B)[p]_q\lambda_1.$$

But

$$\begin{aligned}\left[[p+1]_q(1+B)\left(\lambda_1+\alpha[p]_q\right) + (1-\alpha)(1+A)[p]_q\lambda_1\right] \sum_{n=1}^{\infty}|a_{n+p}| &\leqq \sum_{n=1}^{\infty}\left[\lambda_2(1+B)\left(\lambda_1+\alpha[p+n-1]_q\right) + (1-\alpha)(1+A)[p]_q\lambda_1\right]|a_{n+p}| \\ &\leqq (1-\alpha)(A-B)[p]_q[p-1]_q.\end{aligned}$$

Hence

$$\left[[p+1]_q(1+B)\left(\lambda_1+\alpha[p]_q\right) + (1-\alpha)(1+A)[p]_q\lambda_1\right] \sum_{n=1}^{\infty}|a_{n+p}| \leqq (1-\alpha)(A-B)[p]_q[p-1]_q,$$

which gives

$$\sum_{n=1}^{\infty}|a_{n+p}| \leqq \frac{(1-\alpha)(A-B)[p]_q\lambda_1}{[p+1]_q(1+B)\left(\lambda_1+\alpha[p]_q\right) + (1-\alpha)(1+A)[p]_q\lambda_1}.$$

Now, by substituting from the above inequality into (14) and (15), we get the required result asserted by Theorem 4. □

Now before starting radii problems let us remaind the definition of important classes of multivalent starlike and convex functions.

A function $f \in \mathcal{A}_p$, is said to be multivalent starlike functions of order σ if it satisfies the following inequality

$$\Re\left\{\frac{zf'(z)}{pf(z)}\right\} > \sigma,\ z \in \mathbb{D},\ (0 \leq \sigma \leq p,\ p \in \mathbb{N}),$$

and we denoted this class by $\mathcal{S}_p^*(\sigma)$.

Furthermore, by $\mathcal{C}_p(\sigma)$ we mean the class of multivalent convex functions, that is a function $f \in \mathcal{A}_p$ and satisfies the inequality below

$$\Re\left\{\frac{z\left(zf'(z)\right)'}{p^2 f(z)}\right\} > \sigma,\ z \in \mathbb{D},\ (0 \leq \sigma \leq p,\ p \in \mathbb{N}).$$

Theorem 5. *Let $f \in \mathcal{S}_q(p,\alpha,A,B)$. Then $f \in \mathcal{C}_p(\sigma)$ for $|z| < r_1$, where*

$$r_1 = \left\{\inf\left(\frac{p(p-\sigma)\left((1+B)\lambda_2\left(\lambda_1+\alpha[p+n-1]_q\right) + (1-\alpha)(1+A)[p]_q\lambda_1\right)}{(p+n)(n+p-\sigma)(1-\alpha)(A-B)[p]_q\lambda_1}\right)^{\frac{1}{n}} : n \in \mathbb{N}\right\}.$$

Proof. Suppose that $f \in \mathcal{S}_q(p, \alpha, A, B)$. Then, in order to prove that $f \in \mathcal{C}_p(\sigma)$, we only need to show that

$$\left| \frac{zf''(z) - (p-1)f'(z)}{zf''(z) + (1 - 2\sigma + p)f'(z)} \right| < 1.$$

Using (1) followed by some simplifications, we have

$$\sum_{n=1}^{\infty} \frac{(p+n)(n+p-\sigma)}{p(p-\sigma)} |a_{n+p}||z|^n < 1. \tag{16}$$

Now, from (7), we can easily see that

$$\sum_{n=1}^{\infty} \left(\lambda_2(1+B)\left(\lambda_1 + \alpha[p+n-1]_q\right) + (1-\alpha)(1+A)[p]_q\lambda_1 \right) |a_{n+p}| \\ \leqq (1-\alpha)(A-B)[p]_q\lambda_1,$$

which implies that

$$\sum_{n=1}^{\infty} \frac{\lambda_2(1+B)\left(\lambda_1 + \alpha[p+n-1]_q\right) + (1-\alpha)(1+A)[p]\lambda_1}{(1-\alpha)(A-B)[p]_q\lambda_1} |a_{n+p}| < 1.$$

The inequality in (16) will be true, if the following condition holds true:

$$\sum_{n=1}^{\infty} \frac{(p+n)(n+p-\sigma)}{p(p-\sigma)} |a_{n+p}||z|^n \\ < \sum_{n=1}^{\infty} \frac{\lambda_2(1+B)\left(\lambda_1 + \alpha[p+n-1]_q\right) + (1-\alpha)(1+A)[p]_q\lambda_1}{(1-\alpha)(A-B)[p]_q\lambda_1} |a_{n+p}|,$$

which implies that

$$|z|^n < \frac{p(p-\sigma)\left(\lambda_2(1+B)\left(\lambda_1 + \alpha[p+n-1]_q\right) + (1-\alpha)(1+A)[p]_q\lambda_1\right)}{(p+n)(n+p-\sigma)(1-\alpha)(A-B)[p]_q\lambda_1},$$

or, equivalently, that

$$|z| < \left(\frac{p(p-\sigma)\left(\lambda_2(1+B)\left(\lambda_1 + \alpha[p+n-1]_q\right) + (1-\alpha)(1+A)[p]_q\lambda_1\right)}{(p+n)(n+p-\sigma)(1-\alpha)(A-B)[p]_q\lambda_1} \right)^{\frac{1}{n}} \\ = r_1.$$

This completes the proof of Theorem 5. □

Theorem 6. *Let $f \in \mathcal{S}_q(p, \alpha, A, B)$. Then $f \in \mathcal{S}_p^*(\sigma)$ for $|z| < r_2$, where*

$$r_2 = \left\{ \inf \left(\frac{(p-\sigma)\left(\lambda_2(1+B)\left(\lambda_1 + \alpha[p+n-1]_q\right) + (1-\alpha)(1+A)[p]_q\lambda_1\right)}{(n+p-\sigma)(1-\alpha)(A-B)[p]_q\lambda_1} \right)^{\frac{1}{n}} : n \in \mathbb{N} \right\}.$$

Proof. We know that $f \in \mathcal{S}_p^*(\alpha)$, if and only if

$$\left| \frac{zf'(z) - pf(z)}{zf'(z) + (p-2\sigma)f(z)} \right| \leqq 1.$$

Using (1) and upon simplification, we get

$$\sum_{n=1}^{\infty}\left(\frac{n+p-\sigma}{p-\sigma}\right)|a_{n+p}||z|^{n}<1. \tag{17}$$

Now from (7), we can easily find that

$$\sum_{n=1}^{\infty}\frac{\lambda_2(1+B)\left(\lambda_1+\alpha[p+n-1]_q\right)+(1-\alpha)(1+A)[p]_q\lambda_1}{(1-\alpha)(A-B)[p]_q\lambda_1}|a_{n+p}|<1.$$

For the inequality (17) to be true, it will be sufficient to show that

$$\begin{aligned}&\sum_{n=1}^{\infty}\left(\frac{n+p-\sigma}{p-\sigma}\right)|a_{n+p}||z|^{n}\\&\qquad<\sum_{n=1}^{\infty}\frac{\lambda_2(1+B)\left(\lambda_1+\alpha[p+n-1]_q\right)+(1-\alpha)(1+A)[p]_q}{(1-\alpha)(A-B)[p]_q\lambda_1}|a_{n+p}|,\end{aligned}$$

which yields

$$|z|^{n}<\frac{(p-\sigma)\left(\lambda_2(1+B)\left(\lambda_1+\alpha[p+n-1]_q\right)+(1-\alpha)(1+A)[p]_q\lambda_1\right)}{(n+p-\sigma)(1-\alpha)(A-B)[p]_q\lambda_1},$$

and hence

$$\begin{aligned}|z|&<\left(\frac{(p-\sigma)\left(\lambda_2(1+B)\left(\lambda_1+\alpha[p+n-1]_q\right)+(1-\alpha)(1+A)[p]_q\lambda_1\right)}{(n+p-\sigma)(1-\alpha)(A-B)[p]_q\lambda_1}\right)^{\frac{1}{n}}\\&=r_2.\end{aligned}$$

We thus obtain the required result asserted by Theorem 6. □

Let the functions $f_l(z)$ $(l=1,2,3,\cdots,k)$ be defined by

$$f_l(z)=z+\sum_{n=1}^{\infty}a_{n,l}z^{n+p}\qquad(z\in\mathbb{D}). \tag{18}$$

Now we state and prove the following results.

Theorem 7. *The class $\mathcal{S}_q(p,\alpha,A,B)$ is closed under convex combination.*

Proof. Suppose that the functions $f_l(z)$ $(l=1,2)$, given by (18), belong to the class $\mathcal{S}_q(p,\alpha,A,B)$. Then we need to show that the function $\mathfrak{h}(z)$ given by

$$\mathfrak{h}(z)=\nu f_1(z)+(1-\nu)f_2(z)\qquad(0\leqq\lambda\leqq1),$$

is also in the class $\mathcal{S}_q(p,\alpha,A,B)$. Indeed, for $0\leqq\nu\leqq1$, we have

$$\mathfrak{h}(z)=z+\sum_{n=2}^{\infty}[\nu a_{n,1}+(1-\nu)a_{n,2}]z^{n}.$$

Thus, if

$$\lambda_1=[p-1]_q\qquad\text{and}\qquad\lambda_2=[p+n]_q,$$

then we have

$$\begin{aligned}
&\sum_{n=1}^{\infty}\Big[(1+B)\Big(\lambda_1\lambda_2+\alpha\lambda_2[p+n-1]_q\Big)\\
&\qquad+(1-\alpha)(1+A)[p]_q\lambda_1\Big]\left|\nu a_{n+p,1}+(1-\nu)a_{n+p,2}\right|\\
&\quad=\nu\sum_{n=1}^{\infty}\Big[(1+B)\Big(\lambda_1\lambda_2+\alpha\lambda_2[p+n-1]_q\Big)\\
&\qquad+(1-\alpha)(1+A)[p]_q\lambda_1\Big]\left|a_{n+p,1}\right|\\
&\qquad+(1-\nu)\sum_{n=1}^{\infty}\Big[(1+B)\Big(\lambda_1\lambda_2+\alpha\lambda_2[p+n-1]_q\Big)\\
&\qquad+(1-\alpha)(1+A)[p]_q\lambda_1\Big]\left|a_{n+p,2}\right|\\
&\quad\leqq\nu\Big[(1-\alpha)(A-B)[p]_q\lambda_1\Big]+(1-\nu)\Big[(1-\alpha)(A-B)[p]_q\lambda_1\Big]\\
&\quad=(1-\alpha)(A-B)[p]_q\lambda_1.
\end{aligned}$$

Hence, by Theorem 1, $\mathfrak{h}(z)\in\mathcal{S}_q(p,\alpha,A,B)$. The demonstration of Theorem 7 is thus completed. □

Theorem 8. *Let the L functions $f_l(z)\quad(l=1,2,3,\cdots,L)$, defined by (18), be in the class $\mathcal{S}_q(p,\alpha,A,B)$. Then the function $F(z)$, given by*

$$F(z)=\sum_{l=1}^{L}\nu_l f_l(z)\qquad\left(\lambda_l\geqq 0;\ \sum_{l=1}^{L}\lambda_l=1\right),\tag{19}$$

is also in the class $\mathcal{S}_q(p,\alpha,A,B)$,

Proof. The proof of Theorem 8 is fairly straightforward. We, therefore, omit the details involved. □

3. Concluding Remarks and Observations

Applications of the q-calculus have been the focus point in the recent times in various branches of Mathematics and Physics mentioned [7]. In this article, we have introduced a new q-operator for multivalent functions. Then a new subclass of analytic and multivalent functions has been defined and studied systematically. In particular, we have investigated some of its geometric properties such as sufficient conditions, coefficient estimates, distortion Theorems, radii problems, closure-type results, and so on. The idea used in this article can easily be implemented to define several subclasses of analytic and univalent (or multivalent) functions connected with different image domains. This will open up a lot of new opportunities for research in this and related fields.

Basic (or q-) series and basic (or q-) polynomials, especially the basic (or q-) hypergeometric functions and basic (or q-) hypergeometric polynomials are applicable particularly in several diverse areas (see, for example, [46] (pp. 350–351); see also [28,29,47]). Moreover, as we remarked above and in the introductory Section 1, in Srivastava's recently-published survey-cum-expository review article [7], the triviality of the so-called (p,q)-calculus was exposed and it is also mentioned as an obviously inconsequential variation of the classical q-calculus, the additional parameter p being redundant or superfluous (see, for details, [7] (p. 340)). Indeed one can apply Srivastava's observation and exposition in [7] to any attempt to produce the rather trivial and straightforward (p,q)-variations of the q-results which we have presented in this paper.

Author Contributions: Conceptualization, H.M.S., M.G.K., N.K. and B.A.; methodology, H.M.S.; W.K.M. and B.K. software, B.A., N.K., M.G.K. and B.K.; validation, H.M.S., N.K. and B.K.; formal analysis, H.M.S., W.K.M., Q.H. and N.K., investigation, B.A., M.G.K. and B.K.; writing—original draft preparation, B.A., N.K., M.G.K. and B.K.; writing—review and editing, Q.H. and B.K.; visualization, H.M.S., M.G.K., N.K. and B.A; supervision, H.M.S.; project administration, Q.H.; funding acquisition, Q.H.; All authors have read and agreed to the published version of the manuscript.

Funding: This work was supported by The Key Scientific Research Project of the Colleges and Universities in Henan Province (NO. 19A110024), Natural Science Foundation of Henan Province (CN) (NO. 212300410204), (No.212300410211) and National Project Cultivation Foundation of Luoyang Normal University (No.2020-PYJJ-011).

Data Availability Statement: Not applicable.

Acknowledgments: The authors are grateful to the editor and the reviewers for their valuable comments and suggestions.

Conflicts of Interest: The authors declare that they have no conflict of interest.

References

1. Jackson, F.H. On q-definite integrals. *Quart. J. Pure Appl. Math.* **1910**, *41*, 193–203.
2. Jackson, F.H. q-Difference equations. *Am. J. Math.* **1910**, *32*, 305–314. [CrossRef]
3. Anastassiu, G.A.; Gal, S.G. Geometric and approximation properties of generalized singular integrals. *J. Korean Math. Soc.* **2006**, *23*, 425–443. [CrossRef]
4. Aral, A. On the generalized Picard and Gauss Weierstrass singular integrals. *J. Comput. Anal. Appl.* **2006**, *8*, 249–261.
5. Ismail, M.E.-H.; Merkes, E.; Styer, D. A generalization of starlike functions. *Complex Var. Theory Appl.* **1990**, *14*, 77–84. [CrossRef]
6. Srivastava, H.M. Univalent functions, fractional calculus, and associated generalized hypergeometric functions. In *Univalent Functions, Fractional Calculus, and Their Applications*; Srivastava, H.M., Owa, S., Eds.; Halsted Press: New York, NY, USA; Ellis Horwood Limited: Chichester, UK, 1898; pp. 329–354.
7. Srivastava, H.M. Operators of basic (or q-) calculus and fractional q-calculus and their applications in geometric function theory of complex analysis. *Iran. J. Sci. Technol. Trans. A Sci.* **2020**, *44*, 327–344. [CrossRef]
8. Srivastava, H.M.; Bansal, D. Close-to-convexity of a certain family of q-Mittag–Leffler functions. *J. Nonlinear Var. Anal.* **2017**, *1*, 61–69.
9. Kanas, S.; Răducanu, D. Some class of analytic functions related to conic domains. *Math. Slovaca* **2014**, *64*, 1183–1196. [CrossRef]
10. Mahmmod, S.; Sokół, J. New subclass of analytic functions in conical domain associated with Ruscheweyh q-differential operator. *Results Math.* **2017**, *17*, 1345–1357. [CrossRef]
11. Srivastava, H.M.; Tahir, M.; Khan, B.; Ahmad, Q.Z.; Khan, N. Some general classes of q-starlike functions associated with the Janowski functions. *Symmetry* **2019**, *11*, 292. [CrossRef]
12. Srivastava, H.M.; Tahir, M.; Khan, B.; Ahmad, Q.Z.; Khan, N. Some general families of q-starlike functions associated with the Janowski functions. *Filomat* **2019**, *33*, 2613–2626. [CrossRef]
13. Ahmad, B.; Khan, M.G.; Aouf, M.K.; Mashwani, W.K.; Salleh, Z.; Tang, H. Applications of a new q-difference operator in the Janowski-type meromorphic convex functions. *J. Funct. Spaces* **2021**, *2021*, 5534357.
14. Mahmood, S.; Ahmad, Q.Z.; Srivastava, H.M.; Khan, N.; Khan, B.; Tahir, M. A certain subclass of meromorphically q-starlike functions associated with the Janowski functions. *J. Inequal. Appl.* **2019**, *2019*, 88. [CrossRef]
15. Mahmood, S.; Srivastava, H.M.; Khan, N.; Ahmad, Q.Z.; Khan, B.; Ali, I. Upper bound of the third Hankel determinant for a subclass of q-starlike functions. *Symmetry* **2019**, *11*, 347. [CrossRef]
16. Rehman, M.S.U.; Ahmad, Q.Z.; Srivastava, H.M.; Khan, N.; Darus, M.; Khan, B. Applications of higher-order q-derivatives to the subclass of q-starlike functions associated with the Janowski functions. *AIMS Math.* **2021**, *6*, 1110–1125. [CrossRef]
17. Shi, L.; Khan, Q.; Srivastava, G.; Liu, J.-L.; Arif, M. A study of multivalent q-starlike functions connected with circular domain. *Mathematics* **2019**, *7*, 670. [CrossRef]
18. Shi, L.; Khan, M.G.; Ahmad, B. Some geometric properties of a family of analytic functions involving a generalized q -operator. *Symmetry* **2019**, *12*, 291. [CrossRef]
19. Srivastava, H.M.; Khan, B.; Khan, N.; Ahmad, Q.Z. Coefficient inequalities for q-starlike functions associated with the Janowski functions. *Hokkaido Math. J.* **2019**, *48*, 407–425. [CrossRef]
20. Srivastava, H.M.; Khan, B.; Khan, N.; Ahmad, Q.Z.; Tahir, M. A generalized conic domain and its applications to certain subclasses of analytic functions. *Rocky Mt. J. Math.* **2019**, *49*, 2325–2346. [CrossRef]
21. Khan, Q.; Arif, M.; Raza, M.; Srivastava, G.; Tang, H. Some applications of a new integral operator in q-analog for multivalent functions. *Mathematics* **2019**, *7*, 1178. [CrossRef]
22. Khan, B.; Srivastava, H.M.; Arjika, S.; Khan, S.; Khan, N.; Ahmad, Q.Z. A certain q-Ruscheweyh type derivative operator and its applications involving multivalent functions. *Adv. Differ. Equ.* **2021**, *2021*, 1–14. [CrossRef]

23. Khan, B.; Liu, Z.-G.; Srivastava, H.M.; Khan, N.; Darus, M.; Tahir, M. A study of some families of multivalent q-starlike functions involving higher-order q-Derivatives. *Mathematics* **2020**, *8*, 1470. [CrossRef]
24. Khan, B.; Liu, Z.-G.; Srivastava, H.M.; Khan, N.; Tahir, M. Applications of higher-order derivatives to subclasses of multivalent q-starlike functions. *Maejo Internat. J. Sci. Technol.* **2021**, *15*, 61–72.
25. Khan, B.; Srivastava, H.M.; Khan, N.; Darus, M.; Tahir, M.; Ahmad, Q. Coefficient estimates for a subclass of analytic functions associated with a certain leaf-like domain. *Mathematics* **2020**, *8*, 1334. [CrossRef]
26. Khan, B.; Srivastava, H.M.; Tahir, M.; Darus, M.; Ahmad, Q.Z.; Khan, N. Applications of a certain integral operator to the subclasses of analytic and bi-univalent functions. *AIMS Math.* **2021**, *6*, 1024–1039. [CrossRef]
27. Khan, B.; Srivastava, H.M.; Khan, N.; Darus, M.; Ahmad, Q.Z.; Tahir, M. Applications of certain conic domains to a subclass of q-starlike functions associated with the Janowski functions. *Symmetry* **2021**, *13*, 574. [CrossRef]
28. Liu, Z.-G. Two q-difference equations and q-operator identities. *J. Differ. Equ. Appl.* **2010**, *16*, 1293–1307. [CrossRef]
29. Liu, Z.-G. Some operator identities and q-series transformation formulas. *Discret. Math.* **2003**, *256*, 119–139. [CrossRef]
30. Mahmood, S.; Raza, N.; AbuJarad, E.S.; Srivastava, G.; Srivastava, H.M.; Malik, S.N. Geometric properties of certain classes of analytic functions associated with a q-integral operator. *Symmetry* **2019**, *11*, 719. [CrossRef]
31. Raza, M.; Srivastava, H.M.; Arif, M.; Ahmad, K. Coefficient estimates for a certain family of analytic functions involving a q-derivative operator. *Ramanujan J.* **2021**, *55*, 53–71. [CrossRef]
32. Rehman, M.S.U.; Ahmad, Q.Z.; Srivastava, H.M.; Khan, B.; Khan, N. Partial sums of generalized q-Mittag–Leffler functions. *AIMS Math.* **2020**, *5*, 408–420. [CrossRef]
33. Srivastava, H.M.; Ahmad, Q.Z.; Khan, N.; Khan, N.; Khan, B. Hankel and Toeplitz determinants for a subclass of q-starlike functions associated with a general conic domain. *Mathematics* **2019**, *7*, 181. [CrossRef]
34. Srivastava, H.M.; Aouf, M.K.; Mostafa, A.O. Some properties of analytic functions associated with fractional q-calculus operators. *Miskolc Math. Notes* **2019**, *20*,1245–1260. [CrossRef]
35. Srivastava, H.M.; Arif, M.; Raza, M. Convolution properties of meromorphically harmonic functions defined by a generalized convolution q-derivative operator. *AIMS Math.* **2021**, *6*, 869–5885. [CrossRef]
36. Srivastava, H.M.; Khan, B.; Khan, N.; Tahir, M.; Ahmad, S.; Khan, N. Upper bound of the third Hankel determinant for a subclass of q-starlike functions associated with the q-exponential function. *Bull. Sci. Math.* **2021**, *167*, 102942. [CrossRef]
37. Srivastava, H.M.; Raza, N.; AbuJarad, E.S.A.; Srivastava, G.; AbuJarad, M.H. Fekete-Szegö inequality for classes of (p,q)-starlike and (p,q)-convex functions. *Rev. Real Acad. Cienc. Exactas Fís. Natur. Ser. A Mat. (RACSAM)* **2019**, *113*, 3563–3584. [CrossRef]
38. Yan, C.-M.; Srivastava, R.; Liu, J.-L. Properties of certain subclass of meromorphic multivalent functions associated with q-difference operator. *Symmetry* **2021**, *13*, 1035. [CrossRef]
39. Janwoski, W. Some extremal problems for certain families of analytic functions. *Ann. Polon. Math.* **1973**, *28*, 297–326. [CrossRef]
40. Dziok, J.; Murugusundaramoorthy, G.; Sokół, J. On certain class of meromorphic functions with positive coefficients. *Acta Math. Sci. B* **2012**, *32*, 1–16. [CrossRef]
41. Aldweby, H.; Darus, M. Integral operator defined by q-analogue of Liu-Srivastava operator. *Studia Univ. Babeş-Bolyai Ser. Math.* **2013**, *58*, 529–537.
42. Pommerenke, C. On meromorphic starlike functions. *Pac. J. Math.* **1963**, *13*, 221–235. [CrossRef]
43. Seoudy, T.M.; Aouf, M.K. Coefficient estimates of new classes of q-starlike and q-convex functions of complex order. *J. Math. Inequal.* **2016**, *10*, 135–145. [CrossRef]
44. Uralegaddi, B.A.; Somanatha, C. Certain diferential operators for meromorphic functions. *Houst. J. Math.* **1991**, *17*, 279–284.
45. Rogosinski, W. On the coefficients of subordinate functions. *Proc. Lond. Math. Soc. (Ser. 2)* **1843**, *48*, 48–82. [CrossRef]
46. Srivastava, H.M.; Karlsson, P.W. *Multiple Gaussian Hypergeometric Series*; Ellis Horwood Limited: Chichester, UK, 1985.
47. Srivastava, H.M.; Seoudy, T.M.; Aouf, M.K. A generalized conic domain and its applications to certain subclasses of multivalent functions associated with the basic (or q-) calculus. *AIMS Math.* **2021**, *6*, 6580–6602. [CrossRef]

symmetry
MDPI
Article

Properties of Certain Subclass of Meromorphic Multivalent Functions Associated with q-Difference Operator

Cai-Mei Yan [1], Rekha Srivastava [2,*] and Jin-Lin Liu [3]

1 Information Engineering College, Yangzhou University, Yangzhou 225002, China; cmyan@yzu.edu.cn
2 Department of Mathematics and Statistics, University of Victoria, Victoria, BC V8W 3R4, Canada
3 Department of Mathematics, Yangzhou University, Yangzhou 225002, China; jlliu@yzu.edu.cn
* Correspondence: rekhas@math.uvic.ca

Abstract: A new subclass $\Sigma_{p,q}(\alpha, A, B)$ of meromorphic multivalent functions is defined by means of a q-difference operator. Some properties of the functions in this new subclass, such as sufficient and necessary conditions, coefficient estimates, growth and distortion theorems, radius of starlikeness and convexity, partial sums and closure theorems, are investigated.

Keywords: q-difference operator; Janowski function; meromorphic multivalent function; distortion theorem; partial sum; closure theorem

MSC: 2020 Mathematics Subject Classification; Primary 30C45; 05A30; Secondary 11B65; 47B38

Citation: Yan, C.-M.; Srivastava, R.; Liu, J.-L. Properties of Certain Subclass of Meromorphic Multivalent Functions Associated with q-Difference Operator. *Symmetry* **2021**, *13*, 1035. https://doi.org/10.3390/sym13061035

Academic Editor: Clemente Cesarano and Youssef N. Raffoul

Received: 14 April 2021
Accepted: 19 May 2021
Published: 8 June 2021

1. Introduction

In recent years, q-analysis has attracted the interest of scholars because of its numerous applications in mathematics and physics. Jackson [1,2] was the first to consider the certain application of q-calculus and introduced the q-analog of the derivative and integral. Very recently, several authors published a set of articles [3–13] in which they concentrated upon the classes of q-starlike functions related to the Janowski functions [14] from some different aspects. Further, a recently published survey-cum-expository review paper by Srivastava [15] is very useful for scholars working on these topics. In this review paper, Srivastava [15] gave certain mathematical explanation and addressed applications of the fractional q-derivative operator in Geometric Function Theory. In the same survey-cum-expository review paper [15], the trivial and inconsequential (p,q) variations of various known q-results by adding an obviously redundant parameter p were clearly exposed (see, for details, [15] p. 340).

In this article, motivated essentially by the above works, we shall define a new subclass of meromorphic multivalent functions by using the q-difference operator and Janowski functions and study its geometric properties, such as sufficient and necessary conditions, coefficient estimates, growth and distortion theorems, radius of starlikeness and convexity, partial sums and closure theorems.

Let M_p denote the class of meromorphic multivalent functions of the form

$$f(z) = z^{-p} + \sum_{n=1}^{\infty} a_n z^n \quad (p \in \mathbb{N} = \{1,2,3,\cdots\}),$$

which are analytic in the punctured open unit disk $D^* = \{z \in \mathbb{C} : 0 < |z| < 1\} = D \setminus \{0\}$ with a pole of order p at the origin.

A function $f(z) \in M_p$ is said to be the meromorphic p-valent starlike function of order σ if

$$\mathrm{Re}\left\{-\frac{zf'(z)}{f(z)}\right\} > \sigma \qquad (0 \leqq \sigma < p)$$

for all $z \in D^*$. We denote this class by $MS_p^*(\sigma)$.

A function $f(z) \in M_p$ is said to be the meromorphic p-valent convex function of order σ if

$$\mathrm{Re}\left\{-\left(1+\frac{zf''(z)}{f'(z)}\right)\right\} > \sigma \quad (0 \leqq \sigma < p)$$

for all $z \in D^*$. We denote this class by $MC_p(\sigma)$.

For two functions, $f(z)$ and $g(z)$, which are analytic in D, we can say that $g(z)$ is subordinate to $f(z)$ and denote $g(z) \prec f(z)$ $(z \in D)$, if there exists a Schwarz function $w(z)$, analytic in D with $w(0) = 0$ and $|w(z)| < 1$ $(z \in D)$, such that $g(z) = f(w(z))$ $(z \in D)$. Further, if $f(z)$ is univalent in D, then we have the following equivalence:

$$g(z) \prec f(z) \quad (z \in D) \Longleftrightarrow g(0) = f(0) \quad \text{and} \quad g(D) \subset f(D).$$

A function $\varphi(z)$ is said to be in the class $P[A, B]$, if it is analytic in D with $\varphi(0) = 1$ and

$$\varphi(z) \prec \frac{1+Az}{1+Bz} \quad (-1 \leqq B < A \leqq 1),$$

equivalently, we can write

$$\left|\frac{\varphi(z)-1}{A-B\varphi(z)}\right| < 1.$$

Let $q \in (0,1)$ and define the q-number $[\lambda]_q$ by

$$[\lambda]_q = \begin{cases} \frac{1-q^\lambda}{1-q} & (\lambda \in \mathbb{C}) \\ \sum_{k=0}^{n-1} q^k = 1+q+q^2+\cdots+q^{n-1} & (\lambda = n \in \mathbb{N}). \end{cases}$$

Particularly, when $\lambda = 0$, we write $[0]_q = 0$.

Definition 1. *For $q \in (0,1)$, the q-difference operator D_q of a function $f(z)$ is defined by*

$$D_q f(z) = \begin{cases} \frac{f(z)-f(qz)}{(1-q)z} & (z \neq 0) \\ f'(0) & (z = 0), \end{cases}$$

provided that $f'(0)$ exists.

From Definition 1, we observe that

$$\lim_{q \to 1^-} D_q f(z) = \lim_{q \to 1^-} \frac{f(qz)-f(z)}{(q-1)z} = f'(z)$$

for a differentiable function $f(z)$.

For $f(z) = z^{-p} + \sum_{n=1}^{\infty} a_n z^n \in M_p$, we can see that

$$D_q f(z) = [-p]_q z^{-p-1} + \sum_{n=1}^{\infty} [n]_q a_n z^{n-1} \quad (z \neq 0),$$

where $[-p]_q = \frac{1-q^{-p}}{1-q} = -q^{-p}[p]_q$ and $[p]_q = \frac{1-q^{-p}}{1-q} = 1+q+q^2+\cdots+q^{p-1}$.

We now define a new subclass $\Sigma_{p,q}(\alpha, A, B)$ of M_p as the following.

Definition 2. *For $q \in (0,1)$, $\alpha > 1$ and $-1 \leqq B < A \leqq 1$, a function $f(z) \in M_p$ is said to belong to the class $\Sigma_{p,q}(\alpha, A, B)$, if it satisfies*

$$\frac{1}{1-\alpha}\left(\frac{zD_q f(z)}{[-p]_q f(z)} - \alpha \frac{z^2 D_q^2 f(z)}{[-p]_q[-p-1]_q f(z)}\right) \prec \frac{1+Az}{1+Bz},$$

or equivalently

$$\left|\frac{\frac{zD_q f(z)}{[-p]_q f(z)} - \alpha\frac{z^2 D_q^2 f(z)}{[-p]_q[-p-1]_q f(z)} - (1-\alpha)}{(1-\alpha)A - B\left(\frac{zD_q f(z)}{[-p]_q f(z)} - \alpha\frac{z^2 D_q^2 f(z)}{[-p]_q[-p-1]_q f(z)}\right)}\right| < 1 \quad (z \in D). \tag{1}$$

2. Main Results

Theorem 1. *Let* $1 < \alpha \leqq 1 - \frac{1}{[-p]_q}$ *and*

$$f(z) = z^{-p} + \sum_{n=1}^{\infty} a_n z^n \quad (a_n \geqq 0) \in M_p.$$

Then $f(z) \in \Sigma_{p,q}(\alpha, A, B)$ *if*

$$\begin{aligned}&\sum_{n=1}^{\infty}((1-\alpha)(1+A)[-p]_q[-p-1]_q - (1+B)[n]_q([-p-1]_q - \alpha[n-1]_q))a_n \\ &\quad \leqq (1-\alpha)(B-A)[-p]_q[-p-1]_q.\end{aligned} \tag{2}$$

Proof. Suppose that the inequality (2) holds true. Then we have

$$\begin{aligned}&\left|\frac{\frac{zD_q f(z)}{[-p]_q f(z)} - \alpha\frac{z^2 D_q^2 f(z)}{[-p]_q[-p-1]_q f(z)} - (1-\alpha)}{(1-\alpha)A - B\left(\frac{zD_q f(z)}{[-p]_q f(z)} - \alpha\frac{z^2 D_q^2 f(z)}{[-p]_q[-p-1]_q f(z)}\right)}\right| \\
&= \left|\frac{[-p-1]_q zD_q f(z) - \alpha z^2 D_q^2 f(z) - (1-\alpha)[-p]_q[-p-1]_q f(z)}{(1-\alpha)A[-p]_q[-p-1]_q f(z) - B([-p-1]_q zD_q f(z) - \alpha z^2 D_q^2 f(z))}\right| \\
&= \left|\frac{\sum_{n=1}^{\infty}([n]_q([-p-1]_q - \alpha[n-1]_q) - (1-\alpha)[-p]_q[-p-1]_q)a_n z^n}{(1-\alpha)(A-B)[-p]_q[-p-1]_q z^{-p} - \sum_{n=1}^{\infty}(B[n]_q([-p-1]_q - \alpha[n-1]_q) - (1-\alpha)A[-p]_q[-p-1]_q)a_n z^n}\right| \\
&= \left|\frac{\sum_{n=1}^{\infty}([n]_q([-p-1]_q - \alpha[n-1]_q) - (1-\alpha)[-p]_q[-p-1]_q)a_n z^{n+p}}{(1-\alpha)(A-B)[-p]_q[-p-1]_q - \sum_{n=1}^{\infty}(B[n]_q([-p-1]_q - \alpha[n-1]_q) - (1-\alpha)A[-p]_q[-p-1]_q)a_n z^{n+p}}\right| \\
&= \left|\frac{\sum_{n=1}^{\infty}((1-\alpha)[-p]_q[-p-1]_q - [n]_q([-p-1]_q - \alpha[n-1]_q))a_n z^{n+p}}{(1-\alpha)(B-A)[-p]_q[-p-1]_q - \sum_{n=1}^{\infty}((1-\alpha)A[-p]_q[-p-1]_q - B[n]_q([-p-1]_q - \alpha[n-1]_q))a_n z^{n+p}}\right| \\
&< 1.\end{aligned}$$

This shows that $f(z) \in \Sigma_{p,q}(\alpha, A, B)$.

Conversely, let $f(z) = z^{-p} + \sum_{n=1}^{\infty} a_n z^n \quad (a_n \geqq 0) \in \Sigma_{p,q}(\alpha, A, B)$. From (1), we obtain

$$\begin{aligned}&\left|\frac{\frac{zD_q f(z)}{[-p]_q f(z)} - \alpha\frac{z^2 D_q^2 f(z)}{[-p]_q[-p-1]_q f(z)} - (1-\alpha)}{(1-\alpha)A - B\left(\frac{zD_q f(z)}{[-p]_q f(z)} - \alpha\frac{z^2 D_q^2 f(z)}{[-p]_q[-p-1]_q f(z)}\right)}\right| \\
&= \left|\frac{\sum_{n=1}^{\infty}((1-\alpha)[-p]_q[-p-1]_q - [n]_q([-p-1]_q - \alpha[n-1]_q))a_n z^{n+p}}{(1-\alpha)(B-A)[-p]_q[-p-1]_q - \sum_{n=1}^{\infty}((1-\alpha)A[-p]_q[-p-1]_q - B[n]_q([-p-1]_q - \alpha[n-1]_q))a_n z^{n+p}}\right| \\
&< 1.\end{aligned} \tag{3}$$

The inequality (3) is true for all $z \in D^*$. Thus, we choose $z = \text{Re}z \to 1^-$ and obtain the inequality (2). The proof of Theorem 1 is completed. □

From Theorem 1, we can easily obtain the following coefficient estimates.

Corollary 1. *Let* $-1 < B < A \leqq 1$ *and* $1 < \alpha < 1 - \frac{1+B}{(1+A)[-p]_q}$. *If*

$$f(z) = z^{-p} + \sum_{n=1}^{\infty} a_n z^n \quad (a_n \geqq 0) \in \Sigma_{p,q}(\alpha, A, B),$$

then

$$a_n \leqq \frac{(1-\alpha)(B-A)[-p]_q[-p-1]_q}{(1-\alpha)(1+A)[-p]_q[-p-1]_q-(1+B)[n]_q([-p-1]_q-\alpha[n-1]_q)} \quad (n=1,2,\cdots).$$

The results are sharp for the function given by

$$f(z)=z^{-p}+\frac{(1-\alpha)(B-A)[-p]_q[-p-1]_q}{(1-\alpha)(1+A)[-p]_q[-p-1]_q-(1+B)[n]_q([-p-1]_q-\alpha[n-1]_q)}z^n.$$

Theorem 2. *Let* $-1<B<A\leqq 1$ *and* $1<\alpha<1-\frac{1+B}{(1+A)[-p]_q}$. *If*

$$f(z)=z^{-p}+\sum_{n=1}^{\infty}a_nz^n \qquad (a_n\geqq 0)\in\Sigma_{p,q}(\alpha,A,B),$$

then, for $0<|z|=r<1$, *it is asserted that*

$$\frac{1}{r^p}-r\tau_1\leqq|f(z)|\leqq\frac{1}{r^p}+r\tau_1,$$

where

$$\tau_1=\frac{(1-\alpha)(B-A)[-p]_q}{(1-\alpha)(1+A)[-p]_q-(1+B)}. \tag{4}$$

The results are sharp for the function

$$f(z)=z^{-p}+\frac{(1-\alpha)(B-A)[-p]_q}{(1-\alpha)(1+A)[-p]_q-(1+B)}z.$$

Proof. Let

$$f(z)=z^{-p}+\sum_{n=1}^{\infty}a_nz^n \qquad (a_n\geqq 0)\in\Sigma_{p,q}(\alpha,A,B).$$

Then, by applying the triangle inequality, we have

$$|f(z)|=\left|z^{-p}+\sum_{n=1}^{\infty}a_nz^n\right|\leqq\frac{1}{|z|^p}+\sum_{n=1}^{\infty}a_n|z|^n.$$

Since $|z|=r<1$, we can see that $r^n\leqq r$. Thus, we have

$$|f(z)\leqq\frac{1}{r^p}+r\sum_{n=1}^{\infty}a_n \tag{5}$$

and

$$|f(z)\geqq\frac{1}{r^p}-r\sum_{n=1}^{\infty}a_n. \tag{6}$$

From Theorem 1, we know that

$$\sum_{n=1}^{\infty}((1-\alpha)(1+A)[-p]_q[-p-1]_q-(1+B)[n]_q([-p-1]_q-\alpha[n-1]_q))a_n$$
$$\leqq(1-\alpha)(B-A)[-p]_q[-p-1]_q.$$

It is easy to see that the sequence

$$\{(1-\alpha)(1+A)[-p]_q[-p-1]_q-(1+B)[n]_q([-p-1]_q-\alpha[n-1]_q)\}$$

is an increasing sequence with respect to $n(n \geqq 1)$. Thus,

$$
\begin{aligned}
&\left((1-\alpha)(1+A)[-p]_q[-p-1]_q-(1+B)[-p-1]_q\right)\sum_{n=1}^{\infty} a_n \\
&\leqq \sum_{n=1}^{\infty}\left((1-\alpha)(1+A)[-p]_q[-p-1]_q-(1+B)[n]_q([-p-1]_q-\alpha[n-1]_q)\right)a_n \\
&\leqq (1-\alpha)(B-A)[-p]_q[-p-1]_q,
\end{aligned}
$$

which shows that

$$\sum_{n=1}^{\infty} a_n \leqq \frac{(1-\alpha)(B-A)[-p]_q}{(1-\alpha)(1+A)[-p]_q-(1+B)}. \tag{7}$$

Substituting from (7) into the inequalities (5) and (6), we obtain the required results. The proof of Theorem 2 is completed. □

Theorem 3. *Let* $-1 < B < A \leqq 1$ *and* $1 < \alpha < 1 - \frac{1+B}{(1+A)[-p]_q}$. *If*

$$f(z) = z^{-p} + \sum_{n=1}^{\infty} a_n z^n \qquad (a_n \geqq 0) \in \Sigma_{p,q}(\alpha, A, B),$$

then, for $0 < |z| = r < 1$, *it is asserted that*

$$-[-p]_q\frac{1}{r^{p+1}} - \tau_1 \leqq |D_q f(z)| \leqq -[-p]_q\frac{1}{r^{p+1}} + \tau_1,$$

where τ_1 *is given by (4).*

Proof. Let

$$f(z) = z^{-p} + \sum_{n=1}^{\infty} a_n z^n \qquad (a_n \geqq 0) \in \Sigma_{p,q}(\alpha, A, B).$$

Then, from Definition 1, we can write

$$D_q f(z) = [-p]_q z^{-p-1} + \sum_{n=1}^{\infty}[n]_q a_n z^{n-1}.$$

For $|z| = r < 1$, we have

$$|D_q f(z)| = \left|[-p]_q z^{-p-1} + \sum_{n=1}^{\infty}[n]_q a_n z^{n-1}\right| \leqq -[-p]_q\frac{1}{|r|^{p+1}} + \sum_{n=1}^{\infty}[n]_q a_n. \tag{8}$$

Similarly, we obtain

$$|D_q f(z)| \geqq -[-p]_q\frac{1}{r^{p+1}} - \sum_{n=1}^{\infty}[n]_q a_n. \tag{9}$$

Since $f(z) \in \Sigma_{p,q}(\alpha, A, B)$, we know from Theorem 1 that

$$
\begin{aligned}
&\sum_{n=1}^{\infty}\left(\frac{(1-\alpha)(1+A)[-p]_q[-p-1]_q}{[n]_q} - (1+B)([-p-1]_q - \alpha[n-1]_q)\right)[n]_q a_n \\
&\leqq (1-\alpha)(B-A)[-p]_q[-p-1]_q.
\end{aligned}
$$

As we know that the sequence

$$\left\{\frac{(1-\alpha)(1+A)[-p]_q[-p-1]_q}{[n]_q} - (1+B)([-p-1]_q - \alpha[n-1]_q)\right\}$$

is an increasing sequence with respect to n $(n \geqq 1)$. Thus, we have

$$\begin{aligned}
&((1-\alpha)(1+A)[-p]_q[-p-1]_q - (1+B)([-p-1]_q)\sum_{n=1}^{\infty}[n]_q a_n \\
&\leqq \sum_{n=1}^{\infty}\left(\frac{(1-\alpha)(1+A)[-p]_q[-p-1]_q}{[n]_q} - (1+B)([-p-1]_q - \alpha[n-1]_q)\right)[n]_q a_n \\
&\leqq (1-\alpha)(B-A)[-p]_q[-p-1]_q,
\end{aligned}$$

which implies that

$$\sum_{n=1}^{\infty}[n]_q a_n \leqq \frac{(1-\alpha)(B-A)[-p]_q}{(1-\alpha)(1+A)[-p]_q - (1+B)}. \tag{10}$$

Now, the theorem is proven. □

Theorem 4. *Let* $1 < \alpha < 1 - \frac{1+B}{(1+A)[-p]_q}$, $-1 < B < A \leqq 1$ *and* $0 \leqq \sigma < p$. *If*

$$f(z) = z^{-p} + \sum_{n=1}^{\infty} a_n z^n \quad (a_n \geqq 0) \in \Sigma_{p,q}(\alpha, A, B),$$

then $f(z)$ *is meromorphic p-valent starlike function of order* σ *in* $0 < |z| < r_1$, *where*

$$r_1 = \min\left\{\inf_{n\geqq 1}\left(\frac{(p-\sigma)((1-\alpha)(1+A)[-p]_q[-p-1]_q - (1+B)[n]_q([-p-1]_q - \alpha[n-1]_q))}{(n+\sigma)(1-\alpha)(B-A)[-p]_q[-p-1]_q}\right)^{\frac{1}{n+p}}, 1\right\}.$$

Proof. In order to prove that $f(z)$ is the meromorphic p-valent starlike function of order σ in $0 < |z| < r_1$, we need only to show that

$$\frac{\frac{-zf'(z)}{f(z)} - \sigma}{p-\sigma} \prec \frac{1+z}{1-z}, \quad 0 \leqq \sigma < p.$$

The subordination above is equivalent to $\left|\frac{-zf'(z) - pf(z)}{-zf'(z) + (p-2\sigma)f(z)}\right| < 1$. After some calculations and simplifications, we have

$$\sum_{n=1}^{\infty}\frac{n+\sigma}{p-\sigma}a_n|z|^{n+p} < 1. \tag{11}$$

From (2), we can see that

$$\sum_{n=1}^{\infty}\frac{(1-\alpha)(1+A)[-p]_q[-p-1]_q - (1+B)[n]_q([-p-1]_q - \alpha[n-1]_q)}{(1-\alpha)(B-A)[-p]_q[-p-1]_q}a_n < 1.$$

The inequality (11) will be true if

$$\frac{n+\sigma}{p-\sigma}a_n|z|^{n+p} < \frac{(1-\alpha)(1+A)[-p]_q[-p-1]_q - (1+B)[n]_q([-p-1]_q - \alpha[n-1]_q)}{(1-\alpha)(B-A)[-p]_q[-p-1]_q}a_n$$

or

$$|z| < \left(\frac{(p-\sigma)((1-\alpha)(1+A)[-p]_q[-p-1]_q - (1+B)[n]_q([-p-1]_q - \alpha[n-1]_q))}{(n+\sigma)(1-\alpha)(B-A)[-p]_q[-p-1]_q}\right)^{\frac{1}{n+p}}.$$

Let

$$r_1 = \min\left\{\inf_{n\geqq 1}\left(\frac{(p-\sigma)((1-\alpha)(1+A)[-p]_q[-p-1]_q - (1+B)[n]_q([-p-1]_q - \alpha[n-1]_q))}{(n+\sigma)(1-\alpha)(B-A)[-p]_q[-p-1]_q}\right)^{\frac{1}{n+p}}, 1\right\}.$$

Then, clearly, we obtain the required condition. The proof of Theorem 4 is completed. □

Theorem 5. *Let* $1 < \alpha < 1 - \frac{1+B}{(1+A)[-p]_q}$, $-1 < B < A \leqq 1$ *and* $0 \leqq \sigma < p$. *If*

$$f(z) = z^{-p} + \sum_{n=1}^{\infty} a_n z^n \quad (a_n \geqq 0) \in \Sigma_{p,q}(\alpha, A, B),$$

then $f(z)$ *is the meromorphic p-valent convex function of order* σ *in* $0 < |z| < r_2$, *where*

$$r_2 = \min\left\{\inf_{n\geqq 1}\left(\frac{p(p-\sigma)((1-\alpha)(1+A)[-p]_q[-p-1]_q - (1+B)[n]_q([-p-1]_q - \alpha[n-1]_q))}{n(n+\sigma)(1-\alpha)(B-A)[-p]_q[-p-1]_q}\right)^{\frac{1}{n+p}}, 1\right\}.$$

Proof. To prove that $f(z)$ is the meromorphic p-valent convex function of order σ in $0 < |z| < r_2$, we need only to show that

$$\frac{-\left(1 + \frac{zf''(z)}{f'(z)}\right) - \sigma}{p - \sigma} \prec \frac{1+z}{1-z}, \quad 0 \leqq \sigma < p.$$

This subordination relation is equivalent to the inequality $\left|\frac{-zf''(z)-(p+1)f'(z)}{-zf''(z)+(p-1-2\sigma)f'(z)}\right| < 1$. After some calculations and simplifications, we have

$$\sum_{n=1}^{\infty} \frac{n(n+\sigma)}{p(p-\sigma)} a_n |z|^{n+p} < 1. \tag{12}$$

From the inequality (2), we obtain that

$$\sum_{n=1}^{\infty} \frac{(1-\alpha)(1+A)[-p]_q[-p-1]_q - (1+B)[n]_q([-p-1]_q - \alpha[n-1]_q)}{(1-\alpha)(B-A)[-p]_q[-p-1]_q} a_n < 1.$$

The inequality (12) will be true if

$$\frac{n(n+\sigma)}{p(p-\sigma)} a_n |z|^{n+p} < \frac{(1-\alpha)(1+A)[-p]_q[-p-1]_q - (1+B)[n]_q([-p-1]_q - \alpha[n-1]_q)}{(1-\alpha)(B-A)[-p]_q[-p-1]_q} a_n,$$

or

$$|z| < \left(\frac{p(p-\sigma)((1-\alpha)(1+A)[-p]_q[-p-1]_q - (1+B)[n]_q([-p-1]_q - \alpha[n-1]_q))}{n(n+\sigma)(1-\alpha)(B-A)[-p]_q[-p-1]_q}\right)^{\frac{1}{n+p}}.$$

Let

$$r_2 = \min\left\{\inf_{n\geqq 1}\left(\frac{p(p-\sigma)((1-\alpha)(1+A)[-p]_q[-p-1]_q - (1+B)[n]_q([-p-1]_q - \alpha[n-1]_q))}{n(n+\sigma)(1-\alpha)(B-A)[-p]_q[-p-1]_q}\right)^{\frac{1}{n+p}}, 1\right\}.$$

Then, we obtain the required condition. Now, Theorem 5 is proven. □

Theorem 6. *Let* $1 < \alpha \leqq \frac{(1+2A-B)[-p]_q - (1+B)}{(1+2A-B)[-p]_q}$ *and* $-1 < B < A \leqq 1$. *If*

$$f(z) = z^{-p} + \sum_{n=1}^{\infty} a_n z^n \qquad (a_n \geqq 0) \in \Sigma_{p,q}(\alpha, A, B)$$

and

$$f_k(z) = z^{-p} + \sum_{n=1}^{k} a_n z^n \quad (a_n \geqq 0;\ k \geqq 1),$$

then

$$\mathrm{Re}\left(\frac{f(z)}{f_k(z)}\right) \geqq 1 - \frac{1}{\varphi_{k+1}} \tag{13}$$

and

$$\mathrm{Re}\left(\frac{f_k(z)}{f(z)}\right) \geqq \frac{\varphi_{k+1}}{1+\varphi_{k+1}}, \tag{14}$$

where

$$\varphi_{k+1} = \frac{(1-\alpha)(1+A)[-p]_q[-p-1]_q) - (1+B)[k+1]_q([-p-1]_q - \alpha[k]_q)}{(1-\alpha)(B-A)[-p]_q[-p-1]_q}. \tag{15}$$

Proof. In order to prove the inequality (13), we set

$$\varphi_{k+1}\left[\frac{f(z)}{f_k(z)} - \left(1 - \frac{1}{\varphi_{k+1}}\right)\right] = \frac{1 + \sum_{n=1}^{k} a_n z^{n+p} + \varphi_{k+1}\sum_{n=k+1}^{\infty} a_n z^{n+p}}{1 + \sum_{n=1}^{k} a_n z^{n+p}} = \frac{1+w(z)}{1-w(z)}.$$

After some simplifications, we have

$$w(z) = \frac{\varphi_{k+1}\sum_{n=k+1}^{\infty} a_n z^{n+p}}{2 + 2\sum_{n=1}^{k} a_n z^{n+p} + \varphi_{k+1}\sum_{n=k+1}^{\infty} a_n z^{n+p}}$$

and

$$|w(z)| \leqq \frac{\varphi_{k+1}\sum_{n=k+1}^{\infty} a_n}{2 - 2\sum_{n=1}^{k} a_n - \varphi_{k+1}\sum_{n=k+1}^{\infty} a_n}.$$

From (2), we know that $\sum_{n=1}^{\infty} \varphi_n a_n \leqq 1$. The sequence $\{\varphi_n\}$ given by (15) is an increasing sequence with respect to n and $\varphi_n \geqq 1$ $(n = 1, 2, 3, \cdots)$. Therefore,

$$\sum_{n=1}^{k} a_n + \varphi_{k+1}\sum_{n=k+1}^{\infty} a_n \leqq \sum_{n=1}^{k} \varphi_n a_n + \sum_{n=k+1}^{\infty} \varphi_n a_n = \sum_{n=1}^{\infty} \varphi_n a_n \leqq 1.$$

This shows that $|w(z)| < 1 (z \in D)$. Now, the proof of the inequality (13) is completed.

To prove the inequality (14), we put

$$(1+\varphi_{k+1})\left[\frac{f_k(z)}{f(z)} - \frac{\varphi_{k+1}}{1+\varphi_{k+1}}\right] = \frac{1 + \sum_{n=1}^{k} a_n z^{n+p} - \varphi_{k+1}\sum_{n=k+1}^{\infty} a_n z^{n+p}}{1 + \sum_{n=1}^{\infty} a_n z^{n+p}} = \frac{1+w(z)}{1-w(z)}.$$

After some simplifications , we find that

$$w(z) = \frac{-(1+\varphi_{k+1})\sum_{n=k+1}^{\infty} a_n z^{n+p}}{2 + 2\sum_{n=1}^{k} a_n z^{n+p} - (\varphi_{k+1} - 1)\sum_{n=k+1}^{\infty} a_n z^{n+p}}$$

and

$$|w(z)| \leqq \frac{(1+\varphi_{k+1})\sum_{n=k+1}^{\infty} a_n}{2 - 2\sum_{n=1}^{k} a_n - (\varphi_{k+1} - 1)\sum_{n=k+1}^{\infty} a_n}.$$

Now, we can see that $|w(z)| < 1 (z \in D)$ if

$$\sum_{n=1}^{k} a_n + \varphi_{k+1}\sum_{n=k+1}^{\infty} a_n \leqq 1.$$

The proof of Theorem 6 is completed. □

Theorem 7. *Let* $1 < \alpha \leqq 1 - \frac{1}{[-p]_q}$. *If*

$$f_j(z) = z^{-p} + \sum_{n=1}^{\infty} a_{n,j} z^n \qquad (a_{n,j} \geqq 0;\ j = 1,2) \in \Sigma_{p,q}(\alpha, A, B),$$

then, for $0 \leqq \lambda \leqq 1$, *the function* $H(z) = \lambda f_1(z) + (1-\lambda) f_2(z) \in \Sigma_{p,q}(\alpha, A, B)$.

Proof. For $0 \leqq \lambda \leqq 1$, we have

$$H(z) = \lambda f_1(z) + (1-\lambda) f_2(z) = z^{-p} + \sum_{n=1}^{\infty} (\lambda a_{n,1} + (1-\lambda) a_{n,2}) z^n.$$

Since $f_j(z)$ $(j = 1,2) \in \Sigma_{p,q}(\alpha, A, B)$, by Theorem 1, we have

$$\begin{aligned}
&\sum_{n=1}^{\infty} ((1-\alpha)(1+A)[-p]_q[-p-1]_q - (1+B)[n]_q([-p-1]_q - \alpha[n-1]_q))(\lambda a_{n,1} + (1-\lambda) a_{n,2}) \\
&= \lambda \sum_{n=1}^{\infty} ((1-\alpha)(1+A)[-p]_q[-p-1]_q - (1+B)[n]_q([-p-1]_q - \alpha[n-1]_q)) a_{n,1} \\
&+ (1-\lambda) \sum_{n=1}^{\infty} ((1-\alpha)(1+A)[-p]_q[-p-1]_q - (1+B)[n]_q([-p-1]_q - \alpha[n-1]_q)) a_{n,2} \\
&\leqq \lambda(1-\alpha)(B-A)[-p]_q[-p-1]_q + (1-\lambda)(1-\alpha)(B-A)[-p]_q[-p-1]_q \\
&= (1-\alpha)(B-A)[-p]_q[-p-1]_q.
\end{aligned}$$

This shows that $H(z) \in \Sigma_{p,q}(\alpha, A, B)$. The theorem is provem. □

Corollary 2. *Let* $1 < \alpha \leqq 1 - \frac{1}{[-p]_q}$. *If*

$$f_j(z) = z^{-p} + \sum_{n=1}^{\infty} a_{n,j} z^n \qquad (a_{n,j} \geqq 0;\ j = 1,2,\cdots,t) \in \Sigma_{p,q}(\alpha, A, B),$$

then the function

$$F(z) = \sum_{j=1}^{t} \lambda_j f_j(z) \in \Sigma_{p,q}(\alpha, A, B),$$

where $\lambda_j \geqq 0$ *and* $\sum_{j=1}^{t} \lambda_j = 1$.

Theorem 8. *Let* $1 < \alpha \leqq 1 - \frac{1}{[-p]_q}$. *If*

$$f_j(z) = z^{-p} + \sum_{n=1}^{\infty} a_{n,j} z^n \qquad (a_{n,j} \geqq 0;\ j = 1,2) \in \Sigma_{p,q}(\alpha, A, B),$$

then, for $-1 \leqq m \leqq 1$, *the function*

$$Q_m(z) = \frac{(1-m) f_1(z) + (1+m) f_2(z)}{2} \in \Sigma_{p,q}(\alpha, A, B).$$

Proof. For $-1 \leqq m \leqq 1$, we have

$$Q_m(z) = \frac{(1-m) f_1(z) + (1+m) f_2(z)}{2} = z^{-p} + \sum_{n=1}^{\infty} \left(\frac{1-m}{2} a_{n,1} + \frac{1+m}{2} a_{n,2} \right) z^n.$$

In view of $f_1(z), f_2(z) \in \Sigma_{p,q}(\alpha, A, B)$, by Theorem 1, we obtain

$$\begin{aligned}
&\sum_{n=1}^{\infty}((1-\alpha)(1+A)[-p]_q[-p-1]_q-(1+B)[n]_q([-p-1]_q-\alpha[n-1]_q))\left(\frac{1-m}{2}a_{n,1}+\frac{1+m}{2}a_{n,2}\right)\\
&=\frac{1-m}{2}\sum_{n=1}^{\infty}((1-\alpha)(1+A)[-p]_q[-p-1]_q-(1+B)[n]_q([-p-1]_q-\alpha[n-1]_q))a_{n,1}\\
&\quad+\frac{1+m}{2}\sum_{n=1}^{\infty}((1-\alpha)(1+A)[-p]_q[-p-1]_q-(1+B)[n]_q([-p-1]_q-\alpha[n-1]_q))a_{n,2}\\
&\leq\frac{1-m}{2}(1-\alpha)(B-A)[-p]_q[-p-1]_q+\frac{1+m}{2}(1-\alpha)(B-A)[-p]_q[-p-1]_q\\
&=(1-\alpha)(B-A)[-p]_q[-p-1]_q,
\end{aligned}$$

which shows that $Q_m(z) \in \Sigma_{p,q}(\alpha, A, B)$. The proof of the theorem is completed. □

3. Conclusions

In this article, we introduce a new subclass $\Sigma_{p,q}(\alpha, A, B)$ of meromorphic multivalent functions by using the q-difference operator and Janowski functions. Some geometric properties of functions in $\Sigma_{p,q}(\alpha, A, B)$, such as sufficient and necessary conditions, coefficient estimates, growth and distortion theorems, radius of starlikeness and convexity, partial sums and closure theorems, are studied.

Author Contributions: Every author's contribution is equal. All authors have read and agreed to the published version of the manuscript.

Funding: This work is supported by the Natural Science Foundation of China (Grant No. 11571299).

Institutional Review Board Statement: Not applicable.

Informed Consent Statement: Not applicable.

Data Availability Statement: Not applicable.

Acknowledgments: The authors would like to express their sincere thanks to the referee for their careful reading and suggestions, which helped us to improve the paper.

Conflicts of Interest: The authors declare no conflict of interest.

References

1. Jackson, F.H. On q-functions and a certain difference operator. *Trans. R. Soc. Edinb.* **1908**, *46*, 253–281. [CrossRef]
2. Jackson, F.H. On q-definite integrals. *Q. J. Pure Appl. Math.* **1910**, *41*, 193–203.
3. Ahmad, B.; Khan, M.G.; Frasin, B.A.; Aouf, M.K.; Abdeljawad, T.; Mashwani, W.K.; Arif, M. On q-analogue of meromorphic multivalent functions in lemniscate of Bernoulli domain. *AIMS Math.* **2021**, *6*, 3037–3052. [CrossRef]
4. Khan, B.; Liu, Z.-G.; Srivastava, H.M.; Khan, N.; Tahir, M. Applications of higher-order derivatives to subclasses of multivalent q-starlike functions. *Maejo Int. J. Sci. Technol.* **2021**, *45*, 61–72.
5. Mahmood, S.; Raza, N.; Abujarad, E.S.A.; Srivastava, G.; Srivastava, H.M.; Malik, S.N. Geomotric properties of certain classes of analytic functions associated with a q-integral operator. *Symmetry* **2019**, *11*, 719. [CrossRef]
6. Mahmmod, S.; Sokół, J. New subclass of analytic functions in conical domain associated with Ruscheweyh q-differential operator. *Results Math.* **2017**, *71*, 1345–1357. [CrossRef]
7. Raza, M.; Srivastava, H.M.; Arif, M.; Ahmad, K. Coefficient estimates for a certain family of analytic functions involving a q-derivative operator. *Ramanujan J.* **2021**, *54*, 501–519.
8. Rehman, M.S.U.; Ahmad, Q.Z.; Srivastava, H.M.; Khan, N.; Darus, M.; Khan, B. Applications of higher-order q-derivatives to the subclass of q-starlike functions associated with the Janowski functions. *AIMS Math.* **2021**, *6*, 1110–1125. [CrossRef]
9. Srivastava, H.M.; Bansal, D. Close-to-convexity of a certain family of q-Mittag-Leffler functions. *J. Nonlinear Var. Anal.* **2017**, *1*, 61–69.
10. Srivastava, H.M.; Khan, B.; Khan, N.; Ahmad, Q.Z. Coefficient inequalities for q-starlike functions associated with the Janowski functions. *Hokkaido Math. J.* **2019**, *48*, 407–425. [CrossRef]
11. Srivastava, H.M.; Raza, N.; AbuJarad, E.S.A.; Srivastava, G.; AbuJarad, M.H. Fekete-Szegö inequality for classes of (p,q)-starlike and (p,q)-convex functions. *Rev. Real Acad. Cienc. Exactas Fís. Nat. Ser. A Mat.* **2019**, *113*, 3563–3584. [CrossRef]

12. Srivastava, H.M.; Khan, B.; Khan, N.; Tahir, M.; Ahmad, S.; Khan, N. Upper bound of the third Hankel determinant for a subclassof q-starlike functions associated with the q-exponential function. *Bull. Sci. Math.* **2021**, *167*, 102942. [CrossRef]
13. Seoudy, T.M.; Aouf, M.K. Coefficient estimates of new classes of q-starlike and q-convex functions of complex order. *J. Math. Inequal.* **2016**, *10*, 135–145. [CrossRef]
14. Janowski, W. Some extremal problems for certain families of analytic functions. *Ann. Polon. Math.* **1973**, *28*, 297–326. [CrossRef]
15. Srivastava, H.M. Operators of basic (or q-) calculus and fractional q-calculus and their applications in geometric function theory of complex analysis. *Iran. J. Sci. Technol. Trans. A Sci.* **2020**, *44*, 327–344. [CrossRef]

symmetry

MDPI

Article

A Spectral Calculus for Lorentz Invariant Measures on Minkowski Space

John Mashford

School of Mathematics and Statistics, University of Melbourne, Victoria 3010, Australia; mashford@unimelb.edu.au

Received: 4 August 2020; Accepted: 10 October 2020; Published: 15 October 2020

Abstract: This paper presents a spectral calculus for computing the spectra of causal Lorentz invariant Borel complex measures on Minkowski space, thereby enabling one to compute their densities with respect to Lebesque measure. The spectra of certain elementary convolutions involving Feynman propagators of scalar particles are computed. It is proved that the convolution of arbitrary causal Lorentz invariant Borel complex measures exists and the product of such measures exists in a wide class of cases. Techniques for their computation in terms of their spectral representation are presented.

Keywords: Lorentz invariant complex measures; Minkowski space; spectral decomposition; measure convolution; measure product; Feynman propagator

MSC: 28XX; 81XX; 22E15; 22E70

1. Introduction

Let $\mathcal{B}(\mathbf{R}^4)$ denote the Borel algebra of $\mathbf{R}^4$ (with respect to the Euclidean topology) [1] and let

$$\mathcal{B}_0(\mathbf{R}^4) = \{\Gamma \in \mathcal{B}(\mathbf{R}^4) : \Gamma \text{ is relatively compact}\}. \tag{1}$$

Let

$$H_m^+ = \{p \in \mathbf{R}^4 : p^2 = m^2, p^0 > 0\}, \tag{2}$$

and

$$H_m^- = \{p \in \mathbf{R}^4 : p^2 = m^2, p^0 < 0\}, \tag{3}$$

be the mass shells (cones) corresponding to mass $m > 0$ ($m = 0$) and let

$$H_{im}^+ = \{p \in \mathbf{R}^4 : p^2 = -m^2, p^0 > 0\}, \tag{4}$$

and

$$H_{im}^- = \{p \in \mathbf{R}^4 : p^2 = -m^2, p^0 < 0\}, \tag{5}$$

be the positive time (negative time) imaginary mass hyperboloids corresponding to mass $m > 0$.

Define measures on these hyperboloids (cones) by

$$\Omega_m^{\pm}(\Gamma) = \int_{\pi(H_m^{\pm}\cap\Gamma)} \frac{d\vec{p}}{\omega_m(\vec{p})}, m \geq 0, \tag{6}$$

$$\Omega_{im}^{\pm}(\Gamma) = \int_{\pi(H_{im}^{\pm}\cap\Gamma)} \frac{d\vec{p}}{\omega_{im}(\vec{p})}, m > 0, \tag{7}$$

and

$$\Omega_{im} = \Omega_{im}^{+} + \Omega_{im}^{-}, \tag{8}$$

where $\pi : \mathbf{R}^4 \to \mathbf{R}^3$ is defined by

$$\pi(p) = \pi(p^0, p^1, p^2, p^3) = \vec{p} = (p^1, p^2, p^3), \tag{9}$$

and $\omega_m : \mathbf{R}^3 \to [0, \infty)$ is defined by

$$\omega_m(\vec{p}) = (m^2 + |\vec{p}|^2)^{\frac{1}{2}}. \tag{10}$$

An equivalent set of definitions to these definitions of the measures $\Omega_m^{\pm}$ and $\Omega_{im}^{\pm}$ is to specify the effect of applying the measures to measurable functions $\psi : \mathbf{R}^4 \to \mathbf{C}$. In fact

$$< \Omega_m^{\pm}, \psi >= \int \psi(\pm\omega_m(\vec{p}), \vec{p}) \frac{d\vec{p}}{\omega_m(\vec{p})}, m \geq 0, \tag{11}$$

$$< \Omega_{im}^{\pm}, \psi >= \int \psi(\pm\omega_{im}(\vec{p}), \vec{p}) \frac{d\vec{p}}{\omega_{im}(\vec{p})}, m > 0. \tag{12}$$

Consider the following general form of a complex measure $\mu : \mathcal{B}_0(\mathbf{R}^4) \to \mathbf{C}$ on Minkowski space.

$$\mu(\Gamma) = c\delta(\Gamma) + \int_{m=0}^{\infty} \Omega_m^{+}(\Gamma)\, \sigma_1(dm) + \int_{m=0}^{\infty} \Omega_m^{-}(\Gamma)\, \sigma_2(dm) + \int_{m=0}^{\infty} \Omega_{im}(\Gamma)\, \sigma_3(dm), \tag{13}$$

where $c \in \mathbf{C}$ (the complex numbers), δ is the Dirac delta function (measure), $\sigma_1, \sigma_2, \sigma_3 : \mathcal{B}_0([0, \infty)) \to \mathbf{C}$ are Borel complex measures. Then μ is a Lorentz invariant Borel complex measure. Conversely [2] leads to the following.

Theorem 1. *The Spectral Theorem. Let $\mu : \mathcal{B}_0(\mathbf{R}^4) \to \mathbf{C}$ be a Lorentz invariant Borel complex measure. Then μ has the form of Equation (13) for some $c \in \mathbf{C}$ and Borel spectral measures σ_1, σ_2 and σ_3.*

(More generally, from [2] Lorentz invariant distributions in $\mathcal{D}^*(\mathbf{R}^4)$ are of the form of Equation (13) where $c \in \mathbf{C}$, $\sigma_1, \sigma_2, \sigma_3 \in \mathcal{D}^*(\mathbf{R})$ with the possible addition of a distribution supported at the origin of the form $P(\Box)\delta$ where $\Box$ is the wave operator and P is some polynomial.)

In Section 3, we show how the Feynman scalar propagator in momentum space can be identified with the causal Lorentz invariant measure Ω_m. In Section 4, we will present a spectral calculus whereby the spectrum of a causal Lorentz invariant Borel complex measure on Minkowski space can be calculated, whereby causal is meant that the support of the measure is contained in the closed future null cone of the origin.

In Section 5 of the paper, we use the spectral calculus and other methods to compute the spectrum of the measure $\Omega_m * \Omega_m$ which is the convolution of the standard Lorentz invariant measure on the mass m mass shell (i.e., the Feynman scalar propagator corresponding to mass m on the space of positive energy functions) with itself, where $m > 0$. In Section 7, we use general arguments to compute the spectrum of $\Omega_{im} * \Omega_{im}, m > 0$. In Section 8, we will show how the density with respect to Lebesque measure associated with a causal Lorentz invariant Borel complex measure can be determined from its spectrum and in Section 9 we will show how the convolution and product of such measures can be computed.

Some of the work of this paper may be compared to the work of Scharf and others, dating back to the paper of Epstein and Glaser [3] on forming products of causal distributions.

The concept of spectral representation in quantum field theory (QFT) dates back to the work of Källén [4] and Lehmann [5] who, independently, proposed the representation

$$< 0|[\phi(x), \phi^{\dagger}(y)]|0 >= i \int_0^{\infty} dm'^2 \sigma(m'^2) \Delta_{m'}(x-y), \tag{14}$$

for the commutator of interacting fields where $\Delta_{m'}$ is the Feynman propagator corresponding to mass m'. Itzykson and Zuber [6] state, with respect to σ, "In general this is a positive measure with δ-function singularities." While Källén, Lehmann and others propose and use this decomposition they do not present a way to compute the spectral measure σ. As mentioned above one of the main results of the present paper is a presentation of a spectral calculus that enables one to compute the spectral function of a causal Lorentz invariant Borel complex measure on Minkowski space. This spectral calculus is quite easy to use in practice but it is somewhat tedious to prove rigorously its validity. This use in practice involves a general form of argument which is exemplified by the argument used in the case of the computation of the spectrum of the convolution $\Omega_m * \Omega_m$ which we call Argument 1. The validity of Argument 1 is proved in Section 6.

2. Related Work

Our work has some connection with the spectral theory of hyperbolic surfaces [7,8] and its multivarious ramifications in quantum physics, number theory, and discrete groups as the hyperboloid H_m is a higher dimensional hyperbolic space and the standard measure Ω_m on H_m is a fundamental solution of the Klein-Gordon equation on Minkowski space (whose solutions are eigenfunctions of the wave operator) whereas the spectral theory of hyperbolic surfaces is concerned with eigenfunctions of the Laplace operator.

Bollini et al. [9] describe how the convolution of two ultradistributions of exponential type (UET) can exist. They then define the product of two UETs in terms of the convolution of their Fourier transforms. They obtain expressions for the Fourier transform of Lorentz invariant UETs (generalizing Bochner's theorem). Kamiński and Mincheva-Kaminska [10] present results concerning the convolution of distributions such as the existence of the convolution of tempered distributions whose supports are polynomially compatible sets. Ortner and Wagner [11] consider the Fourier transform of $O(p,q)$ invariant distributions. They present a condition under which two Lorentz invariant tempered distributions are convolvable and a formula for their convolution.

Zinoviev [12] considers Lorentz invariant tempered distributions on $(\mathbf{R}^4)^k$ supported on the product of closed future light cones. Soloviev [13] discusses the theory of Lorentz covariant distributions, ultradistributions and hyperfunctions.

Harish-Chandra [14] realized the fruitfulness of regarding the space of invariant distributions as a module for the algebra of polynomial differential operators. In this context Kolk and Varadarajan [15] consider Lorentz invariant distributions supported on the boundary of the cone representing the causal future of the origin.

Our work does not consider the complexities and partial results of the general theory of Lorentz invariant distributions, ultradistributions and other such spaces but restricts attention to Lorentz invariant Borel complex measures. There are two reasons for this. Firstly one can obtain complete, unencumbered and "elegant" results. Secondly, many the distributional objects of interest in QFT (such as correlation functions) can, through Wick's theorem, or else the operator product expansion, be represented in terms of Feynman propagators and the propagators are Lorentz invariant measures (or else K invariant matrix-valued measures whose trace is Lorentz invariant) [16].

3. The Feynman Scalar Field Propagator as a Tempered Measure

In this section, we give a well-defined definition of the Feynman scalar propagator of QFT in terms of tempered measures and distributions. The propagator is viewed as being a complex tempered distribution. It is constructed from the Fourier transform of the tempered measure Ω_m.

Consider the Feynman scalar field propagator. It is written as ([6], p. 35)

$$\triangle_F(x) = -(2\pi)^{-4} \int \frac{e^{-ip.x}}{p^2 - m^2 + i\epsilon}\, dp. \tag{15}$$

This is to be understood with respect to the $i-$epsilon procedure described in Mandl and Shaw ([17], p. 57), and the dot product $p.x$ is given by

$$p.x = \eta_{\alpha\beta} p^\alpha x^\beta,$$

where $\eta = \mathrm{diag}(1, -1, -1, -1)$. Therefore $\triangle_F(x)$ is written as

$$\triangle_F(x) = -(2\pi)^{-4} \int_{\mathbf{R}^3} \int_{C_F} \frac{e^{-ip.x}}{p^2 - m^2}\, dp^0 d\overrightarrow{p}, \tag{16}$$

where C_F is the standard Feynman propagator contour. Thus $\triangle_F(x)$ is written as

$$\triangle_F(x) = -(2\pi)^{-4} \int_{\mathbf{R}^3} I(\overrightarrow{p}, x)\, d\overrightarrow{p}, \tag{17}$$

where

$$I(\overrightarrow{p}, x) = \int_{C_F} \frac{e^{-ip.x}}{(p^0)^2 - \omega_m(\overrightarrow{p})^2}\, dp^0, \tag{18}$$

and

$$\omega_m(\overrightarrow{p}) = (|\overrightarrow{p}|^2 + m^2)^{\frac{1}{2}}, m \geq 0. \tag{19}$$

The contour integral Equation (18) exists for $\omega_m(\overrightarrow{p}) \neq 0$ and is given by

$$I(\overrightarrow{p}, x) = -\frac{\pi i}{\omega_m(\overrightarrow{p})} \begin{cases} e^{-i(\omega_m(\overrightarrow{p})x^0 - \overrightarrow{p}.\overrightarrow{x})} & \text{if } x^0 > 0, \\ e^{-i(-\omega_m(\overrightarrow{p})x^0 - \overrightarrow{p}.\overrightarrow{x})} & \text{if } x^0 < 0. \end{cases} \tag{20}$$

To prove this consider the contour $C_1(R)$ given by

$$C_1(R) = \{Re^{it} : 0 \leq t \leq \pi\}.$$

We will show that

$$I_1(R) = \int_{C_1(R)} \frac{e^{-ip.x}}{(p^0)^2 - \omega_m(\overrightarrow{p})^2}\, dp^0 \to 0 \text{ as } R \to \infty,$$

as long as $x^0 < 0$. To this effect we note that

$$\begin{aligned} |I_1(R)| &= \left| \int_{t=0}^{\pi} \frac{e^{-iRe^{it}x^0 + i\overrightarrow{p}.\overrightarrow{x}}}{(Re^{it})^2 - \omega_m(\overrightarrow{p})^2} iRe^{it}\, dt \right| \\ &\leq \int_{t=0}^{\pi} \left| \frac{e^{R\sin t x^0}}{(Re^{it})^2 - \omega_m(\overrightarrow{p})^2} \right| R\, dt \\ &\leq \int_{t=0}^{\pi} \frac{1}{|R^2 - \omega_m(\overrightarrow{p})^2|} R\, dt \\ &= \frac{\pi R}{|R^2 - \omega_m(\overrightarrow{p})^2|}, \\ &\to 0 \text{ as } R \to \infty, \text{ if } x^0 < 0. \end{aligned} \tag{21}$$

Therefore, for $x^0 < 0$,

$$I(\vec{p},x) = 2\pi i \mathrm{res}(p^0 \mapsto \frac{e^{-ip.x}}{(p^0)^2 - \omega_m(\vec{p})^2}, -\omega_m(p)).$$

Now

$$\frac{e^{-ip.x}}{(p^0)^2 - \omega_m(\vec{p})^2} = \frac{e^{-ip.x}}{(p^0 - \omega_m(\vec{p}))(p^0 + \omega_m(\vec{p}))}.$$

Thus

$$I(\vec{p},x) = -\pi i \frac{e^{-i(-\omega_m(\vec{p})x^0 - \vec{p}.\vec{x})}}{\omega_m(\vec{p})}.$$

Similarly, if $x^0 > 0$, then

$$I(\vec{p},x) = -\pi i \frac{e^{-i(\omega_m(\vec{p})x^0 - \vec{p}.\vec{x})}}{\omega_m(\vec{p})}.$$

Hence

$$\int_{\mathbf{R}^3} |I(\vec{p},x)|\, d\vec{p} = \pi \int_{\mathbf{R}^3} \frac{1}{\omega_m(\vec{p})} d\vec{p} = \infty, \tag{22}$$

and so the integral Equation (17) defining $\triangle_F(x)$ does not exist as a Lebesgue integral.

We would like to give a well-defined interpretation of the propagator $\triangle_F$. $H_m^{\pm}$ for $m \geq 0$ are orbits of the action of the Lorentz group on Minkowski space (these orbits correspond to real mass orbits, there are also "imaginary mass" hyperboloid orbits H_{im}). $\Omega_m^{\pm}$ are Lorentz invariant measures for Minkowski space supported on $H_m^{\pm}$ ([18], p. 157). $\Omega_m^{\pm}$ is locally finite for $m \geq 0$. Now, for any non-negative measurable function $\psi : \mathbf{R}^4 \to [0, \infty]$,

$$\int_{\mathbf{R}^4} \psi(p)\, \Omega_m^{\pm}(dp) = \int_{\mathbf{R}^3} \psi(\pm\omega_m(\vec{p}), \vec{p}) \frac{d\vec{p}}{\omega_m(\vec{p})}. \tag{23}$$

Here, and for the rest of the section, the symbol ψ stands for a test function in Minkowski space. It follows from Equations (17), (20) and (23) that one may write

$$\triangle_F(x) = \begin{cases} (2\pi)^{-4}\pi i \int e^{-ip.x}\, \Omega_m^{+}(dp), & \text{if } x^0 > 0, \\ (2\pi)^{-4}\pi i \int e^{-ip.x}\, \Omega_m^{-}(dp), & \text{if } x^0 < 0. \end{cases} \tag{24}$$

Equations (16), (17) and (24) are all integral expressions equivalent to Equation (15) and none of them exist as Lebesgue integrals. However, formally, Equation (24) can be written as

$$\triangle_F(x) = \begin{cases} \pi i \overset{\vee}{\Omega_m^{+}}(-x) & \text{if } x^0 > 0, \\ \pi i \overset{\vee}{\Omega_m^{-}}(-x) & \text{if } x^0 < 0, \end{cases} \tag{25}$$

where $^{\vee}$ denotes the inverse Fourier transform operator (and we use the "physics"convention for the definition of the Fourier transform). Since Ω_m^{+} and Ω_m^{-} are tempered distributions their inverse Fourier transforms exist and are tempered distributions. Let $\mathcal{S}^{\pm}(\mathbf{R}^4) \subset \mathcal{S}(\mathbf{R}^4)$ be the space of test functions supported in $S^{\pm}$, where

$$S^{+} = \{x \in \mathbf{R}^4 : x^0 > 0\}, S^{-} = \{x \in \mathbf{R}^4 : x^0 < 0\}. \tag{26}$$

Then

$$< \triangle_F, \psi >= \pi i < \overset{\vee}{\Omega_m^{\mp}}, \psi >= \pi i < \Omega_m^{\mp}, \overset{\vee}{\psi} >, \tag{27}$$

for $\psi \in \mathcal{S}^{\pm}(\mathbf{R}^4)$, where $< \omega, \psi >$ denotes the evaluation of a distribution ω on its test function argument ψ. Therefore the momentum space scalar field propagator on $(\mathcal{S}^{\pm}(\mathbf{R}^4))^{\wedge}$ is

$$\overset{\wedge}{\triangle}_F = \pi i \Omega_m^{\mp}. \tag{28}$$

$(\mathcal{S}^{+}(\mathbf{R}^4))^{\wedge}$ is the space of wave functions with only positive frequency components while $(\mathcal{S}^{-}(\mathbf{R}^4))^{\wedge}$ is the space of wave functions with only negative frequency components.

This measure is a tempered measure, i.e., it is a tempered distribution as well as being a measure. Equations (15) and (25) lead to the *ansatz*

$$\frac{1}{p^2 - m^2 + i\epsilon} \to -\pi i \Omega_m^{\pm}(p). \tag{29}$$

4. A Spectral Calculus for Lorentz Invariant Measures

Suppose that μ is a Lorentz invariant Borel complex measure on Minkowski space. Then by the spectral theorem, it must have the form of Equation (13). If $\sigma_2 = \sigma_3 = 0$ then μ will be said to be *causal* or a type I measure. If $\sigma_1 = \sigma_3 = 0$ then μ will be said to be a type II measure and if $c = 0$ and $\sigma_1 = \sigma_2 = 0$ then μ will be said to be a type III measure. Thus any Lorentz invariant measure is a sum of a type I measure, a type II measure and a type III measure. In particular, any measure of the form

$$\mu(\Gamma) = \int_{m=0}^{\infty} \sigma(m) \Omega_m^{+}(\Gamma)\, dm, \tag{30}$$

where σ is locally integrable function and the integration is carried out with respect to the Lebesgue measure, is a causal Lorentz invariant Borel complex measure. If σ is polynomially bounded then μ is a tempered measure.

The spectral calculus that we will now explain is a way to compute the spectrum σ of a Lorentz invariant measure μ if we know that μ can be written in the form of Equation (30) and σ is continuous.

For $m > 0$ and $\epsilon > 0$ let $S(m, \epsilon)$ be the hyperbolic (hyper-)disc defined by

$$S(m, \epsilon) = \{p \in \mathbf{R}^4 : p^2 = m^2, |\vec{p}| < \epsilon, p^0 > 0\}, \tag{31}$$

where $\vec{p} = \pi(p) = \pi(p^0, p^1, p^2, p^3) = (p^1, p^2, p^3)$. For $a, b \in \mathbf{R}$ with $0 < a < b$ let $\Gamma(a, b, \epsilon)$ be the hyperbolic cylinder defined by

$$\Gamma(a, b, \epsilon) = \bigcup_{m \in (a,b)} S(m, \epsilon). \tag{32}$$

Now suppose that we have a measure in the form of Equation (30) where σ is continuous. Then we can write

$$\begin{aligned} \mu(\Gamma(a, b, \epsilon)) &= \int_{m=0}^{\infty} \sigma(m) \Omega_m(\Gamma(a, b, \epsilon))\, dm \\ &= \int_{m=0}^{\infty} \sigma(m) \int_{\pi(\Gamma(a,b,\epsilon) \cap H_m^{+})} \frac{d\vec{p}}{\omega_m(\vec{p})}\, dm \\ &= \int_a^b \sigma(m) \int_{B_\epsilon(\vec{0})} \frac{d\vec{p}}{\omega_m(\vec{p})}\, dm \\ &\approx \frac{4}{3}\pi\epsilon^3 \int_a^b \frac{\sigma(m)}{m}\, dm. \end{aligned} \tag{33}$$

where

$$\omega_m(\vec{p}) = (\vec{p}^{\,2} + m^2)^{\frac{1}{2}}, \tag{34}$$

and $B_\epsilon(\vec{0}) = \{\vec{p} \in \mathbf{R}^3 : |\vec{p}| < \epsilon\}$.

The approximation $\approx$ in the last line comes about because ω_m is not constant over $B_\epsilon(\vec{0})$. Thus if we define

$$g_a(b) = g(a,b) = \lim_{\epsilon \to 0} \epsilon^{-3} \mu(\Gamma(a,b,\epsilon)), \tag{35}$$

then we can retreive σ using the formula

$$\sigma(b) = \frac{3}{4\pi} b g_a'(b). \tag{36}$$

Thus we have proved the following fundamental theorem of the spectral calculus of causal Lorentz invariant measures.

Theorem 2. *Suppose that μ is a causal Lorentz invariant measure with continuous spectrum σ. Then σ can be calculated from the formula*

$$\sigma(b) = \frac{3}{4\pi} b g_a'(b), \tag{37}$$

where, for $a, b \in \mathbf{R}, 0 < a < b, g_a : (a, \infty) \to \mathbf{R}$ is given by Equation (35).

To make the proof of this theorem rigorous we prove the following.

Lemma 1. *Let $a, b \in \mathbf{R}, 0 < a < b$. Then*

$$\lim_{\epsilon \to 0} \epsilon^{-3} \int_{B_\epsilon(0)} \frac{d\vec{p}}{\omega_m(\vec{p})} = \frac{4\pi}{3} \frac{1}{m}, \tag{38}$$

uniformly for $m \in [a, b]$.

Proof. Define

$$I = I(m, \epsilon) = \int_{B_\epsilon(0)} \frac{d\vec{p}}{\omega_m(\vec{p})}. \tag{39}$$

Then

$$I = \int_{r=0}^{\epsilon} \frac{4\pi r^2 \, dr}{(r^2 + m^2)^{\frac{1}{2}}}. \tag{40}$$

Now

$$I_1 < I < I_2,$$

where

$$I_1 = \int_{r=0}^{\epsilon} \frac{4\pi r^2 \, dr}{(\epsilon^2 + m^2)^{\frac{1}{2}}} = \frac{4\pi}{(\epsilon^2 + m^2)^{\frac{1}{2}}} \frac{1}{3} \epsilon^3,$$

$$I_2 = \int_{r=0}^{\epsilon} \frac{4\pi r^2 \, dr}{m} = \frac{4\pi}{m} \frac{1}{3} \epsilon^3.$$

Therefore

$$\frac{4\pi}{3(\epsilon^2 + m^2)^{\frac{1}{2}}} < \epsilon^{-3} I < \frac{4\pi}{3m}.$$

Thus

$$\frac{4\pi}{3m} - \frac{4\pi}{3(\epsilon^2 + m^2)^{\frac{1}{2}}} > \frac{4\pi}{3m} - \epsilon^{-3} I > 0.$$

Hence

$$\left| \epsilon^{-3} I - \frac{4\pi}{3m} \right| < \frac{4\pi}{3m} - \frac{4\pi}{3(\epsilon^2 + m^2)^{\frac{1}{2}}}. \tag{41}$$

We have

$$\begin{aligned}\frac{4\pi}{3m}-\frac{4\pi}{3(\epsilon^2+m^2)^{\frac{1}{2}}} &= \frac{4\pi}{3}\frac{(\epsilon^2+m^2)^{\frac{1}{2}}-m}{m(\epsilon^2+m^2)^{\frac{1}{2}}}\\ &= \frac{4\pi}{3}\frac{\epsilon^2}{m(\epsilon^2+m^2)^{\frac{1}{2}}((\epsilon^2+m^2)^{\frac{1}{2}}+m)}\\ &< \frac{4\pi}{3}\frac{\epsilon^2}{2m^3}\\ &\leq \frac{4\pi}{3}\frac{\epsilon^2}{2a^3},\ \text{for all } m\in[a,b].\end{aligned}$$

Therefore

$$\left|\epsilon^{-3}I-\frac{4\pi}{3m}\right|<\frac{2\pi\epsilon^2}{3a^3}, \tag{42}$$

for all $m \in [a,b]$ □

This lemma justifies the step of taking the limit under the integral sign (indicated by the symbol $\approx$) in the proof of Theorem 2.

More generally, suppose that $\mu : \mathcal{B}_0(\mathbf{R}^4) \to \mathbf{C}$ is a causal Lorentz invariant Borel measure on Minkowski space with spectrum σ. Then, by the Lebesgue decomposition theorem there exist unique measures $\sigma_c, \sigma_s : \mathcal{B}_0([0,\infty)) \to \mathbf{C}$ such that $\sigma = \sigma_c + \sigma_s$ where σ_c, the continuous part of the spectrum of μ, is absolutely continuous with respect to Lebesque measure and σ_s, the singular part of the spectrum of μ, is singular with respect to σ_c.

It is straightforward to prove the following.

Theorem 3. *Suppose that $a', b' \in \mathbf{R}$ are such that $0 < a' < b'$, $\sigma_c|_{(a',b')}$ is continuous. Then for all $a, b \in \mathbf{R}$ with $a' < a < b < b'$, $g_a(b)$ defined by Equation (35) exists and is continuously differentiable. Furthermore $\sigma_c|_{(a',b')}$ can be computed using the formula*

$$\sigma_c(b) = \frac{3}{4\pi} b g_a'(b), \tag{43}$$

and

$$\sigma_s(E) = 0, \forall \text{ Borel } E \subset (a',b'). \tag{44}$$

Conversely suppose that $a', b' \in \mathbf{R}$ are such that $0 < a' < b'$ and for all $a, b \in \mathbf{R}$ with $a' < a < b < b'$, $g_a(b)$ defined by Equation (35) exists and is continuously differentiable. Then $\sigma_c|_{(a',b')}$ is continuous and can be retrieved using the formula of Equation (43).

5. Investigation of the Measure Defined by the Convolution $\Omega_m * \Omega_m$

5.1. Determination of Some Properties of $\Omega_m * \Omega_m$

Consider the measure defined by

$$\mu(\Gamma) = (\Omega_m * \Omega_m)(\Gamma) = \int \chi_\Gamma(p+q)\,\Omega_m(dp)\,\Omega_m(dq), \tag{45}$$

where, for any set Γ, χ_Γ denotes the characteristic function of Γ defined by

$$\chi_\Gamma(p) = \begin{cases} 1 \text{ if } p \in \Gamma \\ 0 \text{ otherwise.} \end{cases} \tag{46}$$

μ exists as a Borel measure because as $|p|, |q| \to \infty$ with $p, q \in H_m^+$, $(p+q)^0 \to \infty$ and so $p + q$ is eventually $\notin \Gamma$ for any compact set $\Gamma \subset \mathbf{R}^4$. Now

$$\begin{aligned}\mu(\Lambda(\Gamma)) &= \int \chi_{\Lambda(\Gamma)}(p+q)\,\Omega_m(dp)\,\Omega_m(dq)\\ &= \int \chi_\Gamma(\Lambda^{-1}p+\Lambda^{-1}q)\,\Omega_m(dp)\,\Omega_m(dq)\\ &= \int \chi_\Gamma(p+q)\,\Omega_m(dp)\,\Omega_m(dq)\\ &= \mu(\Gamma),\end{aligned} \tag{47}$$

for all $\Lambda \in O(1,3)^{+\uparrow}, \Gamma \in \mathcal{B}_0(\mathbf{R}^4)$. Thus μ is a Lorentz invariant measure.

We will now show that μ is concentrated in the set

$$C_{2m} = \{p \in \mathbf{R}^4 : p^2 \geq 4m^2, p^0 > 0\}, \tag{48}$$

and therefore, that μ is causal. Let $U \subset \mathbf{R}^4$ be open. Then

$$\mu(U) = \int_{\mathbf{R}^3}\int_{\mathbf{R}^3} \chi_U(\omega_m(\vec{p}) + \omega_m(\vec{q}), \vec{p} + \vec{q})\,\frac{d\vec{p}}{\omega_m(\vec{p})}\,\frac{d\vec{q}}{\omega_m(\vec{q})}. \tag{49}$$

Therefore, using continuity, it follows that

$$\mu(U) > 0 \quad \Leftrightarrow \quad (\exists \vec{q}_1, \vec{q}_2 \in \mathbf{R}^3)\ (\omega_m(\vec{q}_1) + \omega_m(\vec{q}_2), \vec{q}_1 + \vec{q}_2) \in U.$$

Suppose that $p \in \mathrm{supp}(\mu)$ (the support of the measure μ) i.e., p is such that $\mu(U) > 0$ for all open neighborhoods U of p. Let U be an open neighborhood of p. Then, as $\mu(U) > 0$, there exists $q \in U, \vec{q}_1, \vec{q}_2 \in \mathbf{R}^3$ such that $q = (\omega_m(\vec{q}_1) + \omega_m(\vec{q}_2), \vec{q}_1 + \vec{q}_2)$. Clearly $q^0 \geq 2m$. As this is true for all neighborhoods U of p it follows that $p^0 \geq 2m$. By Lorentz invariance we may assume without loss of generality that $\vec{p} = 0$. Therefore $p^2 \geq 4m^2$. Thus $\mathrm{supp}(\mu) \subset C_{2m}$.

For the converse, let $p = (\omega_m(\vec{p}), \vec{p}), q = (\omega_m(\vec{p}), -\vec{p}) \in H_m^+$ for $\vec{p} \in \mathbf{R}^3$. As $\vec{p}$ ranges over $\mathbf{R}^3$, $p + q = (2\omega_m(\vec{p}), \vec{0})$ ranges over $\{(m', \vec{0}) : m' \geq 2m\}$. It follows using Lorentz invariance that $\mathrm{supp}(\mu) \supset C_{2m}$.

Therefore the support $\mathrm{supp}(\mu)$ of μ is C_{2m}. Therefore by the spectral theorem μ has a spectral representation of the form

$$\mu(\Gamma) = \int_{m'=2m}^{\infty} \Omega_{m'}(\Gamma)\,\sigma(dm'), \tag{50}$$

for some Borel measure $\sigma : \mathcal{B}_0([2m, \infty)) \to \mathbf{C}$.

5.2. Computation of the Spectrum of $\Omega_m * \Omega_m$ Using the Spectral Calculus

Let $a, b \in \mathbf{R}$ with $0 < a < b$. Let

$$g_a(b, \epsilon) = \mu(\Gamma(a, b, \epsilon)) \text{ for } \epsilon > 0. \tag{51}$$

We would like to calculate

$$g_a(b) = \lim_{\epsilon \to 0} \epsilon^{-3} g_a(b, \epsilon), \tag{52}$$

and then retreive the spectral function as

$$\sigma(b) = \frac{3}{4\pi} b g'(b). \tag{53}$$

To this effect we calculate

$$\begin{aligned}
g(a,b,\epsilon) &= \mu(\Gamma(a,b,\epsilon)) \\
&= \int \chi_{\Gamma(a,b,\epsilon)}(p+q)\,\Omega_m(dp)\,\Omega_m(dq) \\
&\approx \int \chi_{(a,b)\times B_\epsilon(\vec{0})}(p+q)\,\Omega_m(dp)\,\Omega_m(dq) \\
&= \int \chi_{(a,b)}(\omega_m(\vec{p})+\omega_m(\vec{q}))\chi_{B_\epsilon(\vec{0})}(\vec{p}+\vec{q})\frac{d\vec{p}}{\omega_m(\vec{p})}\frac{d\vec{q}}{\omega_m(\vec{q})} \\
&= \int \chi_{(a,b)}(\omega_m(\vec{p})+\omega_m(\vec{q}))\chi_{B_\epsilon(\vec{0})-\vec{q}}(\vec{p})\frac{d\vec{p}}{\omega_m(\vec{p})}\frac{d\vec{q}}{\omega_m(\vec{q})} \\
&\approx \int \chi_{(a,b)}(2\omega_m(\vec{q}))\frac{\frac{4}{3}\pi\epsilon^3}{\omega_m(\vec{q})^2}\,d\vec{q}.
\end{aligned}$$

We will call this argument Argument 1. This argument is intuitively reasonable but it needs to be justified rigorously. It is proved in the proof of Theorem 4 of the next section.

Given that Argument 1 is valid we now proceed to compute the spectrum of μ. We have

$$\begin{aligned}
a < 2\omega_m(\vec{q}) < b \quad &\Leftrightarrow \quad \left(\frac{a}{2}\right)^2 - m^2 < \vec{q}^{\,2} < \left(\frac{b}{2}\right)^2 - m^2 \\
&\Leftrightarrow \quad mZ(a) < |\vec{q}| < mZ(b),
\end{aligned}$$

where

$$Z(m') = \left(\frac{m'^2}{4m^2} - 1\right)^{\frac{1}{2}}, \text{ for } m' \geq 2m. \tag{54}$$

Thus

$$g(a,b,\epsilon) \approx \frac{16\pi^2}{3}\epsilon^3 \int_{r=mZ(a)}^{mZ(b)} \frac{r^2}{m^2+r^2}\,dr. \tag{55}$$

Hence

$$g_a(b) = \frac{16\pi^2}{3} \int_{r=mZ(a)}^{mZ(b)} \frac{r^2}{m^2+r^2}\,dr. \tag{56}$$

Therefore g_a is continuously differentiable and so Theorem 3 applies. Using the Leibniz integral rule

$$g_a'(b) = \frac{16\pi^2}{3}\frac{m^2Z^2(b)}{m^2+m^2Z^2(b)}mZ'(b) = \frac{16\pi^2}{3}\frac{mZ(b)}{b}. \tag{57}$$

Therefore we compute the spectrum σ of μ as

$$\sigma(b) = \begin{cases} 4\pi m Z(b) \text{ for } b \geq 2m \\ 0 \text{ otherwise.} \end{cases} \tag{58}$$

6. Proof of the Validity of Argument 1

The following theorem establishes that Argument 1 is justified.

Theorem 4. *Let $g(a,b,\epsilon)$ be defined by $g(a,b,\epsilon) = \mu(\Gamma(a,b,\epsilon))$ for $a,b \in \mathbf{R}, a < b, \epsilon > 0$, where $\mu = \Omega_m * \Omega_m$. Then the following formal argument (Argument 1)*

$$\begin{aligned} g(a,b,\epsilon) &= \mu(\Gamma(a,b,\epsilon)) \\ &= \int \chi_{\Gamma(a,b,\epsilon)}(p+q)\,\Omega_m(dp)\,\Omega_m(dq) \\ &\approx \int \chi_{(a,b)\times B_\epsilon(0)}(p+q)\,\Omega_m(dp)\,\Omega_m(dq) \\ &= \int \chi_{(a,b)}(\omega_m(\vec{p})+\omega_m(\vec{q}))\chi_{B_\epsilon(0)}(\vec{p}+\vec{q})\frac{d\vec{p}}{\omega_m(\vec{p})}\frac{d\vec{q}}{\omega_m(\vec{q})} \\ &= \int \chi_{(a,b)}(\omega_m(\vec{p})+\omega_m(\vec{q}))\chi_{B_\epsilon(0)-\vec{q}}(\vec{p})\frac{d\vec{p}}{\omega_m(\vec{p})}\frac{d\vec{q}}{\omega_m(\vec{q})} \\ &\approx \int \chi_{(a,b)}(2\omega_m(\vec{q}))\frac{\frac{4}{3}\pi\epsilon^3}{\omega_m(\vec{q})^2}\,d\vec{q}, \end{aligned}$$

is justified in the sense that

$$\lim_{\epsilon\to 0}\epsilon^{-3}g(a,b,\epsilon) = \frac{4}{3}\pi\int \chi_{(a,b)}(2\omega_m(\vec{q}))\frac{1}{\omega_m(\vec{q})^2}\,d\vec{q}. \tag{59}$$

Proof. There are 2 $\approx$ signs that we have to consider. The first is in line 3 and arises because we are approximating the hyperbolic cylinder of radius ϵ between a and b with an ordinary cylinder of radius ϵ. We will show that the error is of order greater than ϵ^3. Let $\Gamma = \Gamma(a,b,\epsilon)$ be the aforementioned hyperbolic cylinder. Then

$$\Gamma = \bigcup_{m'\in(a,b)} S(m',\epsilon). \tag{60}$$

Let

$$\begin{aligned} \Gamma' &= \bigcup_{m'\in(a,b)} \{m'\}\times B_\epsilon(\vec{0}) = (a,b)\times B_\epsilon(\vec{0}) \\ \Gamma'^- &= \bigcup_{m'\in(a,a^+)} \{(m',\vec{p}) : \vec{p}^2 > m'^2 - a^2\} \\ &\subset \bigcup_{m'\in(a,a^+)} (\{m'\}\times B_\epsilon(\vec{0})) = (a,a^+)\times B_\epsilon(\vec{0}) \\ \Gamma'^+ &= \bigcup_{m'\in(b,b^+)} \{(m',\vec{p}) : \vec{p}^2 > m'^2 - b^2\} \\ &\subset \bigcup_{m'\in(b,b^+)} (\{m'\}\times B_\epsilon(\vec{0})) = (b,b^+)\times B_\epsilon(\vec{0}), \end{aligned}$$

in which

$$a^+ = (a^2+\epsilon^2)^{\frac{1}{2}}, b^+ = (b^2+\epsilon^2)^{\frac{1}{2}}. \tag{61}$$

Then Γ differs from $(\Gamma' \sim \Gamma'^-)\cup\Gamma'^+$ on a set of measure zero,

It is straightforward to show that if $\Gamma_1,\Gamma_2 \in \mathcal{B}_0(\mathbf{R}^4), \Gamma_1\cap\Gamma_2 = \emptyset$ then

$$\begin{aligned} \int \chi_{\Gamma_1\cup\Gamma_2}(p+q)\,\Omega_m(dp)\,\Omega_m(dq) &= \int \chi_{\Gamma_1}(p+q)\,\Omega_m(dp)\,\Omega_m(dq) + \\ &\quad \int \chi_{\Gamma_2}(p+q)\,\Omega_m(dp)\,\Omega_m(dq). \end{aligned}$$

Therefore

$$|\int \chi_\Gamma(p+q)\,\Omega_m(dp)\,\Omega_m(dq) - \int \chi_{\Gamma'}(p+q)\,\Omega_m(dp)\,\Omega_m(dq)| \leq$$
$$\int \chi_{\Gamma'^-}(p+q)\,\Omega_m(dp)\,\Omega_m(dq) + \int \chi_{\Gamma'^+}(p+q)\,\Omega_m(dp)\,\Omega_m(dq).$$

We will show that

$$\lim_{\epsilon\to 0}(\epsilon^{-3}\int \chi_{\Gamma'^\pm}(p+q)\,\Omega_m(dp)\Omega_m(dq)) = 0. \tag{62}$$

It suffices to consider the $-$ case. We have

$$\begin{aligned}\int \chi_{\Gamma'^-}(p+q)\,\Omega_m(dp)\Omega_m(dq) &\leq \int \chi_{(a,a^+)\times B_\epsilon(0)}(p+q)\,\Omega_m(dp)\,\Omega_m(dq)\\ &= \int \chi_{(a,a^+)}(\omega_m(\vec{p})+\omega_m(\vec{q}))\chi_{B_\epsilon(0)-\vec{q}}(\vec{p})\,\frac{d\vec{p}}{\omega_m(\vec{p})}\\ &\frac{d\vec{q}}{\omega_m(\vec{q})}.\end{aligned} \tag{63}$$

We will come back to this equation later but will now return to the general argument Argument 1 and consider the second and final $\approx$. This $\approx$ arises because we are approximating $\vec{p}$ by $-\vec{q}$ since $\vec{p}$ ranges over a ball of radius ϵ centred on $-\vec{q}$.

Suppose that $\vec{p}$ and $\vec{q}$ are such that $\chi_{B_\epsilon(0)-\vec{q}}(\vec{p}) = 1$. Then $|\vec{p}+\vec{q}| < \epsilon$. Thus $||\vec{p}|-|\vec{q}|| < \epsilon$. Hence

$$\begin{aligned}|\omega_m(\vec{p})-\omega_m(\vec{q})| &= |(\vec{p}^{\,2}+m^2)^{\frac{1}{2}} - (\vec{q}^{\,2}+m^2)^{\frac{1}{2}})|\\ &= \left|\frac{\vec{p}^{\,2}-\vec{q}^{\,2}}{(\vec{p}^{\,2}+m^2)^{\frac{1}{2}}+(\vec{q}^{\,2}+m^2)^{\frac{1}{2}})}\right|\\ &\leq \frac{|\vec{p}^{\,2}-\vec{q}^{\,2}|}{2m}\\ &= \frac{||\vec{p}|-|\vec{q}||(|\vec{p}|+|\vec{q}|)}{2m}\\ &< \frac{\epsilon}{2m}(|\vec{p}|+|\vec{q}|).\end{aligned}$$

We have $|\vec{p}| \in (|\vec{q}|-\epsilon, |\vec{q}|+\epsilon)$. Therefore $|\vec{p}|+|\vec{q}| < 2|\vec{q}|+\epsilon$. Thus

$$|\omega_m(\vec{p})-\omega_m(\vec{q})| < \frac{\epsilon}{2m}(2|\vec{q}|+\epsilon).$$

Therefore

$$\begin{aligned}\omega_m(\vec{p})+\omega_m(\vec{q}) &= \omega_m(\vec{p})-\omega_m(\vec{q})+\omega_m(\vec{q})+\omega_m(\vec{q})\\ &\leq |\omega_m(\vec{p})-\omega_m(\vec{q})|+2\omega_m(\vec{q})\\ &< 2\omega_m(\vec{q})+\frac{\epsilon}{2m}(2|\vec{q}|+\epsilon).\end{aligned} \tag{64}$$

Now let

$$\begin{aligned} I(\epsilon) &= \int \chi_{(a,b)}(\omega_m(\vec{p}) + \omega_m(\vec{q}))\chi_{B_\epsilon(\vec{0})-\vec{q}}(\vec{p}) \frac{d\vec{p}}{\omega_m(\vec{p})} \frac{d\vec{q}}{\omega_m(\vec{q})} \\ J(\epsilon) &= \int \chi_{(a,b)}(2\omega_m(\vec{q}))\chi_{B_\epsilon(\vec{0})-\vec{q}}(\vec{p}) \frac{d\vec{p}}{\omega_m(\vec{p})} \frac{d\vec{q}}{\omega_m(\vec{q})} \\ K(\epsilon) &= \int \chi_{(a,b)}(2\omega_m(\vec{q})) \frac{\frac{4}{3}\pi\epsilon^3}{\omega_m(\vec{q})^2} d\vec{q}. \end{aligned}$$

We will show that

$$\lim_{\epsilon\to 0} \epsilon^{-3}(I(\epsilon) - J(\epsilon)) = 0, \text{ and } \lim_{\epsilon\to 0} \epsilon^{-3}(J(\epsilon) - K(\epsilon)) = 0. \tag{65}$$

Concerning the first limit we note that $\chi_{(a,b)}(\omega_m(\vec{p}) + \omega_m(\vec{q}))$ differs from $\chi_{(a,b)}(2\omega_m(\vec{q}))$ if and only if

1. $\omega_m(\vec{p}) + \omega_m(\vec{q}) \in (a,b)$ but $2\omega_m(\vec{q}) \leq a$ or
2. $\omega_m(\vec{p}) + \omega_m(\vec{q}) \in (a,b)$ but $2\omega_m(\vec{q}) \geq b$ or
3. $2\omega_m(\vec{q}) \in (a,b)$ but $\omega_m(\vec{p}) + \omega_m(\vec{q}) \leq a$ or
4. $2\omega_m(\vec{q}) \in (a,b)$ but $\omega_m(\vec{p}) + \omega_m(\vec{q}) \geq b$.

Thus

$$|I(\epsilon) - J(\epsilon)| = I_1(\epsilon) + I_2(\epsilon) + I_3(\epsilon) + I_4(\epsilon), \tag{66}$$

where

$$\begin{aligned} I_1(\epsilon) = \int \chi_{(a,b)}(\omega_m(\vec{p}) + \omega_m(\vec{q}))\chi_{(-\infty,a]}(2\omega_m(\vec{q}))\chi_{B_\epsilon(\vec{0})-\vec{q}}(\vec{p}) \\ \frac{d\vec{p}}{\omega_m(\vec{p})} \frac{d\vec{q}}{\omega_m(\vec{q})}, \end{aligned} \tag{67}$$

and I_2, I_3, I_4 are defined similarly. We will show that

$$\lim_{\epsilon\to 0} \epsilon^{-3} I_1(\epsilon) = 0. \tag{68}$$

I_2, I_3 and I_4 can be dealt with similarly.

Using Equation (64)

$$\omega_m(\vec{p}) + \omega_m(\vec{q}) \in (a,b) \text{ and } 2\omega_m(\vec{q}) \leq a \Rightarrow a - \frac{\epsilon}{2m}(2|\vec{q}| + \epsilon) < 2\omega_m(\vec{q}) \leq a.$$

Therefore

$$\begin{aligned} I_1(\epsilon) &\leq \int \chi_{(a-(2|\vec{q}|+\epsilon)\epsilon/(2m),a]}(2\omega_m(\vec{q}))\chi_{B_\epsilon(\vec{0})-\vec{q}}(\vec{p}) \frac{1}{m^2} d\vec{p}\, d\vec{q} \\ &= \frac{4}{3}\pi\epsilon^3 \int \chi_{(a-(2|\vec{q}|+\epsilon)\epsilon/(2m),a]}(2\omega_m(\vec{q})) \frac{1}{m^2} d\vec{q}. \end{aligned}$$

Hence

$$\epsilon^{-3} I_1(\epsilon) \leq \frac{4}{3}\pi \int \chi_{(a-(2|\vec{q}|+\epsilon)\epsilon/(2m),a]}(2\omega_m(\vec{q})) \frac{1}{m^2} d\vec{q}. \tag{69}$$

The integrand is integrable for all $\epsilon > 0$, vanishes outside the compact set

$$C = \{\vec{q} \in \mathbf{R}^3 : 2\omega_m(\vec{q}) \leq a\},$$

is dominated by the integrable function

$$g(\vec{q}) = \frac{1}{m^2}\chi_{[0,a]}(2\omega_m(\vec{q})),$$

and converges pointwise to 0 everywhere on $\mathbf{R}^3$ as $\epsilon \to 0$ except on the set $\partial C = \{\vec{q} \in \mathbf{R}^3 : 2\omega_m(\vec{q}) = a\}$ which is a set of measure 0. Therefore by the dominated convergence theeorem

$$\lim_{\epsilon \to 0} \epsilon^{-3} I_1(\epsilon) = 0, \tag{70}$$

as required.

Now regarding the second limit in Equation (65) consider the function $f : [0, \infty) \to (0, m^{-1}]$ defined by

$$f(p) = (m^2 + p^2)^{-\frac{1}{2}}. \tag{71}$$

f is analytic. Therefore by Taylor's theorem for all $q, p \geq 0$

$$f(p) = f(q) + f'(q)(p-q) + \frac{1}{2}f''(\xi)(p-q)^2, \tag{72}$$

for some ξ between q and p. Now

$$\begin{aligned} f'(p) &= -p(m^2 + p^2)^{-\frac{3}{2}} \\ f''(p) &= (m^2 + p^2)^{-\frac{5}{2}}(2p^2 - m^2). \end{aligned}$$

Therefore

$$\begin{aligned} |f''(\xi)| &= (m^2 + \xi^2)^{-\frac{5}{2}}|2\xi^2 - m^2| \\ &\leq m^{-5}(2(q+\epsilon)^2 + m^2), \end{aligned}$$

as long as $|p - q| < \epsilon$. Thus

$$\begin{aligned} |f(p) - f(q)| &= |f'(q)(p-q) + \frac{1}{2}f''(\xi)(p-q)^2| \\ &< q(m^2 + q^2)^{-\frac{3}{2}}\epsilon + \frac{1}{2}m^{-5}(2(q+\epsilon)^2 + m^2)\epsilon^2 \\ &< m^{-1}\epsilon + \frac{1}{2}m^{-5}(2(q+\epsilon)^2 + m^2)\epsilon^2, \end{aligned}$$

as long as $|p - q| < \epsilon$. Hence

$$|J(\epsilon) - K(\epsilon)| = |\int \chi_{(a,b)}(2\omega_m(\vec{q}))(\int \chi_{B_\epsilon(\vec{0})-\vec{q}}(\vec{p})(\frac{1}{\omega_m(\vec{p})} - \frac{1}{\omega_m(\vec{q})})\, d\vec{p})\frac{d\vec{q}}{\omega_m(\vec{q})}|$$
$$\leq \int \chi_{(a,b)}(2\omega_m(\vec{q}))(\int \chi_{B_\epsilon(\vec{0})-\vec{q}}(\vec{p})(|f(|\vec{p}|) - f(|\vec{q}|)|)\, d\vec{p})\frac{d\vec{q}}{\omega_m(\vec{q})}$$
$$\leq \int \chi_{(a,b)}(2\omega_m(\vec{q})) \int \chi_{B_\epsilon(\vec{0})-\vec{q}}(\vec{p})(m^{-1}\epsilon + \frac{1}{2}m^{-5}(2(|\vec{q}| + \epsilon)^2 + m^2)\epsilon^2)$$
$$d\vec{p}\frac{d\vec{q}}{\omega_m(\vec{q})}$$
$$= \frac{4}{3}\pi\epsilon^3 \int \chi_{(a,b)}(2\omega_m(\vec{q}))(m^{-1}\epsilon + \frac{1}{2}m^{-5}(2(|\vec{q}| + \epsilon)^2 + m^2)\epsilon^2)\frac{d\vec{q}}{\omega_m(\vec{q})}.$$

Therefore

$$\lim_{\epsilon\to 0} \epsilon^{-3}|J(\epsilon) - K(\epsilon)| = \lim_{\epsilon\to 0} \frac{4}{3}\pi \int \chi(a,b)(2\omega_m(\vec{q}))(m^{-1}\epsilon + \frac{1}{2}m^{-5}(2(|\vec{q}| + \epsilon)^2 + m^2)\epsilon^2)$$
$$\frac{d\vec{q}}{\omega_m(\vec{q})}$$
$$= 0,$$

as required. We have therefore dealt with the second $\approx$ in Argument 1.

To finish dealing with the first $\approx$ suppose that $\epsilon_1 > 0$ is given. Choose $c \in (a, b)$ such that

$$\frac{16}{3}\pi^2(Z(c) - Z(a)) < \frac{\epsilon_1}{2}. \tag{73}$$

Now choose $\delta_1 > 0$ such that if $0 < \epsilon < \delta_1$ then

$$|\epsilon^{-3} \int \chi_{(a,c)}(\omega_m(\vec{p}) + \omega_m(\vec{q}))\chi_{B_\epsilon(\vec{0})-\vec{q}}(\vec{p})\frac{d\vec{p}}{\omega_m(\vec{p})}\frac{d\vec{q}}{\omega_m(\vec{q})} -$$
$$\frac{4}{3}\pi \int \chi_{(a,c)}(2\omega_m(\vec{q}))\frac{d\vec{q}}{\omega_m(\vec{q})^2}| < \frac{\epsilon_1}{2}.$$

(We can do this because of the validity of the second $\approx$.) Choose $\delta_2 > 0$ such that if $0 < \epsilon < \delta_2$ then $a^+ = a^+(\epsilon) < c$. Let $\delta = \min(\delta_1, \delta_2)$.

Then if $\epsilon < \delta$ then

$$
\begin{aligned}
&|\epsilon^{-3}\int \chi_{(a,a^+)}(\omega_m(\vec{p})+\omega_m(\vec{q}))\chi_{B_\epsilon(\vec{0})-\vec{q}}(\vec{p})\frac{d\vec{p}}{\omega_m(\vec{p})}\frac{d\vec{q}}{\omega_m(\vec{q})}| \\
\leq\ &|\epsilon^{-3}\int \chi_{(a,c)}(\omega_m(\vec{p})+\omega_m(\vec{q}))\chi_{B_\epsilon(\vec{0})-\vec{q}}(\vec{p})\frac{d\vec{p}}{\omega_m(\vec{p})}\frac{d\vec{q}}{\omega_m(\vec{q})}| \\
\leq\ &|\epsilon^{-3}\int \chi_{(a,c)}(\omega_m(\vec{p})+\omega_m(\vec{q}))\chi_{B_\epsilon(\vec{0})-\vec{q}}(\vec{p})\frac{d\vec{p}}{\omega_m(\vec{p})}\frac{d\vec{q}}{\omega_m(\vec{q})}- \\
&\frac{4}{3}\pi\int \chi_{(a,c)}(2\omega_m(\vec{q}))\frac{d\vec{q}}{\omega_m(\vec{q})^2}|+|\frac{4}{3}\pi\int \chi_{(a,c)}(2\omega_m(\vec{q}))\frac{d\vec{q}}{\omega_m(\vec{q})^2}| \\
<\ &\frac{\epsilon_1}{2}+\frac{4}{3}\pi\int \chi_{(a,c)}(2\omega_m(\vec{q}))\frac{d\vec{q}}{\omega_m(\vec{q})^2} \\
=\ &\frac{\epsilon_1}{2}+\frac{4}{3}\pi\int_{Z(a)}^{Z(c)}\frac{4\pi r^2}{m^2+r^2}dr \\
\leq\ &\frac{\epsilon_1}{2}+\frac{16}{3}\pi^2(Z(c)-Z(a)) \\
<\ &\frac{\epsilon_1}{2}+\frac{\epsilon_1}{2} \\
=\ &\epsilon_1.
\end{aligned}
$$

Thus

$$\lim_{\epsilon\to 0}\epsilon^{-3}\int \chi_{(a,a^+)}(\omega_m(\vec{p})+\omega_m(\vec{q}))\chi_{B_\epsilon(\vec{0})-\vec{q}}(\vec{p})\frac{d\vec{p}}{\omega_m(\vec{p})}\frac{d\vec{q}}{\omega_m(\vec{q})}=0, \tag{74}$$

thereby completing the proof of the validity of the first $\approx$ and therefore the validity of Argument 1. □

7. Investigation of the Measure Defined by the Convolution $\Omega_{im} * \Omega_{im}$

The measure Ω^+_{im} is defined by

$$\Omega^+_{im}(\Gamma)=\int_{\pi(\Gamma\cap H^+_{im})}\frac{d\vec{p}}{\omega_{im}(\vec{p})}\ \text{for}\ \Gamma\in\mathcal{B}_0(\mathbf{R}^4), \tag{75}$$

where

$$H^+_{im}=\{p\in\mathbf{R}^4 : p^2=-m^2, p^0\geq 0\}. \tag{76}$$

Ω^+_{im} is a measure concentrated on the positive time imaginary mass hyperboloid H^+_{im} corresponding to mass im. There is also a measure Ω^-_{im} on H^-_{im} and we may define $\Omega_{im}=\Omega^+_{im}+\Omega^-_{im}$, for $m>0$. Ω_{im} is a Lorentz invariant measure on $H_{im}=\{p\in\mathbf{R}^4 : p^2=-m^2\}$.

Define, for $m\in\mathbf{C}$

$$J^+_m=\{p\in\mathbf{C}^4 : p^2=m^2, \mathrm{Re}(p^0)\geq 0, \mathrm{Im}(p^0)\geq 0\}, \tag{77}$$

where $p^2=\eta_{\mu\nu}p^\mu p^\nu$ (in which $\eta_{\mu\nu}$ is the Minkowski space metric tensor). Then, for $m>0$,

$$J^+_m\cap\mathbf{R}^4=\{p\in\mathbf{R}^4 : p^2=m^2, p^0\geq 0\}=H^+_m, \tag{78}$$

$$\begin{aligned} J_m^+ \cap (i\mathbf{R}^4) &= \{p \in i\mathbf{R}^4 : p^2 = m^2, \mathrm{Re}(p^0) \geq 0, \mathrm{Im}(p^0) \geq 0\} \\ &= \{iq : q \in \mathbf{R}^4, q^2 = -m^2, q^0 \geq 0\} \\ &= iH_{im}^+. \end{aligned} \tag{79}$$

Now if $\vec{p} \in \mathbf{R}^3, m > 0$, we may write

$$\omega_{im}(\vec{p}) = ((im)^2 + \vec{p}^{\,2})^{\frac{1}{2}} = (-m^2 + \vec{p}^{\,2})^{\frac{1}{2}} = (-(m^2 + (i\vec{p})^2))^{\frac{1}{2}} = i(m^2 + (i\vec{p})^2)^{\frac{1}{2}} = i\omega_m(i\vec{p}).$$

One may consider the measure Ω_m^+ to be defined on $i\mathbf{R}^4$ as well as $\mathbf{R}^4$ and for all $m \in \mathbf{R}$ or $m \in i\mathbf{R}$ according to

$$\Omega_m^+(\Gamma) = \int_{\pi(\Gamma \cap J_m^+)} \frac{d\vec{p}}{\omega_m(\vec{p})}. \tag{80}$$

Then from Equation (79)

$$\Omega_m^+(i\Gamma) = \int_{i\pi(\Gamma \cap H_{im}^+)} \frac{d\vec{p}}{\omega_m(\vec{p})}. \tag{81}$$

Now make the substitution $\vec{p} = i\vec{q}$. Then $d\vec{p} = -id\vec{q}$. Thus

$$\Omega_m^+(i\Gamma) = \int_{\pi(\Gamma \cap H_{im}^+)} \frac{-id\vec{q}}{-i\omega_{im}(\vec{q})} = \Omega_{im}^+(\Gamma). \tag{82}$$

Now suppose that

$$\psi = \sum_k c_k \chi_{E_k}, \tag{83}$$

where $c_i \in \mathbf{C}$ and $E_k \in \mathcal{B}_0(\mathbf{R}^4)$, is a simple function. Then

$$\begin{aligned} \int_{\mathbf{R}^4} \psi(p)\, \Omega_{im}^+(dp) &= \sum_k c_k \Omega_{im}^+(E_k) \\ &= \sum_k c_k \Omega_m^+(iE_k) \\ &= \sum_k c_k \int_{i\mathbf{R}^4} \chi_{iE_k}(p)\, \Omega_m^+(dp) \\ &= \sum_k c_k \int_{i\mathbf{R}^4} \chi_{E_k}(\frac{p}{i})\, \Omega_m^+(dp) \\ &= \int_{i\mathbf{R}^4} \psi(\frac{p}{i})\, \Omega_m^+(dp). \end{aligned} \tag{84}$$

As this is true for every such simple function ψ it follows that

$$\int_{\mathbf{R}^4} \psi(p)\, \Omega_{im}^+(dp) = \int_{i\mathbf{R}^4} \psi(\frac{p}{i})\, \Omega_m^+(dp), \tag{85}$$

for every function ψ which is integrable with respect to Ω_{im}^+. Therefore

$$\begin{aligned} (\Omega_{im}^+ * \Omega_{im}^+)(\Gamma) &= \int_{(\mathbf{R}^4)^2} \chi_\Gamma(p+q)\, \Omega_{im}^+(dp)\, \Omega_{im}^+(dq) \\ &= \int_{(i\mathbf{R}^4)^2} \chi_\Gamma\left(\frac{p+q}{i}\right) \Omega_m^+(dp)\, \Omega_m^+(dq) \\ &= \int_{(i\mathbf{R}^4)^2} \chi_{i\Gamma}(p+q) \Omega_m^+(dp) \Omega_m^+(dq) \\ &= (\Omega_m^+ * \Omega_m^+)(i\Gamma), \end{aligned} \tag{86}$$

for all $\Gamma \in \mathcal{B}_0(\mathbf{R}^4)$.

Now in general, suppose that a measure μ has a causal spectral representation of the form

$$\mu(\Gamma) = \int_{m'=0}^{\infty} \Omega_{m'}^{+}(\Gamma)\,\sigma(m'), \tag{87}$$

for some Borel spectral measure $\sigma : \mathcal{B}_0([0,\infty)) \to \mathbf{C}$. Then μ extends to a measure defined on $i\mathbf{R}^4$ by

$$\mu(i\Gamma) = \int_{m'=0}^{\infty} \Omega_{m'}^{+}(i\Gamma)\,\sigma(dm') = \int_{m'=0}^{\infty} \Omega_{im'}^{+}(\Gamma)\,\sigma(dm'), \tag{88}$$

for $\Gamma \in \mathcal{B}_0(\mathbf{R}^4)$. Therefore since, as we have determined above, $\Omega_m^+ * \Omega_m^+$ is a causal spectral measure with spectrum

$$\sigma(m') = \begin{cases} 4\pi m Z(m') \text{ for } m' \geq 2m \\ 0 \text{ otherwise,} \end{cases} \tag{89}$$

it follows that

$$(\Omega_m^+ * \Omega_m^+)(i\Gamma) = \int_{m'=0}^{\infty} \Omega_{im'}^{+}(\Gamma)\,\sigma(dm'). \tag{90}$$

Therefore using Equation (86) $\Omega_{im}^+ * \Omega_{im}^+$ is a measure with spectral representation

$$(\Omega_{im}^+ * \Omega_{im}^+)(\Gamma) = \int_{m'=0}^{\infty} \Omega_{im'}^{+}(\Gamma)\,\sigma(m')\,dm', \tag{91}$$

where σ is the spectral function given by Equation (89). Note that $\Omega_{im}^+ * \Omega_{im}^+$ is not causal, it is a type III measure, and

$$\operatorname{supp}(\Omega_{im}^+ * \Omega_{im}^+) = \{p \in \mathbf{R}^4 : p^2 \leq -4m^2, p^0 \geq 0\}. \tag{92}$$

8. Determination of the Density Defining a Causal Lorentz Invariant Borel Measure from Its Spectrum

Suppose that μ is of the form of Equation (30) where σ is a well behaved (e.g., locally integrable) function. We would like to see if μ can be defined by a density with respect to the Lebesgue measure, i.e., if there exists a function $g : \mathbf{R}^4 \to \mathbf{C}$ such that

$$\mu(\Gamma) = \int_{\Gamma} g(p)\,dp. \tag{93}$$

Well we have that

$$\mu(\Gamma) = \int_{m=0}^{\infty} \sigma(m)\Omega_m^+(\Gamma)\,dm = \int_{m=0}^{\infty} \sigma(m) \int_{\pi(\Gamma\cap H_m^+)} \frac{d\vec{p}}{\omega_m(\vec{p})}\,dm. \tag{94}$$

Now

$$\begin{aligned} \vec{p} \in \pi(\Gamma \cap H_m^+) &\Leftrightarrow (\exists p \in \mathbf{R}^4)\,\vec{p} = \pi(p), p \in H_m^+, p \in \Gamma \\ &\Leftrightarrow (\omega_m(\vec{p}), \vec{p}) \in \Gamma \\ &\Leftrightarrow \chi_\Gamma(\omega_m(\vec{p}), \vec{p}) = 1. \end{aligned}$$

Therefore

$$\mu(\Gamma) = \int_{m=0}^{\infty} \sigma(m) \int_{\mathbf{R}^3} \chi_\Gamma(\omega_m(\vec{p}), \vec{p}) \frac{1}{\omega_m(\vec{p})}\,d\vec{p}\,dm. \tag{95}$$

Now consider the transformation defined by the function $h : (0, \infty) \times \mathbf{R}^3 \to \mathbf{R}^4$ given by

$$h(m, \vec{p}) = (\omega_m(\vec{p}), \vec{p}). \tag{96}$$

Let

$$q = h(m, \vec{p}) = (\omega_m(\vec{p}), \vec{p}) = ((m^2 + \vec{p}^{\,2})^{\frac{1}{2}}, \vec{p}). \tag{97}$$

Then

$$\frac{\partial q^0}{\partial m} = m\omega_m(\vec{p})^{-1}, \frac{\partial q^0}{\partial p^j} = p^j \omega_m(\vec{p})^{-1}, \frac{\partial q^i}{\partial m} = 0, \frac{\partial q^i}{\partial p^j} = \delta_{ij}, \tag{98}$$

for $i, j = 1, 2, 3$. Thus the Jacobian of the transformation is

$$J(m, \vec{p}) = m\omega_m(\vec{p})^{-1}. \tag{99}$$

Now $q = (\omega_m(\vec{p}), \vec{p})$. Therefore $q^2 = \omega_m(\vec{p})^2 - \vec{p}^{\,2} = m^2$. So $m = (q^2)^{\frac{1}{2}}, q^2 > 0$. Thus

$$\begin{aligned} \mu(\Gamma) &= \int_{q \in \mathbf{R}^4, q^2 > 0, q^0 > 0} \chi_\Gamma(q) \frac{\sigma(m)}{\omega_m(\vec{p})} \frac{dq}{J(m, \vec{p})} \\ &= \int_{q^2 > 0, q^0 > 0} \chi_\Gamma(q) \frac{\sigma(m)}{m} \, dq. \end{aligned} \tag{100}$$

Hence

$$\begin{aligned} \mu(\Gamma) &= \int_{q^2 > 0, q^0 > 0} \chi_\Gamma(q) \frac{\sigma((q^2)^{\frac{1}{2}})}{(q^2)^{\frac{1}{2}}} \, dq \\ &= \int_\Gamma g(q) \, dq, \end{aligned}$$

where $g : \mathbf{R}^4 \to \mathbf{C}$ is defined by

$$g(q) = \begin{cases} (q^2)^{-\frac{1}{2}} \sigma((q^2)^{\frac{1}{2}}) \text{ if } q^2 > 0, q^0 > 0 \\ 0 \text{ otherwise.} \end{cases} \tag{101}$$

We have therefore shown how, given a spectral representation of a causal Lorentz invariant Borel complex measure in which the spectrum is a complex function, one can obtain and equivalent representation of the measure in terms of a density with respect to Lebesgue measure.

9. Convolutions and Products of Causal Lorentz Invariant Borel Measures

9.1. Convolution of Measures

Let μ and ν be causal Lorentz invariant Borel complex measures. Then (up to possible atoms at the origin which can be dealt with in a straightforward way) there exist Borel spectral measures $\sigma, \rho : \mathcal{B}_0([0, \infty)) \to \mathbf{C}$ such that

$$\begin{aligned} \mu &= \int_{m=0}^\infty \Omega_m \, \sigma(dm), \\ \nu &= \int_{m=0}^\infty \Omega_m \, \rho(dm). \end{aligned} \tag{102}$$

We will assume, without loss of generality, that σ and ρ are complex measures, i.e. $\sigma, \rho : \mathcal{B}([0, \infty)) \to \mathbf{C}$ and are countably additive. The convolution of μ and ν, if it exists, is given by

$$(\mu * \nu)(\Gamma) = \int \chi_\Gamma(p + q) \, \mu(dp) \, \nu(dq). \tag{103}$$

Now let $\psi = \sum_i c_i \chi_{E_i}$ with $c_i \in \mathbf{C}, E_i \in \mathcal{B}_0(\mathbf{R}^4)$ be a simple function. Then

$$\begin{aligned}\int \psi(p)\,\mu(dp) &= \int \sum_i c_i \chi_{E_i}\,\mu(dp) \\ &= \sum_i c_i \mu(E_i) \\ &= \sum_i c_i \int_{m=0}^{\infty} \Omega_m(E_i)\,\sigma(dm) \\ &= \sum_i c_i \int_{m=0}^{\infty} \int_{\mathbf{R}^4} \chi_{E_i}(p)\,\Omega_m(dp)\,\sigma(dm) \\ &= \int_{m=0}^{\infty} \int_{\mathbf{R}^4} \psi(p)\,\Omega_m(dp)\,\sigma(dm).\end{aligned}$$

Therefore for any sufficiently well behaved measurable function $\psi : \mathbf{R}^4 \to \mathbf{C}$ (e.g. bounded measurable functions of compact support)

$$\int \psi(p)\mu(dp) = \int \psi(p)\,\Omega_m(dp)\,\sigma(dm). \tag{104}$$

(Note that the integral exists because σ is a Borel measure.) Hence for all $\Gamma \in \mathcal{B}_0(\mathbf{R}^4)$

$$\begin{aligned}(\mu * \nu)(\Gamma) &= \int \chi_\Gamma(p+q)\,\mu(dp)\,\nu(dq) \\ &= \int \chi_\Gamma(p+q)\,\Omega_m(dp)\,\sigma(dm)\,\Omega_{m'}(dq)\,\rho(dm') \\ &= \int \chi_\Gamma(p+q)\,\Omega_m(dp)\,\Omega_{m'}(dq)\sigma(dm)\,\rho(dm'),\end{aligned} \tag{105}$$

by Fubini's theorem, as long as

$$\int \chi_\Gamma(p+q)\,\Omega_m(dp)\,\Omega_{m'}(dq)|\sigma|(dm) < \infty, \forall m' \in [0,\infty), \tag{106}$$

where $|\sigma|$ is the total variations of the measure σ.

Suppose that $\Gamma \in \mathcal{B}_0(\mathbf{R}^4)$. Then there exists $a, R \in (0,\infty)$ such that $\Gamma \subset (-a,a) \times B_R(\vec{0})$, where $B_R(\vec{0}) = \{\vec{p} \in \mathbf{R}^3 : |\vec{p}| < R\}$. Now

$$\int \chi_\Gamma(p+q)\,\Omega_m(dp) = \int_{\Gamma \quad q} \Omega_m(dp) = \Omega_m(\Gamma - q) < \infty, \tag{107}$$

for all $q \in \mathbf{R}^4$ because Ω_m is Borel and Γ is compact.

Now suppose that $m, m' > a$. Then

$$p \in H_m^+, q \in H_{m'}^+ \Rightarrow (p+q)^0 = p^0 + q^0 \geq m + m' > 2a \Rightarrow (p+q) \notin \Gamma. \tag{108}$$

Thus

$$\int \chi_\Gamma(p+q)\,\Omega_m(dp)\,\Omega_{m'}(dq) = 0. \tag{109}$$

Therefore since σ and ρ are Borel, $(\mu * \nu)(\Gamma)$ exists, is finite and is given by Equation (105).

Now let $\Lambda \in O(1,3)^{+\uparrow}$, $\psi : \mathbf{R}^4 \to \mathbf{C}$ be a measurable function of compact support. Then

$$\begin{aligned}< \mu * \nu, \Lambda\psi > &= \int \psi(\Lambda^{-1}(p+q))\,\Omega_m(dp)\,\Omega_{m'}(dq)\,\sigma(dm)\,\rho(dm') \\ &= \int \psi(p+q)\,\Omega_m(dp)\,\Omega_{m'}(dq)\,\sigma(dm)\,\rho(dm'). \\ &= < \mu * \nu, \psi >\end{aligned}$$

Therefore $\mu * \nu$ is Lorentz invariant. It can be shown, by an argument similar to that used for the case $\Omega_m * \Omega_m$ that $\mu * \nu$ is causal.

We have therefore shown that the convolution of two causal Lorentz invariant Borel complex measures exists and is a causal Lorentz invariant Borel complex measure.

9.2. Product of measures

We now turn to the problem of computing the product of two causal Lorentz invariant Borel complex measures. The problem of computing the product of measures or distributions is difficult in general and has attracted a large amount of research [10,19,20]. In such work one generally seeks a definition of the product of measures or distributions which agrees with the ordinary product when the measures or distributions are functions (i.e., densities with respect to Lebesgue measure). The most common approach is to use the fact that, for Schwartz functions $f, g \in \mathcal{S}(\mathbf{R}^4)$ multiplication in the spatial domain corresponds to convolution in the frequency domain, i.e., $(fg)^\wedge = f^\wedge * g^\wedge$ (where $\wedge$ denotes the Fourier transform operator). Thus one defines the product of measures or distributions μ, ν as

$$\mu\nu = (\mu^\wedge * \nu^\wedge)^\vee. \tag{110}$$

However, this definition is only successful when the convolution that it involves exists which may not be the case in general. If μ, ν are tempered measures then $\mu^\wedge$ and $\nu^\wedge$ exist as tempered distributions, however, they are generally not causal, even if μ, ν are causal.

We will therefore not use the "frequency space" approach to define the product of measures but will use a different approach. Our approach is just as valid as the frequency space approach because our product will coincide with the usual function product when the measures are defined by densities. Furthermore, our approach is useful for the requirements of QFT because measures and distributions in QFT are frequently Lorentz invariant and causal.

Let $\text{int}(C) = \{p \in \mathbf{R}^4 : p^2 > 0, p^0 > 0\}$. Suppose that $f : \text{int}(C) \to \mathbf{C}$ is a Lorentz invariant locally integrable function. Then it defines a causal Lorentz invariant Borel measure μ_f which, by the spectral theorem, must have a representation of the form

$$\mu_f(\Gamma) = \int_\Gamma f(p)\, dp = \int_{m=0}^{\infty} \Omega_m(\Gamma)\, \sigma(dm), \tag{111}$$

for some spectral measure $\sigma : \mathcal{B}_0([0,\infty)) \to \mathbf{C}$. As μ_f is absolutely continuous with respect to Lebesgue measure it follows that σ must be non-singular, i.e., a function. By the result of the previous section a density defining μ_f is $\tilde{f} : \text{int}(C) \to \mathbf{C}$ defined by

$$\tilde{f}(p) = (p^2)^{-\frac{1}{2}} \sigma((p^2)^{\frac{1}{2}}), p \in \text{int}(C). \tag{112}$$

We must have that $\tilde{f} = f$ (almost everywhere). Therefore (almost everywhere on $\text{int}(C)$)

$$f(p) = (p^2)^{-\frac{1}{2}} \sigma((p^2)^{\frac{1}{2}}). \tag{113}$$

Without loss of generality, it can be assumed that equality holds everywhere in Equation (113). $f(p)$ depends only on p^2. Therefore for all $m > 0$, $\sigma(m) = mf(p)$ for all $p \in \text{int}(C)$ such that $p^2 = m^2$. In particular

$$\sigma(m) = mf((m, \vec{0})^T), \forall m > 0. \tag{114}$$

Now we are seeking a definition of product which has useful properties. Two such properties would be that it is distributive with respect to generalized sums such as integrals and also that it agrees with the ordinary product when the measures are defined by functions. Suppose that we had such a product.

Let $f, g : \mathrm{int}(C) \to \mathbf{C}$ be Lorentz invariant locallly integrable functions. Let $\mu, \nu : \mathcal{B}_0(\mathrm{int}(C)) \to \mathbf{C}$ be the associated measures with spectra σ, ρ. Then

$$\begin{aligned} \mu\nu &= \int_{m=0}^{\infty} \Omega_m\, \sigma(dm) \int_{m'=0}^{\infty} \Omega_{m'}\, \rho(dm') \\ &= \int_{m=0}^{\infty} \Omega_m\, mf((m,\vec{0})^T)\, dm \int_{m'=0}^{\infty} \Omega_{m'}\, m'g((m',\vec{0})^T)\, dm' \\ &= \int_{m=0}^{\infty}\int_{m'=0}^{\infty} \Omega_m\Omega_{m'} mf((m,\vec{0})^T)m'g((m',\vec{0})^T)\, dm\, dm'. \end{aligned}$$

Now we want this to be equal to

$$\int_{m=0}^{\infty} \Omega_m m(fg)((m,\vec{0})^T)\, dm \tag{115}$$

This will be the case (formally) if we have

$$\Omega_m\Omega_{m'} = \frac{1}{m}\delta(m-m')\Omega_m, \forall m, m' > 0. \tag{116}$$

Physicists will be familiar with such a formula (e.g., the equal time commutation relations). Rather than attempting to define its meaning in a rigorous way, we will simply carry out the following formal computation for general Lorentz invariant Borel measures μ, ν with spectra σ, ρ

$$\begin{aligned} \mu\nu &= \int_{m=0}^{\infty} \Omega_m\, \sigma(dm) \int_{m'=0}^{\infty} \Omega_{m'}\, \rho(dm') \\ &= \int_{m=0}^{\infty}\int_{m'=0}^{\infty} \Omega_m\Omega_{m'}\sigma(m)\rho(m')\, dm\, dm' \\ &= \int_{m=0}^{\infty}\int_{m'=0}^{\infty} \frac{1}{m}\Omega_m\delta(m-m')\sigma(m)\rho(m')\, dm'\, dm \\ &= \int_{m=0}^{\infty} \frac{1}{m}\Omega_m\sigma(m)\rho(m)\, dm. \end{aligned}$$

Therefore we can simply define the product $\mu\nu$ in general by

$$\mu\nu = \int_{m=0}^{\infty} \frac{1}{m}\Omega_m\, (\sigma\rho)(dm), \tag{117}$$

i.e.,

$$(\mu\nu)(\Gamma) = \int_{m=0}^{\infty} \frac{1}{m}\Omega_m(\Gamma)\, (\sigma\rho)(dm), \tag{118}$$

for $\Gamma \in \mathcal{B}_0(\mathbf{R}^4)$.

We have therefore reduced the problem of computing the product of measures on $\mathrm{int}(C)$ to the problem of computing the product of their 1D spectral measures. The problem of multiplying 1D measures is somewhat less problematic than the problem of multiplying 4D measures. A large class of 1D measures is made up of measures which are of the form of a function plus a finite number of "atoms" (singularities of the form $c\delta_a$ where $c \in \mathbf{C}\backslash\{0\}, a \in [0,\infty)$, where δ_a is the Dirac delta function (measure) concentrated at a). There are other pathological types of the 1D measure but these may not be of interest for physical applications.

In the general non-pathological case, if μ, ν are causal Lorentz invariant Borel measures with spectra $\sigma(m) = \xi(m) + \sum_{i=1}^{k} c_i\delta(m-a_i), \rho(m) = \zeta(m) + \sum_{j=1}^{l} d_j\delta(m-b_j)$ where $\xi, \zeta : [0,\infty) \to \mathbf{C}$ are locally integrable functions, $c_i, d_j \in \mathbf{C}\backslash\{0\}, k, l \geq 0, a_i, b_j \in [0,\infty)$ are such that $a_i \neq b_j, \forall i, j$ then we may define the product of μ and ν to be the causal Lorentz invariant measure $\mu\nu$ given by

$$\mu\nu = \int_{m=0}^{\infty} \Omega_m \tau(dm), \tag{119}$$

where

$$\tau(m) = \frac{1}{m}(\xi(m)\zeta(m) + \zeta(m)\sum_{i=1}^{k} c_i\delta(m-a_i) + \xi(m)\sum_{j=1}^{l} d_j\delta(m-b_j))$$
$$= \frac{1}{m}(\xi(m)\zeta(m) + \sum_{i=1}^{k} \zeta(a_j)c_i\delta(m-a_i) + \sum_{j=1}^{l} \xi(b_j)d_j\delta(m-b_j)),$$

for $m > 0$.

10. Conclusions

We have defined a spectral calculus that enables one to compute the spectrum of any causal Lorentz invariant Borel complex measure on Minkowski space whose spectrum is a continuous function. This calculus can be used in many applications in QFT and leads to a method called spectral regularization [21].

We have computed the spectra associated with certain elementary convolutions involving Feynman propagators of mass m scalar particles. It has been shown how one can compute the density associated with a causal Lorentz invariant Borel complex measure from its spectrum.

We have shown that the convolution of arbitrary measures of the prescribed type exists and how their product exists in a wide class of cases of physical interest. Methods for the computation of these objects from the spectra of their components have been presented.

The spectral calculus can be used to compute the spectrum, and hence density, associated with the contraction of the vacuum polarization tensor [21]. A generalization of the spectral calculus to Lorentz invariant tensor valued measures on Minkowski space can be used to compute the form of the vacuum polarization tensor and therefore to compute the vacuum polarization function. This function is shown to have a close agreement, up to finite renormalization, with the vacuum polarization function obtained using dimensional regularization /renormalization. This can be used to compute the Uehling potential function without using renormalization from which the Uehling contribution to the Lamb shift for the H atom can be computed exactly.

Funding: This research received no external funding.

Conflicts of Interest: The author declares no conflict of interest.

References

1. Halmos, P.R. *Measure Theory*; Springer: New York, NY, USA, 1988.
2. Garding, L.; Lions, J.L. Functional Analysis. *Nuovo C.* **1959**, *14* (Suppl. 1), 9–66. [CrossRef]
3. Epstein, H.; Glaser, V. Role of locality in perturbation theory. *Ann. l'Institut Henri Poincaré Sect. A Phys. Théorique* **1973**, *19*, 211–295.
4. Källén, G. On the definition of the renormalization constants in quantumelectrodynamics. *Helv. Phys. Acta* **1952**, *25*, 417–434.
5. Lehmann, H. Über eigenshaften von ausbreitungsfunkionen und renormierungskonstanten quantisierter felder. *Il Nuovo C.* **1954**, *11*, 342–357. [CrossRef]
6. Itzykson, C.; Zuber, J.-B. *Quantum Field Theory*; McGraw-Hill: New York, NY, USA, 1980.
7. Borthwick, D. *Spectral Theory of Infinite-Area Hyperbolic Surfaces*; Birkhauser: Basel, Switzerland, 2007.
8. Sarnak, P. Spectra of hyperbolic surfaces. *Bull. Am. Math. Soc.* **2003**, *40*, 441–478. [CrossRef]
9. Bollini, C.G.; Marchiano, P.; Rocca, M.C. Convolution of ultradistributions, field theory, Lorentz invariance and resonances. *Int. J. Theor. Phys.* **2007**, *46*, 3030–3059. [CrossRef]
10. Kamiński, A.; Mincheva-Kaminska, S. Compatibility conditions and the convolution of functions and generalized functions. *J. Funct. Spaces Appl.* **2013**, 356724. [CrossRef]
11. Ortner, N.; Wagner, P. Applications of $O(p,q)$ invariant distributions. *Math. Nachrichten* **2017**, *290*, 2995–3005. [CrossRef]

12. Zinoviev, Y.M. Lorentz covariant distributions with spectral conditions. *Theor. Math. Phys.* **2008**, *156*, 1247–1267. [CrossRef]
13. Soloviev, M.A. Lorentz-covariant ultradistributions, hyperfunctions and analytic functionals. *Theoret. Mat. Phys.* **2001**, *128*, 1252–1270. [CrossRef]
14. Harish-Chandra. Invariant distributions on Lie algebras. *Am. J. Math.* **1964**, *86*, 271–309. [CrossRef]
15. Kolk, J.A.C.; Varadarajan, V.S. Lorentz invariant distributions supported on the forward light cone. *Compos. Math.* **1992**, *81*, 61–106.
16. Mashford J. Second quantized quantum field theory based on invariance properties of locally conformally flat space-times. *arXiv* **2017**, arXiv:1709.09226,
17. Mandl, F.; Shaw, G. *Quantum Field Theory*; Wiley: Chichester, UK, 1991.
18. Bogolubov, N.N.; Logunov, A.A.; Todorov, I.T. *Introduction to Axiomatic Quantum Field Theory*; W. A. Benjamin, Inc.: Reading, MA, USA, 1975.
19. Colombeau, J.F. *New Generalized Functions and Multiplication of Distributions*; Elsevier: Amsterdam, The Netherlands, 1984.
20. Oberguggenberger, M. *Multiplication of Distributions and Applications to Partial Differential Equations*; Longman: New York, NY, USA, 1992.
21. Mashford, J. An introduction to spectral regularization for quantum field theory. In *Springer Proceedings in Mathematics and Statistics, Proceedings of the XIII International Workshop on Lie Theory and Its Applications in Physics, Varna, Bulgaria, 17–23 June 2019*; Dobrev, V., Ed.; Springer: Berlin/Heidelberg, Germany, 2020; Volume 335.

Publisher's Note: MDPI stays neutral with regard to jurisdictional claims in published maps and institutional affiliations.

symmetry

MDPI

Article

An Upper Bound of the Third Hankel Determinant for a Subclass of q-Starlike Functions Associated with k-Fibonacci Numbers

Muhammad Shafiq [1], Hari M. Srivastava [2,3,4], Nazar Khan [1,*], Qazi Zahoor Ahmad [5], Maslina Darus [6] and Samiha Kiran [1]

1 Department of Mathematics, Abbottabad University of Science and Technology, Abbottabad 22010, Pakistan; shafiqiqbal19@gmail.com (M.S.); samiiikhan26@gmail.com (S.K.)
2 Department of Mathematics and Statistics, University of Victoria, Victoria, BC V8W 3R4, Canada; harimsri@math.uvic.ca
3 Department of Medical Research, China Medical University Hospital, China Medical University, Taichung 40402, Taiwan
4 Department of Mathematics and Informatics, Azerbaijan University, 71 Jeyhun Hajibeyli Street, Baku AZ1007, Azerbaijan
5 Government Akhtar Nawaz Khan (Shaheed) Degree College KTS, Haripur 22620, Pakistan; zahoorqazi5@gmail.com
6 Department of Mathematical Sciences, Faculty of Science and Technology, Universiti Kebangsaan Malaysia, Bangi 43600, Selangor, Malaysia; maslina@ukm.edu.my
* Correspondence: nazarmaths@aust.edu.pk

Received: 15 May 2020; Accepted: 16 June 2020; Published: 22 June 2020

Abstract: In this paper, we use q-derivative operator to define a new class of q-starlike functions associated with k-Fibonacci numbers. This newly defined class is a subclass of class $\mathcal{A}$ of normalized analytic functions, where class $\mathcal{A}$ is invariant (or symmetric) under rotations. For this function class we obtain an upper bound of the third Hankel determinant.

Keywords: starlike functions; subordination; q-Differential operator; k-Fibonacci numbers

MSC: Primary 05A30, 30C45; Secondary 11B65, 47B38

1. Introduction and Definitions

The calculus without the notion of limits is called quantum calculus; it is usually called q-calculus or q-analysis. By applying q-calculus, univalent functions theory can be extended. Moreover, the q-derivative, such as the q-calculus operators (or the q-difference) operator, are used to developed a number of subclasses of analytic functions (see, for details, the survey-cum-expository review article by Srivastava [1]; see also a recent article [2] which appeared in this journal, *Symmetry*).

Ismail et al. [3] instigated the generalization of starlike functions by defining the class of q-starilke functions. A firm footing of the usage of the q-calculus in the context of Geometric Functions Theory was actually provided and the basic (or q-) hypergeometric functions were first used in Geometric Function Theory by Srivastava (see, for details [4]). Raghavendar and Swaminathan [5] studied certain basic concepts of close-to-convex functions. Janteng et al. [6] published a paper in which the (q) generalization of some subclasses of analytic functions have studied. Further, q-hypergeometric functions, the q-operators were studied in many recent works (see, for example, [7–9]). The q-calculus applications in operator theory could be found in [4,10]. The coefficient inequality for q-starlike and q-close-to-convex functions with respect to Janowski functions were studied by Srivastava et al. [8,11] recently, (see also [12]). Further development on this subject could be seen in [7,9,13,14]. For a comprehensive review of the theory and applications of the q-derivative (or the q-difference) operator and related literature, we refer the reader to the above-mentioned work [1].

We denote by $\mathcal{A}$ the class of functions which are analytic and having the form:

$$f(z) = z + \sum_{n=2}^{\infty} a_n z^n \tag{1}$$

in the open unit disk $\mathbb{U}$ given by

$$\mathbb{U} = \{z : z \in \mathbb{C} \quad \text{and} \quad |z| < 1\}$$

and normalized by the following conditions:

$$f(0) = 0 = f'(0) - 1.$$

The subordinate between two functions f and g in $\mathbb{U}$, given by:

$$f \prec g \quad \text{or} \quad f(z) \prec g(z),$$

if an analytic Schwarz function w exists in such way that

$$w(0) = 0 \qquad \text{and} \qquad |w(z)| < 1,$$

so that

$$f(z) = g(w(z)).$$

In particular, the following equivalence also holds for the univalent function g

$$f(z) \prec g(z) \quad (z \in \mathbb{U}) \implies f(0) = g(0) \text{ and } f(\mathbb{U}) \subset g(\mathbb{U}).$$

Next by the $\mathcal{P}$ class of analytic functions, $p(z)$ in $\mathbb{U}$ is denoted, in which normalization conditions are given as follow:

$$p(z) = 1 + \sum_{n=1}^{\infty} c_n z^n \tag{2}$$

such that

$$\Re(p(z)) > 0 \qquad (\forall\, z \in \mathbb{U}).$$

Let k be any positive real number, then we define the k-Fibonacci number sequence $\{F_{k,n}\}_{n=0}^{\infty}$ recursively by

$$F_{k,0} = 0,\ F_{k,1} = 1 \text{ and } F_{k,n+1} = kF_{k,n} + F_{k,n-1} \text{ for } n \geqq 1. \tag{3}$$

The nth k-Fibonacci number is given by

$$F_{k,n} = \frac{(k - \mathcal{T}_k)^n - \mathcal{T}_k^n}{\sqrt{k^2 + 4}},$$

where

$$\mathcal{T}_k = \frac{k - \sqrt{k^2 + 4}}{2}. \tag{4}$$

If

$$\tilde{p}_k(z) = 1 + \sum_{n=1}^{\infty} \tilde{p}_{k,n} z^n,$$

then we have (see also [15])

$$\tilde{p}_{k,n} = (F_{k,n-1} + F_{k,n+1})\mathcal{T}_k^n \qquad (n \in \mathbb{N};\ \mathbb{N} := \{1,2,3,\cdots\}). \tag{5}$$

Definition 1. *Let $q \in (0,1)$ then the q-number $[\lambda]_q$ is given by*

$$[\lambda]_q = \begin{cases} \dfrac{1-q^\lambda}{1-q} & (\lambda \in \mathbb{C}) \\ \sum\limits_{k=0}^{n-1} q^k = 1 + q + q^2 + \cdots + q^{n-1} & (\lambda = n \in \mathbb{N}). \end{cases}$$

Definition 2. *The q-difference (or the q-derivative) $\mathcal{D}_q$ operator of any given function f is defined, in a given subset of $\mathbb{C}$, of complex numbers by*

$$(\mathcal{D}_q f)(z) = \begin{cases} \dfrac{f(z) - f(qz)}{(1-q)z} & (z \neq 0) \\ f'(0) & (z = 0), \end{cases}$$

led to the existence of the derivative $f'(0)$.

From Definitions 1 and 2, we have

$$\lim_{q \to 1-} (\mathcal{D}_q f)(z) = \lim_{q \to 1-} \frac{f(z) - f(qz)}{(1-q)z} = f'(z)$$

for a differentiable function f. In addition, from (1) and (2), we observe that

$$(\mathcal{D}_q f)(z) = 1 + \sum_{n=2}^{\infty} [n]_q a_n z^{n-1}. \tag{6}$$

In the year 1976, it was Noonan and Thomas [16] who concentrated on the function f given in (1) and gave the $\mathfrak{q}$th Hankel determinant as follows.

Let $n \geqq 0$ and $\mathfrak{q} \in \mathbb{N}$. Than the $\mathfrak{q}$th Hankel determinant is defined by

$$H_{\mathfrak{q}}(n) = \begin{vmatrix} a_n & a_{n+1} & \cdot & \cdot & \cdot & a_{n+\mathfrak{q}-1} \\ a_{n+1} & \cdot & & & & \cdot \\ \cdot & \cdot & & & & \cdot \\ \cdot & \cdot & & & & \cdot \\ \cdot & \cdot & & & & \cdot \\ a_{n+\mathfrak{q}-1} & \cdot & & \cdot & \cdot & \cdot & a_{n+2(\mathfrak{q}-1)} \end{vmatrix}$$

Several authors studied the determinant $H_{\mathfrak{q}}(n)$. In particular, sharp upper bounds on $H_2(2)$ were obtained in such earlier works as, for example, in [17,18] for various subclasses of the normalized analytic function class $\mathcal{A}$. It is well-known for the Fekete-Szegö functional $|a_3 - a_2^2|$ that

$$\left|a_3 - a_2^2\right| = H_2(1).$$

Its worth mentioning that, for a parameter μ which is real or complex, the generalization the functional $|a_3 - \mu a_2^2|$ is given in aspects. In particular, Babalola [19] studied the Hankel determinant $H_3(1)$ for some subclasses of $\mathcal{A}$.

In 2017, Güney et al. [20] explored the third Hankel determinant in some subclasses of $\mathcal{A}$ connected with the above-defined k-Fibonacci numbers. A derivation of the sharp coefficient bound for the third Hankel determinant and the conjecture for the sharp upper bound of the second Hankel determinant is also derived by them, which is employed to solve the related problems to the third Hankel determinant and to present an upper bound for this determinant.

Motivated and inspired by the above-mentioned work and also by the recent works of Güney et al. [20] and Uçar [12], we will now define a new subclass $\mathcal{SL}(k,q)$ of starlike functions associated with the k-Fibonacci numbers. We will then find the Hankel determinant $H_3(1)$ for the newly-defined functions class $\mathcal{SL}(k,q)$.

Definition 3. *Let $\mathcal{P}(\beta)$ $(0 \leqq \beta < 1)$ denote the class of analytic functions p in $\mathbb{U}$ with*

$$p(0) = 1 \quad \text{and} \quad \Re(p(z)) > \beta.$$

Definition 4. *Let the function p be said to belong to the class k-$\tilde{\mathcal{P}}_q(z)$ and let k be any positive real number if*

$$p(z) \prec \frac{2\tilde{p}_k(z)}{(1+q)+(1-q)\tilde{p}_k(z)}, \tag{7}$$

where $\tilde{p}_k(z)$ is given by

$$\tilde{p}_k(z) = \frac{1+\mathcal{T}_k^2 z^2}{1-k\mathcal{T}_k z-\mathcal{T}_k^2 z^2}, \tag{8}$$

and $\mathcal{T}_k$ is given in (4).

Remark 1. For $q = 1$, it is easily seen that

$$p(z) \prec \tilde{p}_k(z).$$

Definition 5. *Let k be any positive real number. Then the function f be in the functions class $\mathcal{SL}(k,q)$ if and only if*

$$\frac{z}{f(z)}(\mathcal{D}_q f)z \prec \frac{2\tilde{p}_k(z)}{(1+q)+(1-q)\tilde{p}_k(z)}, \tag{9}$$

where $\tilde{p}_k(z)$ is given in (8).

Remark 2. For $q = 1$, we have

$$\frac{zf'(z)}{f(z)} \prec \tilde{p}_k(z).$$

We recall that when the f belongs to the class $\mathcal{A}$ of analytic function then it is invariant (or symmetric) under rotations if and only if the function $f_\varsigma(z)$ given by

$$f_\varsigma(z) = e^{-i\varsigma} f(ze^{i\varsigma}) \quad (\varsigma \in \mathbb{R})$$

is also in $\mathcal{A}$. A functional $\mathcal{I}(f)$ defined for functions f is in $\mathcal{A}$ is called invariant under rotations in $\mathcal{A}$ if $f_\varsigma \in \mathcal{A}$ and $\mathcal{I}(f) = \mathcal{I}(f_\varsigma)$ for all $\varsigma \in \mathbb{R}$. It can be easily checked that the functionals $|a_2a_3 - a_4|$, $|H_{2,1}|$ and $|H_{3,1}|$ considered for the class $\mathcal{SL}(k,q)$ satisfy the above definitions.

Lemma 1 (see [21]). *Let*

$$p(z) = 1 + c_1 z + c_2 z^2 + \ldots$$

be in the class $\mathcal{P}$ of functions with positive real part in $\mathbb{U}$. Then

$$|c_k| \leqq 2 \qquad (k \in \mathbb{N}). \tag{10}$$

If $|c_1| = 2$, *then*

$$p(z) \cong p_1(z) \cong \frac{1+xz}{1-xz} \qquad \left(x = \frac{c_1}{2}\right).$$

Conversely, if $p(z) \cong p_1(z)$ *for some* $|x| = 1$, *then* $c_1 = 2x$ *and*

$$\left|c_2 - \frac{c_1^2}{2}\right| \leqq 2 - \frac{|c_1^2|}{2}. \tag{11}$$

Lemma 2 (see [22]). *Let* $p \in \mathcal{P}$ *with its coefficients* c_k *as in Lemma 1, then*

$$\left|c_3 - 2c_1c_2 + c_1^3\right| \leqq 2. \tag{12}$$

Lemma 3 (see [23]). *Let* $p \in \mathcal{P}$ *with its coefficients* c_k *as in Lemma 1, then*

$$|c_1c_2 - c_3| \leqq 2. \tag{13}$$

Lemma 4 (see [20]). *If the function* f *given in the form (1) belongs to class* $\mathcal{SL}^k$, *then*

$$|a_n| \leqq |\mathcal{T}_k|^{n-1} F_{k,n}, \tag{14}$$

where $\mathcal{T}_k$ *is given in (4). Equality holds true in (14) for the function* g *given by*

$$\begin{aligned} g_k(z) &= \frac{z}{1 - k\mathcal{T}_k z - \mathcal{T}_k^2 z} \\ &= \sum_{n=1}^{\infty} \mathcal{T}_k^{n-1} F_{k,n} z^n, \end{aligned}$$

which can be written as follows:

$$g_k(z) = z + \mathcal{T}_k z^2 + \left(k^2 + 1\right)(\mathcal{T}_k k + 1) z^3 + \cdots. \tag{15}$$

2. Main Results

Here, we investigate the sharp bounds for the second Hankel determinant and the third Hankel determinant. We also find sharp bounds for the Fekete-Szegö functional $|a_3 - \lambda a_2^2|$ for a real number λ. Throughout our discussion, we will assume that $q \in (0,1)$.

Theorem 1. *Let the function* $f \in \mathcal{A}$ *given in (1) belong to the class* $\mathcal{SL}(k,q)$. *Then*

$$\left|a_2a_4 - a_3^2\right| \leqq \frac{1}{q^3(q+1)^2 \mathcal{Q}} \left\{\mathcal{Q}(q+1)^2 + \left(|\mathcal{B}_q| k^2 + |\mathcal{C}_q|\right) 16k^2\right\} \mathcal{T}_k^2, \tag{16}$$

where

$$\mathcal{Q} = \left(q + q^2 + q^3\right) \tag{17}$$

$$\mathcal{B}_q = \frac{1}{64}(q+1)^4 \left\{\frac{1}{(q+1)^2}\mathcal{Q}\left(q^2 + 6q - 3\right) - \frac{1}{4}q^2(q-1)(2q-3)\right\} \tag{18}$$

$$\mathcal{C}_q = \frac{1}{16}(q+1)^2 \left[(2q-1) - \mathcal{Q}\left(3 + \frac{1}{2}q^2(q-1)\right)(q+1)^2\right] \tag{19}$$

and $\mathcal{T}_k$ *is given in (4).*

Proof. If $f \in \mathcal{SL}(k,q)$, then it follows from the definition that

$$\frac{z\left(\mathcal{D}_q f\right)(z)}{f(z)} \prec \tilde{q}(z),$$

where

$$\tilde{q}(z) = \frac{2\breve{p}_k(z)}{(1+q)+(1-q)\,\breve{p}(z)}.$$

For a given $f \in \mathcal{SL}(k,q)$, we find for the function $p(z)$, where

$$p(z) = 1 + p_1 z + p_2 z^2 + \cdots,$$

that

$$\frac{z\left(\mathcal{D}_q f\right)(z)}{f(z)} = p(z) := 1 + p_1 z + p_2 z^2 + \cdots,$$

where

$$p \prec \tilde{q}(z).$$

If

$$p(z) \prec \tilde{q}(z),$$

then there is an analytic function w such that

$$|w(z)| \leqq |z| \quad \text{in} \quad \mathbb{U}$$

and

$$p(z) = \tilde{q}(w(z)).$$

Therefore, the function $g(z)$, given by

$$g(z) = \frac{1+w(z)}{1-w(z)} = 1 + c_1 z + c_2 z + \cdots \qquad (\forall\, z \in \mathbb{U}), \tag{20}$$

is in the class $\mathcal{P}$. It follows that

$$w(z) = \left(\frac{c_1}{2}\right) z + \left(c_2 - \frac{c_1^2}{2}\right)\frac{z^2}{2} + \cdots$$

and

$$\begin{aligned}
\tilde{q}(w(z)) &= 1 + \frac{1}{4}(q+1)\,\tilde{p}_{k,1} c_1 z \\
&\quad + \left[\frac{1}{4}(q+1)\,\tilde{p}_{k,1}\left(c_2 - \frac{c_1^2}{2}\right) z^2 + \frac{c_1^2}{16}(q+1)\left[(q-1)\tilde{p}_{k,1}^2 + 2\tilde{p}_{k,2}\right]\right] z^2 \\
&\quad + \left[\frac{1}{4}(q+1)\,\tilde{p}_{k,1}\left(c_3 - c_1 c_2 + \frac{c_1^3}{4}\right) + \frac{1}{8}(q+1)\left\{(q-1)\tilde{p}_{k,1}^2 + 2\tilde{p}_{k,2}\right\} c_1\right. \\
&\quad \cdot \left(c_2 - \frac{c_1^2}{2}\right) + \frac{1}{64}(q+1)\left\{(q-1)^2\,\tilde{p}_{k,1}^3 \quad + 4\tilde{p}_{k,2}\tilde{p}_{k,1}(q-1) + 4\tilde{p}_{k,3}\right\} c_1^3] z^3 + \cdots \\
&= p(z).
\end{aligned} \tag{21}$$

From (5), we find the coefficient $\tilde{p}_{k,n}$ of the function $\tilde{q}$ given by

$$\tilde{p}_{k,n} = (F_{k,n-1} + F_{k,n+1})\mathcal{T}_k^n.$$

This shows the following relevant connection $\tilde{q}$ with the sequence of k-Fibonacci numbers:

$$\tilde{q}(w(z)) = 1 + \frac{1}{4}(q+1)\,k\mathcal{T}_k c_1 z + \left[\frac{1}{4}(q+1)\,k\mathcal{T}_k\left(c_2 - \frac{c_1^2}{2}\right) + \frac{c_1^2}{16}(q+1)\right.$$
$$\cdot\left((q-1)k^2 + 2\left(2+k^2\right)\right)\mathcal{T}_k^2\Big] z^2 + \left[\frac{1}{4}(q+1)\,k\mathcal{T}_k\left(c_3 - c_1c_2 + \frac{c_1^3}{4}\right)\right.$$
$$+\frac{1}{8}(q+1)\left\{(q-1)k^2 + 2\left(2+k^2\right)\right\}\mathcal{T}_k^2 c_1\left(c_2 - \frac{c_1^2}{2}\right) + \frac{1}{64}(q+1)$$
$$\left.\cdot\left\{(q-1)^2k^2 + 4\left(2+k^2\right)(q-1) + 4\left(k^2+3\right)\right\}k\mathcal{T}_k^3 c_1^3\right] z^3 + \cdots \tag{22}$$

If

$$p(z) = 1 + p_1 z + p_2 z^2 + \cdots,$$

then, by (21) and (22), we find that

$$p_1 = \left(\frac{q+1}{2}\right)\frac{k\mathcal{T}_k c_1}{2} \tag{23}$$

$$p_2 = \frac{1}{4}(q+1)\left(k\mathcal{T}_k\left(c_2 - \frac{c_1^2}{2}\right) + \frac{c_1^4}{4}\right)\left\{(q-1)\,k^2 + 2(2+k^2)\right\}\mathcal{T}_k^2 \tag{24}$$

$$p_3 = (q+1)\left[\frac{k\mathcal{T}_k}{2}\left(c_3 - c_1c_2 + \frac{c_1^3}{4}\right) + \left\{(q-1)\,k^2\mathcal{T}_k^2 + 2\left(2+k^2\right)\mathcal{T}_k^2\right\}\right.$$
$$\cdot\left(c_2 - \frac{c_1^2}{2}\right)\frac{c_1}{8} + \frac{c_1^3}{64}\left\{(q-1)^2k^2 + 4\left(2+k^2\right)(q-1)\right.$$
$$\left.\left.+4\left(k^2+3\right)\right\}k\mathcal{T}_k^3\right]. \tag{25}$$

Moreover, we have

$$\frac{z\left(\mathcal{D}_q f\right)(z)}{f(z)} = 1 + qa_2 z + q\left\{(1+q)\,a_3 - a_2^2\right\}z^2$$
$$+\left\{\mathcal{Q}a_4 - q\,(2+q)\,a_2a_3 + qa_2^3\right\}z^3 + \cdots$$
$$= 1 + p_1 z + p_2 z^2 + \cdots$$

and

$$a_2 = \frac{p_1}{q}$$
$$a_3 = \frac{qp_2 + p_1^2}{q^2(q+1)},$$
$$a_4 = \frac{q^2(q+1)\,p_3 - p_1^3(q+1) + (2+q)\left(p_1p_2q + p_1^3\right)}{q^3(q+1)\,\mathcal{Q}}$$

Therefore, we obtain

$$\begin{aligned}
\left|a_2a_4 - a_3^2\right| = \left|\frac{\mathcal{T}_k^2}{q^3\,(q+1)\,\mathcal{Q}}\right| \Bigg| &\left(\frac{(q+1)\,c_1}{2}\right)^2 \left\{ \frac{\mathcal{Q}c_1^2}{16}\,(q+1)^2 \right. \\
&+ \left\{\mathcal{Q} - q^2\,(q+1)^2\right\} \left(c_2 - \frac{c_1^2}{4}\right) \left.\frac{(2+k^2)}{4} \right\} \frac{k\,\mathcal{T}_k^n}{F_{k,n}} \\
&- \left(\frac{(q+1)\,c_1}{2}\right)^2 \left\{ \frac{\mathcal{Q}c_1^2}{16}\,(q+1)^2 \left\{\mathcal{Q} - q^2\,(q+1)^2\right\} \right. \\
&\cdot \left(c_2 - \frac{c_1^2}{4}\right) \left.\frac{(2+k^2)}{4} \right\} x_{k,n} + \left(\frac{(q+1)\,k}{2}\right)^2 \\
&\cdot \left\{ q^2 \left(\frac{q+1}{2}\right)^2 c_1(c_1c_2 - c_3) + \frac{\mathcal{Q}}{4}\left(c_2 - \frac{c_1^2}{2}\right)^2 \right\} \\
&+ \left\{ \frac{\mathcal{Q}}{8}\,(q+1) - \frac{1}{4}q^2k^2 \right\} \left(\frac{q+1}{2}\right) c_1^4 \\
&+ \left\{ \mathcal{E}_q k^3 \frac{c_1^2}{2}\left(c_2 - \frac{c_1^2}{2}\right) + \mathcal{B}_q k^5 c_1^4 + \mathcal{C}_q k^3 c_1^4 \right\} \frac{\mathcal{T}_k^n}{F_{k,n}} \\
&- \left\{ \mathcal{E}_q k^3 \frac{c_1^2}{2}\left(c_2 - \frac{c_1^2}{2}\right) + \mathcal{B}_q k^5 c_1^4 + \mathcal{C}_q k^3 c_1^4 \right\} x_{k,n} + \mathcal{B}_q k^4 c_1^4 + \mathcal{C}_q k^2 c_1^4 \Bigg|,
\end{aligned}$$

where

$$\mathcal{E}_q = \frac{1}{16}\,(q+1)^2\,(q-1)\left\{\mathcal{Q} - q^2\,(q+1)^2\right\}.$$

This can be written as follows:

$$\begin{aligned}
\left|a_2a_4 - a_3^2\right| = \left|\frac{\mathcal{T}_k^2}{q^2\,(q+1)^2\,\mathcal{Q}}\right| \Bigg| &\left\{ \mathcal{Q}\left(\frac{q+1}{2}\right)^4 \frac{c_1^4}{4} + \left(\frac{q+1}{2}\right)^2 \right. \\
&\cdot \left\{\mathcal{Q} - q^2\,(q+1)^2\right\} c_1^2 \left(c_2 - \frac{c_1^2}{4}\right) \left.\frac{(2+k^2)}{4} \right\} \frac{k\mathcal{T}_k^n}{F_{k,n}} \\
&+ q^2 \left(\frac{q+1}{2}\right)^4 k^2 c_1(c_1c_2 - c_3) - \frac{q^2}{64}\,(q+1)^4 k^2 c_1^4 + \frac{\mathcal{Q}}{16}\,(q+1)^2 c_1^4 \\
&- \mathcal{Q}\left(\frac{q+1}{2}\right)^4 \frac{c_1^4}{4} k x_{k,n} + \frac{\mathcal{Q}k^2}{4}\left(\frac{q+1}{2}\right)^2 c_2 \left(c_2 - \frac{c_1^2}{2}\right) \\
&+ \frac{3}{8}\,(q+1)^2 \left\{ \frac{1}{6}\left(k^2+2\right)\left\{\mathcal{Q} - q^2\,(q+1)^2\right\} k x_{k,n} - \frac{k^2}{4} \right\} \\
&\cdot c_1^2 \left(c_2 - \frac{c_1^2}{2}\right) + \left\{ \mathcal{E}_q k^3 \frac{c_1^2}{2}\left(c_2 - \frac{c_1^2}{2}\right) + \mathcal{B}_q k^5 c_1^4 + \mathcal{C}_q k^3 c_1^4 \right\} \frac{\mathcal{T}_k^n}{F_{k,n}} \\
&- \left\{ \mathcal{E}_q k^3 \frac{c_1^2}{2}\left(c_2 - \frac{c_1^2}{2}\right) + \mathcal{B}_q k^5 c_1^4 + \mathcal{C}_q k^3 c_1^4 \right\} x_{k,n} + \mathcal{B}_q k^4 c_1^4 + \mathcal{C}_q k^2 c_1^4 \Bigg|.
\end{aligned} \tag{26}$$

It is known that

$$\forall\, n \in \mathbb{N},\ \mathcal{T}_k = \frac{\mathcal{T}_k^n}{F_{k,n}} - x_{k,n}, \quad x_{k,n} = \frac{F_{k,n-1}}{F_{k,n}}, \quad \lim_{n\to\infty} \frac{F_{k,n-1}}{F_{k,n}} = |\mathcal{T}_k|\,. \tag{27}$$

Applying (27) together with (11)–(13), we get

$$
\begin{aligned}
\left|a_2a_4-a_3^2\right| \leqq & \left|\frac{\mathcal{T}_k^2}{q^2\left(q+1\right)^2\mathcal{Q}}\right|\left|\left\{\mathcal{Q}\left(\frac{q+1}{2}\right)^4\frac{c_1^4}{4}+\left(\frac{q+1}{2}\right)^2\right.\right.\\
&\left.\cdot\{\mathcal{Q}-q^2\left(q+1\right)^2\}c_1^2\left(c_2-\frac{c_1^2}{4}\right)\frac{(2+k^2)}{4}\right\}\left|\frac{k\mathcal{T}_k^n}{F_{k,n}}\right.\\
&+\left|q^2\left(\frac{q+1}{2}\right)^4k^2\right||c_1||c_1c_2-c_3|-\left|\frac{1}{4}q^2\left(\frac{q+1}{2}\right)k^2\right|\left|c_1^4\right|\\
&+\frac{\mathcal{Q}}{4}\left(\frac{q+1}{2}\right)^2\left|c_1^4\right|-\mathcal{Q}\left(\frac{q+1}{2}\right)^4\frac{|c_1^4|}{4}kx_{k,n}\\
&+\left|\frac{\mathcal{Q}k^2}{4}\left(\frac{q+1}{2}\right)^2\right||c_2|\left|c_2-\frac{c_1^2}{2}\right|+\left|\frac{3}{8}\left\{2\left(\frac{q+1}{2}\right)^2\right.\right.\\
&\left.\left.\cdot\{\mathcal{Q}-q^2\left(q+1\right)^2\}kx_{k,n}-\left(\frac{q+1}{2}\right)^2k^2\right\}\right||c_1|^2\\
&\cdot\left|c_2-\frac{c_1^2}{2}\right|+\left\{\mathcal{E}_qk^3\frac{c_1^2}{2}\left(c_2-\frac{c_1^2}{2}\right)+\mathcal{B}_qk^5c_1^4+\mathcal{C}_qk^3c_1^4\right\}\frac{\mathcal{T}_k^n}{F_{k,n}}\\
&-\mathcal{E}_qk^3\frac{|c_1^2|}{2}\left|2-\frac{c_1^2}{2}\right|x_{k,n}-\mathcal{B}_qk^5|c_1|^4x_{k,n}-\mathcal{C}_qk^3|c_1|^4x_{k,n}+\mathcal{B}_qk^4|c_1^4|+\mathcal{C}_qk^2|c_1^4|.
\end{aligned}
$$

From (27), we obtain

$$
\left(\frac{q+1}{8}\right)\left\{\mathcal{Q}\left(\frac{q+1}{2}\right)-\mathcal{Q}\left(\frac{q+1}{2}\right)^3kx_{k,n}-q^2k^2\right\}|c_1^4|>0
$$

and

$$
\left(\frac{q+1}{2}\right)^2\left\{\frac{2}{3}\left(k^2+2\right)\left\{\mathcal{Q}-q^2\left(q+1\right)^2\right\}kx_{k,n}-k^2\right\}>0,
$$

which, for sufficiently large n, yields

$$
|c_1|=:y\in[0,2]\,.
$$

After some computations, we can find that

$$
\begin{aligned}
&\max_{y\in[0,2]}\left\{\frac{q^2}{8}\left(q+1\right)^4k^2y+\left|-\frac{q^2}{8}\left(q+1\right)k^2+\mathcal{Q}\frac{(q+1)^2}{16}\right.\right.\\
&\left.-\frac{1}{64}\mathcal{Q}\left(q+1\right)^4kx_{k,n}\right|y^4+\left|\mathcal{Q}k^2\frac{(q+1)^2}{8}\right|\left(2-\frac{y^2}{2}\right)\\
&+\left|\frac{3}{8}\left\{\left(\frac{q+1}{2}\right)^2\left(\frac{2}{3}(k^2+2)\{\mathcal{Q}-q^2\left(q+1\right)^2\}kx_{k,n}-k^2\right)\right\}y^2\left(2-\frac{y^2}{2}\right)\right.\\
&\left.-|\mathcal{E}_q|k^3\frac{y^2}{2}\left(2-\frac{y^2}{2}\right)x_{k,n}-|\mathcal{B}_q|k^5y^4x_{k,n}-|\mathcal{C}_q|k^3y^4x_{k,n}+|\mathcal{B}_q|k^4y^4+|\mathcal{C}_q|k^2y^4\right|\\
&=4\mathcal{Q}\left(\frac{q+1}{2}\right)^2\{1-kx_{k,n}\}+\left(16|\mathcal{B}_q|k^4+16|\mathcal{C}_q|k^2\right)\{1-kx_{k,n}\}.
\end{aligned}
$$

As a result of the following limit formula:

$$\lim_{n\to\infty}\left[\left|\left(\frac{q+1}{2}\right)^2\frac{c_1^2}{4}\left\{\left(\frac{q+1}{2}\right)^2\mathcal{Q}c_1^2+\left\{\mathcal{Q}-q^2(q+1)^2\right\}\left(c_2-\frac{c_1^2}{4}\right)(2+k^2)\right\}\right|\right.$$
$$\left.+\left|\mathcal{E}_qk^3\frac{c_1^2}{2}\left(c_2-\frac{c_1^2}{2}\right)+\mathcal{B}_qk^5c_1^4+\mathcal{C}_qk^3c_1^4\right|\right]\frac{|\mathcal{T}_k^n|}{F_{k,n}}=0,$$

and by using (27), we get

$$\lim_{n\to\infty}\left[\max_{y\in[0,2]}\left\{q^2\left(\frac{q+1}{2}\right)^4 2k^2y+\left|-\frac{1}{8}q^2(q+1)k^2+\frac{\mathcal{Q}}{16}(q+1)^2\right.\right.\right.$$
$$\left.-\frac{1}{64}\mathcal{Q}(q+1)^4kx_{k,n}\right|y^4+\mathcal{Q}k^2\frac{(q+1)^2}{8}\left(2-\frac{y^2}{2}\right)^2+\frac{3}{32}(q+1)^2$$
$$\cdot\left\{\frac{2}{3}(k^2+2)\{\mathcal{Q}-q^2(q+1)^2\}kx_{k,n}-k^2\right\}y^2\left(2-\frac{y^2}{2}\right)-|A|k^3\frac{y^2}{2}$$
$$\left.\cdot\left(2-\frac{y^2}{2}\right)x_{k,n}-|\mathcal{B}_q|k^5y^4x_{k,n}-|\mathcal{C}_q|k^3y^4x_{k,n}+|\mathcal{B}_q|k^4y^4+|\mathcal{C}_q|k^2y^4\right]$$
$$=\mathcal{Q}(q+1)^2\mathcal{T}_k^2+(|\mathcal{B}_q|k^2+|\mathcal{C}_q|)16k^2\mathcal{T}_k^2$$

We thus find that

$$\left|a_2a_4-a_3^2\right|\leqq\frac{\mathcal{T}_k^2}{q^2(q+1)^2\mathcal{Q}}\left\{\mathcal{Q}(q+1)^2+(|\mathcal{B}_q|k^2+|\mathcal{C}_q|)16k^2\right\}\mathcal{T}_k^2.$$

If, in (20), we set

$$g(z)=\frac{1+z}{1-z}=1+2z+2z^2+\cdots,$$

then, by putting $c_1=c_2=c_3=2$ in (26), we obtain

$$\left|a_2a_4-a_3^2\right|=\frac{\mathcal{T}_k^2}{q^2(q+1)^2\mathcal{Q}}\left\{\mathcal{Q}(q+1)^2+(|\mathcal{B}_q|k^2+|\mathcal{C}_q|)16k^2\right\}\mathcal{T}_k^2$$

This completes the proof of Theorem 1. □

Remark 3. In the next result, for simplicity, we take the values of $\mathcal{S}_q$, $\mathcal{L}_q$ and $\mathcal{M}_q$ as given by

$$\mathcal{S}_q=q^3(1+q)\mathcal{Q},$$

$$\mathcal{L}_q=\{q(1+q)-q(2+q)+\mathcal{Q}\}\left(\frac{q+1}{2}\right)^3\left(\frac{k^3c_1^3}{8}\right)-q^3\left(\frac{q+1}{2}\right)^2k^3c_1^3$$
$$+\frac{c_1^3k}{16}\left[\{q\mathcal{Q}-q^2(2+q)\}\left(\frac{q+1}{2}\right)^2\left\{(q-1)k^2+2\left(2+k^2\right)\right\}\right]$$
$$-\frac{c_1^3}{32}q^3\left(\frac{q+1}{2}\right)^2\left\{(q-1)k^3-(q-1)\left(2+k^2\right)k\right\}$$

and

$$\mathcal{M}_q=\left[\{q\mathcal{Q}-q^2(2+q)\}\left(\frac{q+1}{2}\right)^2-q^3\left(\frac{q+1}{2}\right)^3(q-1)k^2\right]\frac{c_1}{2}\left(c_2-\frac{c_1^2}{2}\right).$$

Theorem 2. *Let the function $f \in \mathcal{A}$ given in (1) belong to the class $\mathcal{SL}(k,q)$. Then*

$$|a_2a_3 - a_4| = \frac{2}{\mathcal{S}_q}\left(\frac{3}{4}kq^3(q+1)^2|\mathcal{T}_k^3| + \frac{1}{2}\left|\left\{\mathcal{M}_q + \left(1+k^2\right)\mathcal{L}_q\right\}\right| kx_{k,n} + \frac{1}{2}k|\mathcal{L}_q|\right). \tag{28}$$

Proof. Let $f \in \mathcal{SL}(k,q)$ and let $p \in \mathcal{P}$ be given in (2). Then, from (23)–(25) and

$$\begin{aligned}\frac{z\mathcal{D}_qf(z)}{f(z)} &= 1 + qa_2z + \left\{\left(q+q^2\right)a_3 - qa_2^2\right\}z^2 \\ &\quad + \left\{\mathcal{Q}a_4 - \left(2q+q^2\right)a_2a_3 + qa_2^3\right\}z^3 + \cdots \\ &= 1 + p_1z + p_2z^2 + \cdots,\end{aligned}$$

we have

$$\begin{aligned}a_2a_3 - a_4 = \frac{1}{\mathcal{S}_q}\Big[&\left\{q\mathcal{Q} - q^2(2+q)\right\}p_1p_2 \\ &+ \{q(1+q) - q(2+q) + \mathcal{Q}\}p_1^3 - q^3(1+q)p_3\Big],\end{aligned}$$

which, together with (27), yields

$$\begin{aligned}|a_2a_3 - a_4| = \frac{2}{\mathcal{S}_q}\Bigg|&\frac{q^3}{4}(q+1)^2\left[\frac{1}{4}\left(c_2 - \frac{c_1^3}{2}\right)c_1k^2 + \frac{1}{2}(c_1c_2 - c_3)\right. \\ &\left.- \left(\frac{3k^2+4}{4}\right)c_1c_2\right]\frac{k\mathcal{T}_k}{f_{k,n}} + \frac{1}{8}q^3(q+1)^2\left(c_3 - 2c_1c_2 + c_1^3\right)kx_{k,n} \\ &+ \frac{q^3}{8}(q+1)^2\left\{\frac{1}{2}\left(4-k^2\right)\left(c_2 - \frac{c_1^2}{2}\right) + \left(3k^2+2\right)c_2\right\}kc_1x_{k,n} \\ &+ \frac{q^3}{16}(q+1)^2\left\{\left(4-k^2\right)\left(c_2 - \frac{c_1^2}{2}\right) - 3k^2c_2\right\}c_1 - \frac{q^3}{32}(q+1)^3 \\ &\cdot(q-1)\left(c_2 - \frac{c_1^2}{2}\right)c_1k^2 + \frac{1}{2}\left\{\mathcal{M}_q + \left(1+k^2\right)\mathcal{L}_q\right\}\frac{k\mathcal{T}_k}{f_{k,n}} \\ &- \frac{1}{2}\left\{\mathcal{M}_q + \left(1+k^2\right)\mathcal{L}_q\right\}kx_{k,n} + \frac{1}{2}k\mathcal{L}_q\Bigg|.\end{aligned} \tag{29}$$

Now, applying the triangle inequality in (10)–(13), we get

$$\begin{aligned}|a_2a_3 - a_4| \leq \frac{2}{\mathcal{S}_q}\Bigg|&\frac{q^3}{16}(q+1)^2\left(c_2 - \frac{c_1^3}{2}\right)c_1k^2 + \frac{q^3}{8}(q+1)^2(c_1c_2 - c_3) \\ &- \frac{q^3}{16}(q+1)^2\left(3k^2+4\right)c_1c_2\Bigg|\frac{k\mathcal{T}_k}{f_{k,n}} + \frac{q^3}{4}(q+1)^2kx_{k,n} \\ &+ \frac{q^3}{4}(q+1)^2kx_{k,n} + \frac{q^3}{8}(q+1)^2\left|k\left(4-k^2\right)x_{k,n} - \left(4-k^2\right)\right||c_1| \\ &- \frac{q^3}{32}(q+1)^2\left|k\left(4-k^2\right)x_{k,n} - \left(4-k^2\right)\right|\left|c_1^3\right| + \frac{q^3}{4}(q+1)^2 \\ &\cdot\left|\left(3k^2+2k\right)x_{k,n} - 3k^2\right| - \frac{q^3}{32}(q+1)^3(q-1)\left(2 - \frac{|c_1|^2}{2}\right)k^2|c_1| \\ &+ \frac{1}{2}\left|\left\{\mathcal{M}_q + \left(1+k^2\right)\mathcal{L}_q\right\}\right|\frac{k\mathcal{T}_k}{f_{k,n}} - \frac{1}{2}\left|\left\{\mathcal{M}_q + \left(1+k^2\right)\mathcal{L}_q\right\}\right|kx_{k,n} + \frac{1}{2}k|\mathcal{L}_q|.\end{aligned}$$

In addition, by using (27), we have

$$\frac{q^3}{4}(q+1)^2\left(4-k^2\right)kx_{k,n}-\left(4-k^2\right)<0 \qquad (0<k<2)$$

and

$$\frac{q^3}{4}(q+1)^2(3k+2)\,kx_{k,n}-3k^2>0$$

for $0<k\leqq 1$ and sufficiently large n. Therefore, we have got a function of the variable $|c_1| =: y\in[0,2]$ and, after some computations, we can find that

$$\begin{aligned}&\max_{y\in[0,2]}\left\{\frac{q^3}{4}(q+1)^2\left\{kx_{k,n}+\frac{1}{2}\left(k\left(4-k^2\right)x_{k,n}-\left(4-k^2\right)\right)y\right\}-\frac{q^3}{32}(q+1)^2\right.\\&\quad\left.\cdot\left(k\left(4-k^2\right)x_{k,n}-\left(4-k^2\right)\right)y^3+\left|\frac{1}{2}\left\{\mathcal{M}_q+\left(1+k^2\right)\mathcal{L}_q\right\}\right|kx_{k,n}+\left|\frac{1}{2}k\mathcal{L}_q\right|\right\}\\&=\frac{q^3}{4}(q+1)^2\left(3k^2+3k\right)x_{k,n}-3k^2+\left|\frac{1}{2}\left\{\mathcal{M}_q+\left(1+k^2\right)\mathcal{L}_q\right\}\right|kx_{k,n}+\left|\frac{1}{2}k\mathcal{L}_q\right|.\end{aligned}$$

As a result of the following limit relation:

$$\begin{aligned}&\lim_{n\to\infty}\left[\frac{q^3}{16}(q+1)^2\left(c_2-\frac{c_1^3}{2}\right)c_1k^2+\frac{q^3}{8}(q+1)^2(c_1c_2-c_3)-\frac{q^3}{16}(q+1)^2\right.\\&\quad\left.\cdot\left(3k^2+4\right)c_1c_2+\frac{1}{2}\left\{\mathcal{M}_q+\left(1+k^2\right)\mathcal{L}_q\right\}\right]\frac{k\mathcal{T}_k}{f_{k,n}}=0\end{aligned}$$

and, by means of (27), we have

$$\begin{aligned}&\lim_{n\to\infty}\left[\max_{y\in[0,2]}\left\{\frac{q^3}{4}(q+1)^2\left\{kx_{k,n}+\frac{1}{2}\left(k\left(4-k^2\right)x_{k,n}-\left(4-k^2\right)\right)y\right\}-\frac{q^3}{32}(q+1)^2\right.\right.\\&\quad\left.\cdot\left(k\left(4-k^2\right)x_{k,n}-\left(4-k^2\right)\right)y^3+\frac{q^3}{4}(q+1)^2\left(3k^2+2k\right)x_{k,n}-3k^2\right\}\\&\quad\left.+\frac{1}{2}\left\{\mathcal{M}_q+\left(1+k^2\right)\mathcal{L}_q\right\}kx_{k,n}+\frac{1}{2}k\mathcal{L}_q\right]\\&=q^3\left(\frac{q+1}{2}\right)^2\left\{\left(3k^2+3k\right)\mathcal{T}_k-3k^2\right\}+\frac{1}{2}\left|\left\{\mathcal{M}_q+\left(1+k^2\right)\mathcal{L}_q\right\}\right|kx_{k,n}+\frac{1}{2}k\left|\mathcal{L}_q\right|\\&=q^3\left(\frac{q+1}{2}\right)^2\left\{-3k\left(\left(k^2+1\right)\mathcal{T}_k+k\right)\right\}+\frac{1}{2}\left|\left\{\mathcal{M}_q+\left(1+k^2\right)\mathcal{L}_q\right\}\right|kx_{k,n}+\frac{1}{2}k\left|\mathcal{L}_q\right|\\&=q^3\left(\frac{q+1}{2}\right)^2\left(-3k\mathcal{T}_k^3\right)+\frac{1}{2}\left|\left\{\mathcal{M}_q+\left(1+k^2\right)\mathcal{L}_q\right\}\right|kx_{k,n}+\frac{1}{2}k\left|\mathcal{L}_q\right|\Big)\\&=3q^3\left(\frac{q+1}{2}\right)^2k\left|\mathcal{T}_k^3\right|+\frac{1}{2}\left|\left\{\mathcal{M}_q+\left(1+k^2\right)\mathcal{L}_q\right\}\right|kx_{k,n}+\frac{1}{2}k\left|\mathcal{L}_q\right|\end{aligned}$$

If, in the formula (20) , we set

$$g(z)=\frac{1+z}{1-z}=1+2z+2z^2+\cdots,$$

then, by putting $c_1=c_2=c_3=2$ in (26), we obtain

$$|a_2a_3-a_4|=\frac{k}{2}\left\{\frac{3q^3}{\mathcal{S}_q}(q+1)^2\left|\mathcal{T}_k^3\right|+1\left|\left\{\mathcal{M}_q+\left(1+k^2\right)\mathcal{L}_q\right\}\right|x_{k,n}+\left|\mathcal{L}_q\right|\right\}$$

This completes the proof of Theorem 2. □

Theorem 3. *Let the function $f \in \mathcal{A}$ given in (1) belong to the class $\mathcal{SL}(k,q)$. Then*

$$\left|a_3 - \lambda a_2^2\right| \leq \frac{\mathcal{T}_k^2}{q^2(q+1)}\left[\left\{\left|\mathcal{G}\left(\frac{1+q}{2}\right)^2\right| + \left|\frac{q}{4}\left(q^2-1\right) + q\left(\frac{q+1}{2}\right)\right|\right\}k^2 + q\,(q+1)\right] \tag{30}$$

Proof. Let $f \in \mathcal{SL}(k,q)$ and let $p \in \mathcal{P}$ given in (2). Then, from (23)–(25) and

$$\begin{aligned}\frac{z\mathcal{D}_q f(z)}{f(z)} &= 1 + qa_2 z + \left\{\left(q+q^2\right)a_3 - qa_2^2\right\}z^2 \\ &\quad + \left\{\mathcal{Q}a_4 - \left(2q+q^2\right)a_2a_3 + qa_2^3\right\}z^3 + \cdots \\ &= 1 + p_1 z + p_2 z^2 + \cdots,\end{aligned}$$

we have

$$\left|a_3 - \lambda a_2^2\right| = \frac{1}{q^2(1+q)}\left|\left[1 + \left|\lambda^2\right|(1+q)\right]p_1^2 + qp_2\right|.$$

Therefore, we obtain

$$\begin{aligned}\left|a_3 - \lambda a_2^2\right| &= \frac{1}{q^2(1+q)}\left|(1-\lambda^2(1+q))\left[\left(\frac{1+q}{2}\right)^2\left(\frac{\kappa c_1 \mathcal{T}}{2}\right)^2\right]\right. \\ &\quad \left. + q\left\{\frac{q+1}{4}\kappa\mathcal{T}_k(c_2 - \frac{c_1^2}{2})c_1^2(\frac{q+1}{16})\right\}\left\{(q-1)\kappa^2 + 2(2+k^2)\mathcal{T}_k^2\right\}\right| \\ &= \frac{\mathcal{T}_k}{q^2(1+q)}\left|(1+\left|\lambda^2\right|(1+q)\,(\frac{1+q}{2})^2\frac{k^2c_1^2}{4}\mathcal{T}_k + q\left[\frac{q+1}{4}\kappa\left(c_2 - \frac{c_1^2}{2}\right)\right.\right. \\ &\quad \left.\left. + \frac{q(q+1)c_1^2}{16}(q-1)k^2\mathcal{T}_k + 2(2+k^2)\mathcal{T}_k^2\right]\right|.\end{aligned}$$

Thus, by applying (27), we have

$$\begin{aligned}\left|a_3 - \lambda a_2^2\right| &= \frac{\mathcal{T}_k}{q^2(1+q)}\left|\mathcal{G}\left(\frac{1+q}{2}\right)^2\frac{k^2c_1^2}{4}\mathcal{T}_k + q(\frac{q+1}{4})\frac{c_1^2}{4}\right. \\ &\quad \cdot\left[(q-1)k^2 + 2(2+k^2)\frac{\mathcal{T}_k^n}{f_{k,n}}\right] - \left(\mathcal{G}\left(\frac{1+q}{2}\right)^2\frac{k^2c_1^2}{4} + q\frac{(q+1)}{4}\frac{c_1^2}{4}\right) \\ &\quad \left.\cdot\left[(q-1)k^2 + 2(2+k^2)\right]x_{k,n} + q\left(\frac{q+1}{4}\right)k\left(c_2 - \frac{c_1^2}{2}\right)\right|,\end{aligned}$$

where

$$\mathcal{G} = 1 + \left|\lambda^2\right|(1+q).$$

Now, by applying the triangle inequality in (10)–(13), we have

$$
\begin{aligned}
&\left|\frac{\mathcal{T}_k}{q^2(1+q)}\right|\left|\mathcal{G}\left(\frac{1+q}{2}\right)^2\frac{k^2c_1^2}{4}+q\left(\frac{q+1}{4}\right)\frac{c_1^2}{4}\left[(q-1)k^2+2(2+k^2)\right]\right| \\
&\cdot\frac{\mathcal{T}_k^n}{f_{k,n}}-\left[\left|\mathcal{G}\left(\frac{1+q}{2}\right)^2\frac{k^2}{4}\right||c_1|^2_{x_{k,n}}+q\left(\frac{q^2-1}{4}\right)k^2|c_1|^2_{x_{k,n}}+q\left(\frac{q+1}{4}\right)\right. \\
&\left.\cdot\left|\frac{c_1^2}{2}(2+k^2)x_{k,n}\right|+q\left(\frac{q+1}{4}\right)k\left|c_2-\frac{c_1^2}{2}\right|\right] \\
&=\left|\frac{\mathcal{T}_k}{q^2(1+q)}\right|\left|\frac{\mathcal{G}}{16}(1+q)^2k^2c_1^2+\frac{q}{16}(q+1)c_1^2\left[(q-1)k^2+2(2+k^2)\right]\right| \\
&\cdot\frac{\mathcal{T}_k^n}{f_{k,n}}-\left[\left|\mathcal{G}\left(\frac{1+q}{2}\right)^2\frac{k^2}{4}\right||c_1|^2_{x_{k,n}}+q\left(\frac{q+1}{4}\right)(q-1)k^2|c_1|^2_{x_{k,n}}\right. \\
&\left.+\frac{q(q+1)}{4}\left|\frac{c_1^2}{2}(2+k^2)x_{k,n}\right|+\left|q\left(\frac{q+1}{4}\right)k\right|\left|c_2-\frac{c_1^2}{2}\right|\right],
\end{aligned}
$$

which, after some computations, yields

$$
\begin{aligned}
&\max_{y\epsilon[0,2]}\left|\mathcal{G}\left(\frac{1+q}{2}\right)^2\frac{k^2}{4}\right|y^2x_{k,n}+\left|q\left(\frac{q+1}{4}\right)(q-1)k^2\right|x_{k,n}+\left|q\left(\frac{q+1}{4}\right)(2+k^2)\right|2x_{k,n} \\
&=\left[\left|\mathcal{G}\left(\frac{1+q}{2}\right)^2\right|+\left|q\left(\frac{q^2-1}{4}\right)\right|+\left|q\left(\frac{q+1}{2}\right)\right|\right]k^2x_{k,n}+q(q+1)x_{k,n},
\end{aligned}
$$

in which we have set $y=2$. As a result of the following limit formula:

$$
\lim_{n\to\infty}\left|\mathcal{G}\left(\frac{1+q}{2}\right)^2\frac{k^2c_1^2}{4}+q\left(\frac{q+1}{2}\right)\frac{c_1^2}{4}\left\{(q-1)k^2+2(2+k^2)\right\}\right|\frac{\mathcal{T}_k^n}{f_{x_{k,n}}}=0,
$$

which, by applying (27), yields

$$
\begin{aligned}
\left|a_3-\lambda a_2^2\right| \leqq \frac{\mathcal{T}_k^2}{q^2(q+1)}&\left[\left\{\frac{1}{4}\left|\mathcal{G}(1+q)^2\right|+\left|\frac{q}{4}\left(q^2-1\right)+\frac{q}{2}(q+1)\right|\right\}k^2\right. \\
&\left.+q(q+1)\right].
\end{aligned}
$$

This completes the proof of Theorem 3. □

Theorem 4. *Let the function $f\in\mathcal{A}$ given in (1) belong to the class $\mathcal{SL}(k,q)$. Then*

$$
\begin{aligned}
|H_3(1)| \leqq &\frac{\mathcal{T}_k^6}{q^4(1+q)^3\mathcal{Q}}\left[(2q+(q+1)k^2]\{16\left|\mathcal{B}_q\right|k^4+16\left|\mathcal{C}_q\right|k^2\right\} \\
&+\frac{\mathcal{T}_k^3 2(2+q)k^3+(5q+7)k}{2q^2(1+q)\mathcal{Q}}\left|\{\mathcal{M}_q+(1+k^2)\mathcal{L}_q\}kx_{k,n}+k\mathcal{L}_q\right|.
\end{aligned}
$$

Proof. Let $f\in\mathcal{SL}(k,q)$. Then as we know that

$$
H_3(1)=\begin{vmatrix} a_1 & a_2 & a_3 \\ a_2 & a_3 & a_4 \\ a_3 & a_4 & a_5 \end{vmatrix}=a_3(a_2a_4-a_3^2)-a_4(a_4-a_2a_3)+a_5(a_3-a_2^2),
$$

where $a_1=1$ so, we have

$$
|H_3(1)| \leqq |a_3|\left|a_2a_4-a_3^2\right|+|a_4|\left|a_4-a_2a_3\right|+|a_5|\left|a_3-a_2^2\right| \tag{31}
$$

Thus, by using Lemma 4, Theorems 1–3, as well as the formula (31), we find that

$$|H_3(1)| \leqq 2(\digamma_q k^6 + 4\Psi_q k^4 + 4\Upsilon_q k^2 + \Gamma_q)\mathcal{T}_k^6, \tag{32}$$

where

$$\digamma_q = \frac{\varkappa_{q,\lambda}}{2q^2\mathcal{Q}}\left\{1 + \frac{\chi_q(2+q)}{\mathcal{Q}q^2} - \frac{(q+3)(q+1)^2}{4q^2} + \left(\frac{q+1}{q^3}\right) + \frac{(q+1)^4}{16q^3}\right\}$$

$$\Psi_q = \frac{3(2+q)}{q^3(1+q)^2\mathcal{Q}} + \frac{\varkappa_{q,\lambda}}{2q^2(\mathcal{Q}+1)}\left\{4 + \frac{(5q+7)\chi_q}{2q^2\mathcal{Q}} - \frac{(q+3)(q+1)}{2q} + \frac{4}{q^2}\right\}$$
$$+ \frac{(q+1)}{2q(\mathcal{Q}+1)}\left\{1 + \frac{\chi_q(2+q)}{q^2\mathcal{Q}} - \frac{(q+3)(q+1)^2}{4q^2} + \frac{q+1}{q^3} + q\left(\frac{q+1}{2q}\right)^4\right\}$$

$$\Upsilon_q = \frac{1}{q^2} + \frac{1}{(\mathcal{Q}+1)}\left(4 + \frac{\chi_q(5q+7)}{2q^2\mathcal{Q}}\right) - \frac{q+1}{2q}\left(\frac{(q+1)(q+3)}{2q} + \frac{4}{q^2}\right)$$
$$+ \frac{3(5q+7)}{2q^2\mathcal{Q}} + \frac{\varkappa_{q,\lambda}}{2q^2(\mathcal{Q}+1)}\left(2 + \frac{4}{q(1+q}\right)$$

$$\Gamma_q = \frac{1}{q^4(1+q)^3\mathcal{Q}}\left(2q + (q+1)k^2\right)\left(16\left|\mathcal{B}_q\right|k^4 + 16\left|\mathcal{C}_q\right|k^2\right)$$
$$+ \frac{2(2+q)k^3 + (5q+7)k}{2q^2(1+q)\mathcal{Q}}\left|\{\mathcal{M}_q + (1+k^2)L_q\}kx_{k,n} + kL_q\right|,$$

and

$$\chi_q = q^2 + q + 2.$$

This completes the proof of Theorem 4. □

3. Conclusions

A new subclass of analytic functions associated with k-Fibonacci numbers has been introduced by means of quantum (or q-) calculus. Upper bound of the third Hankel determinant has been derived for this functions class. We have stated and proved our main results as Theorems 1–4 in this article.

Further developments based upon the the q-calculus can be motivated by several recent works which are reported in (for example) [24,25], which dealt essentially with the second and the third Hankel determinants, as well as [26–29], which studied many different aspects of the Fekete-Szegö problem.

Author Contributions: Conceptualization, H.M.S. and Q.Z.A.; methodology, Q.Z.A.; formal analysis, H.M.S. and M.D.; Investigation, M.S.; resources, X.X.; data curation, S.K.; writing—review and editing, N.K.; visualization, N.K.; funding acquisition, M.D. All authors have read and agreed to the published version of the manuscript.

Funding: This research was funded by UKM Grant: FRGS/1/2019/STG06/UKM/01/1.

Acknowledgments: The work here is supported by UKM Grant: FRGS/1/2019/STG06/UKM/01/1.

Conflicts of Interest: The authors declare that they have no conflicts of interest.

References

1. Srivastava, H.M. Operators of basic (or q-) calculus and fractional q-calculus and their applications in Geometric Function Theory of Complex Analysis. *Iran. J. Sci. Technol. Trans. A Sci.* **2020**, *44*, 327–344. [CrossRef]
2. Mahmood, S.; Raza, N.; Abujarad, E.S.A.; Srivastava, G.; Srivastava, H.M.; Malik, S.N. Geometric properties of certain classes of analytic functions associated with a q-integral operator. *Symmetry* **2019**, *11*, 719. [CrossRef]
3. Ismail, M.E.-H.; Merkes, E.; Styer, D. A generalization of starlike functions. *Complex Variables Theory Appl.* **1990**, *14*, 77–84. [CrossRef]
4. Srivastava, H.M. Univalent functions, fractional calculus, and associated generalized hypergeometric functions. In *Univalent Functions, Fractional Calculus and Their Applications*; Srivastava, H.M., Owa, S., Eds.; Ellis Horwood Limited: Chichester, UK; John Wiley and Sons: New York, NY, USA; Chichester, UK; Brisbane, Australia; Toronto, QC, Canada, 1989; pp. 329–354.
5. Raghavendar, K.; Swaminathan, A. Close-to-convexity properties of basis hypergeometric functions using their Taylor coefficients. *J. Math. Appl.* **2012**, *35*, 53–67.
6. Janteng, A.; Abdulhalirn, S.; Darus, M. Coefficient inequality for a function whose derivative has positive real part. *J. Inequal. Pure Appl. Math.* **2006**, *50*, 1–5.
7. Khan, N.; Shafiq, M.; Darus, M.; Khan, B.; Ahmad, Q.Z. Upper bound of the third Hankel determinant for a subclass of q-starlike functions associated with Lemniscate of Bernoulli. *J. Math. Inequal.* **2020**, *14*, 51–63. [CrossRef]
8. Mahmood, S.; Ahmad, Q.Z.; Srivastava, H.M.; Khan, N.; Khan, B.; Tahir, M. A certain subclass of meromorphically q-starlike functions associated with the Janowski functions. *J. Inequal. Appl.* **2019**, *2019*. [CrossRef]
9. Srivastava, H.M.; Khan, B.; Khan, N.; Ahmad, Q.Z. Coefficient inequalities for q-starlike functions associated with the Janowski functions. *Hokkaido Math. J.* **2019**, *48*, 407–425. [CrossRef]
10. Jackson, F.H. q-Difference equations. *Am. J. Math.* **1910**, *32*, 305–314. [CrossRef]
11. Srivastava, H.M.; Tahir, M.; Khan, B.; Ahmad, Q.Z.; Khan, N. Some general families of q-starlike functions associated with the Janowski functions. *Filomat* **2019**, *33*, 2613–2626. [CrossRef]
12. Uçar, H.E.Ö. Coefficient inequality for q-starlike functions. *Appl. Math. Comput.* **2016**, *276*, 122–126.
13. Ahmad, Q.Z.; Khan, N.; Raza, M.; Tahir, M.; Khan, B. Certain q-difference operators and their applications to the subclass of meromorphic q-starlike functions. *Filomat* **2019**, *33*, 3385–3397. [CrossRef]
14. Jackson, F.H. On q-definite integrals. *Quart. J. Pure Appl. Math.* **1910**, *41*, 193–203.
15. Özgür, N.Y.; Sokół, J. On starlike functions connected with k-Fibonacci numbers. *Bull. Malays. Math. Sci. Soc.* **2015**, *38*, 249–258. [CrossRef]
16. Noonan, J.W.; Thomas, D.K. On the second Hankel determinant of areally mean p-valent functions. *Trans. Am. Math. Soc.* **1976**, *223*, 337–346.
17. Mishra, A.K.; Gochhayat, P. Second Hankel determinant for a class of analytic functions defined by fractional derivative. *Internat. J. Math. Math. Sci.* **2008**, *2008*, 153280. [CrossRef]
18. Singh, G.; Singh, G. On the second Hankel determinant for a new subclass of analytic functions. *J. Math. Sci. Appl.* **2014**, *2*, 1–3.
19. Babalola, K.O. On $H_3(1)$ Hankel determinant for some classes of univalent functions. *Inequal. Theory Appl.* **2007**, *6*, 1–7.
20. Güney, H.O.; Ilhan, S.; Sokół, J. An upper bound for third Hankel determinant of starlike functions connected with k-Fibonacci numbers. *Bol. Soc. Mat. Mex.* **2019**, *25*, 117–129. [CrossRef]
21. Pommerenke, C. *Univalent Functions*; Vanderhoeck and Ruprecht: Göttingen, Germany, 1975.
22. Libera, R.J.; Złotkiewicz, E.J. Coefficient bounds for the inverse of a function with derivative in $\mathcal{P}$. *Proc. Am. Math. Soc.* **1983**, *87*, 251–257. [CrossRef]
23. Ravichandran, V.; Verma, S. Bound for the fifth coefficient of certain starlike functions. *C. R. Math. Acad. Sci. Paris* **2015**, *353*, 505–510. [CrossRef]
24. Güney, H.Ö.; Murugusundaramoorthy, G.; Srivastava, H.M. The second Hankel determinant for a certain class of bi-close-to-convex functions. *Results Math.* **2019**, *74*, 93. [CrossRef]

25. Shi, L.; Srivastava, H.M.; Arif, M.; Hussain, S.; Khan, H. An investigation of the third Hankel determinant problem for certain subfamilies of univalent functions involving the exponential function. *Symmetry* **2019**, *11*, 598. [CrossRef]
26. Mahmood, S.; Srivastava, H.M.; Khan, N.; Ahmad, Q.Z.; Khan, B.; Ali, I. Upper bound of the third Hankel determinant for a subclass of q-starlike functions. *Symmetry* **2019**, *11*, 347. [CrossRef]
27. Srivastava, H.M.; Ahmad, Q.Z.; Khan, N.; Khan, N.; Khan, B. Hankel and Toeplitz determinants for a subclass of q-starlike functions associated with a general conic domain. *Mathematics* **2019**, *7*, 181. [CrossRef]
28. Srivastava, H.M.; Hussain, S.; Raziq, A.; Raza, M. The Fekete-Szegö functional for a subclass of analytic functions associated with quasi-subordination. *Carpathian J. Math.* **2018**, *34*, 103–113.
29. Srivastava, H.M.; Mostafa, A.O.; Aouf, M.K.; Zayed, H.M. Basic and fractional q-calculus and associated Fekete-Szegö problem for p-valently q-starlike functions and p-valently q-convex functions of complex order. *Miskolc Math. Notes* **2019**, *20*, 489–509. [CrossRef]

symmetry

MDPI

Article

Difference of Some Positive Linear Approximation Operators for Higher-Order Derivatives

Vijay Gupta [1,†], Ana Maria Acu [2,†] and Hari Mohan Srivastava [3,4,5,*,†]

1 Department of Mathematics, Netaji Subhas University of Technology, Sector 3 Dwarka, New Delhi 110078, India; vijay@nsut.ac.in
2 Department of Mathematics and Informatics, Lucian Blaga University of Sibiu, Str. Dr. I. Ratiu, No. 5-7, R-550012 Sibiu, Romania; anamaria.acu@ulbsibiu.ro
3 Department of Mathematics and Statistics, University of Victoria, Victoria, BC V8W 3R4, Canada
4 Department of Medical Research, China Medical University Hospital, China Medical University, Taichung 40402, Taiwan
5 Department of Mathematics and Informatics, Azerbaijan University, 71 Jeyhun Hajibeyli Street, AZ1007 Baku, Azerbaijan
* Correspondence: harimsri@math.uvic.ca
† All three authors contributed equally to this work.

Received: 21 April 2020; Accepted: 25 May 2020; Published: 2 June 2020

Abstract: In the present paper, we deal with some general estimates for the difference of operators which are associated with different fundamental functions. In order to exemplify the theoretical results presented in (for example) Theorem 2, we provide the estimates of the differences between some of the most representative operators used in Approximation Theory in especially the difference between the Baskakov and the Szász–Mirakyan operators, the difference between the Baskakov and the Szász–Mirakyan–Baskakov operators, the difference of two genuine-Durrmeyer type operators, and the difference of the Durrmeyer operators and the Lupaş–Durrmeyer operators. By means of illustrative numerical examples, we show that, for particular cases, our result improves the estimates obtained by using the classical result of Shisha and Mond. We also provide the symmetry aspects of some of these approximations operators which we have studied in this paper.

Keywords: approximation operators; differences of operators; Szász–Mirakyan–Baskakov operators; Durrmeyer type operators; Bernstein polynomials; modulus of continuity

1. Introduction, Definitions and Preliminary Results

Approximation by positive linear operators is a classical and important topic of research in Approximation Theory and Computer-Aided Geometric Design (CAGD). The basis of the familiar Bernstein operators is an important tool in Computer-Aided Geometric Design. This basis is used in order to construct Bézier curves, which have applications for designing curves for the cars industry and problems involving animations. In addition, the Bézier curves are used in order to control the velocity over time. A class of symmetric Beta-type distributions involving the symmetric Bernstein-type basis function was introduced and studied in [1]. In recent years, the quantum (or the q-) calculus and its variation, the so-called post-quantum or the (p,q)-calculus, which have many applications in quantum physics, attracted the attention of many researchers. For example, some variations of positive linear operators by using the (p,q)-calculus instead of their known forms involving the traditional q-calculus were, in fact, published recently in *Symmetry* itself (see [2]). In this connection, the readers are referred also to a subsequent survey-cum-expository review article by Srivastava [3] in which the above-mentioned variation aspect of the (p,q)-calculus was exposed. Several other applications of the positive linear operators in learning theory can also be found in the literature. For more details about this topic, the reader is referred to the applications of the Bernstein operators and the iterated Boolean sums of operators (see [4]) and the applications of the Durrmeyer operators (see [5]).

The attention of many researchers in the study of the differences of positive linear operators began with the question raised by Lupaş in regard with the possibility to give an estimate for the following commutator:

$$[B_n, \overline{\mathbb{B}}_n] := B_n \circ \overline{\mathbb{B}}_n - \overline{\mathbb{B}}_n \circ B_n,$$

where B_n are the Bernstein operators and $\overline{\mathbb{B}}_n$ are the Beta operators (see, for details, [6]).

In [7], an algebraic structure of positive linear operators, which map $C[0,1]$ into itself, was considered in order to give an inequality for the commutators of certain positive linear operators. In several sequels to this study, Gonska et al. (see, for example, [8–10]) considered an algebraic structure $(S, +, \circ, 0, I)$ which satisfies each of the following conditions:

(i) It is closed under both "$+$" and "$\circ$";
(ii) Both "$+$" and "$\circ$" are associative;
(iii) 0 is the identity for $+$ and I is the identity for "$\circ$";
(iv) 0 is an annihilator for "$\circ$", that is, $A \circ 0 = 0 \circ A = 0$;
(v) "$+$" is commutative;
(vi) "$\circ$" distributes over "$+$", that is, both of the distributive laws hold true.

The set

$$PLO = \{L : C[0,1] \to C[0,1] \quad \text{and} \quad L \text{ is linear and positive}\},$$

which is equipped with the canonical operations of addition and operator composition, is an algebraic structure defined above. The commutator given by

$$[A, B] := AB - BA \qquad (A, B \in PLO)$$

was studied from a quantitative point of view in [7].

A solution of the Lupaş problem was given by Gonska et al. [7] by using the Taylor expansion. The estimates for the differences of two positive linear operators, which have the same moments up to a certain order, were derived in [8–10]. In [11], the differences of certain positive linear operators, which have the same fundamental functions, were studied. These studies of the positive linear operators, which are defined on unbounded interval, become an interesting area of research in Approximation Theory (see [12–15]). Estimates for the differences of these operators in terms of weighted modulus of smoothness were obtained by Aral et al. [16]. The Bernstein polynomials are, by all means, the most investigated polynomials in Approximation Theory and were introduced by Bernstein in order to prove the Weierstrass Theorem. Various new generalizations of these operators were considered in, for example, [17,18]. In [19], estimates of the differences of the Bernstein operators and their derivatives were obtained. Recently, some interesting results on this topic were published in [20–25]. In the present paper, our approach involves positive linear operators which have substantially different fundamental functions. In fact, the results presented in this paper extend the earlier studies in [11] for more general classes of positive linear operators.

We denote by $E(I)$ the space of real-valued continuous functions defined on an interval $I \subseteq \mathbb{R}$, which contains the polynomials. Let

$$||f|| = \sup\{|f(x)| : x \in I\}$$

and

$$E_B(I) := \{f \in E(I) \quad \text{and} \quad ||f|| < \infty\}.$$

Let $e_j(t) := t^j$ $(j = 0, 1, 2, \cdots$. We consider the linear positive functional $F : E(I) \to R$ preserving constant function, namely, $F(e_0) = 1$. We also put

$$\mu_r^F = F((e_1 - \phi^F e_0)^r) := \sum_{i=0}^{r} \binom{r}{i} (-1)^i \, F(e_{r-i})[\phi^F]^i \quad (r \in \mathbb{N}),$$

where $\phi^F := F(e_1)$. For the functional F, the following basic result was obtained in [11].

Lemma 1 (see [11]). *Let $f \in E(I)$ with $f^{(4)} \in E_B(I)$. Then*

$$\left| F(f) - f(\phi^F) - \frac{\mu_2^F}{2!} f^{(2)}(\phi^F) - \frac{\mu_3^F}{3!} f^{(3)}(\phi^F) \right| \leq \frac{\mu_4^F}{4!} \|f^{(4)}\|.$$

Let us now consider the fundamental functions $p_{m,k}, b_{m,k} \geq 0$, $k \in K$, and $p_{m,k}, b_{m,k} \in C(I)$ such that

$$\sum_{k\in K} p_{m,k}(x) = \sum_{k\in K} b_{m,k}(x) = e_0,$$

where K is a set of non-negative integers, that is,

$$K = \mathbb{N}_0 := \mathbb{N} \cup \{0\}.$$

Suppose also that $F_{m,k}, G_{m,k} : E(I) \to R$ are the linear positive functionals such that

$$F_{m,k}(e_0) = G_{m,k}(e_0) = 1$$

and denote

$$D(I) := \left\{ f \in E(I) \,\middle|\, \sum_{k\in K} p_{m,k} F_{m,k}(f) \in C(I) \text{ and } \sum_{k\in K} b_{m,k} G_{m,k}(f) \in C(I) \right\}.$$

Define the positive linear operators $U_m, V_m : D(I) \to C(I)$ as follows:

$$U_m(f,x) := \sum_{k\in K} p_{m,k}(x) F_{m,k}(f) \quad \text{and} \quad V_m(f,x) := \sum_{k\in K} b_{m,k}(x) G_{m,k}(f).$$

In [11], the following result concerning the difference of the operators U_m and V_m was proved.

Theorem 1 (see [11]). *Suppose that*

$$p_{m,k} = b_{m,k} \quad \text{and} \quad \phi^{F_{m,k}} = \phi^{G_{m,k}} \quad k \in K;\ m \in \mathbb{N}.$$

Let $f \in D(I)$ with $f^{(i)} \in E_B(I)$ $(i = 2,3,4)$. Then

$$|(U_m - V_m)(f,x)| \leq \|f^{(2)}\|\gamma(x) + \|f^{(3)}\|\beta(x) + \|f^{(4)}\|\alpha(x) \quad (x \in I),$$

where

$$\gamma(x) := \sum_{k\in K} |\mu_2^{F_{m,k}} - \mu_2^{G_{m,k}}| p_{m,k}(x),$$

$$\beta(x) := \sum_{k\in K} |\mu_3^{F_{m,k}} - \mu_3^{G_{m,k}}| p_{m,k}(x)$$

and

$$\alpha(x) := \sum_{k\in K} (\mu_4^{F_{m,k}} + \mu_4^{G_{m,k}}) p_{m,k}(x).$$

In the series of papers [8–10], the results concerning the estimations of the differences of certain positive linear operators were based upon the fact that the positive linear operators have the same moments up to a certain order. In the recent paper [11], the approach involved the positive linear operators which have the same fundamental functions. The main goal of this paper is to extend the above result for the positive linear operators that have different fundamental functions. Furthermore, the condition $\phi^{F_{m,k}} = \phi^{G_{m,k}}$ of ([11], Theorem 4) is shown to be not necessary in order to obtain an estimate of the differences of the positive linear operators V_m and U_m.

Theorem 2. *Let $f \in D(I)$. If $f^{(i)} \in E_B(I)$ $(i = 2, 3, 4)$, then*

$$|(U_m - V_m)(f, x)| \leq A(x)\|f^{(4)}\| + B(x)\|f^{(3)}\| + C(x)\|f^{(2)}\| + 2\omega_1(f, \delta_1(x)) + 2\omega_1(f, \delta_2(x)) \quad (x \in I),$$

where $\omega_1(f, \cdot)$ is the usual modulus of continuity,

$$A(x) = \frac{1}{4!} \sum_{k\in K} (p_{m,k}(x)\mu_4^{F_{m,k}} + b_{m,k}(x)\mu_4^{G_{m,k}}),$$

$$B(x) = \frac{1}{3!} \left| \sum_{k\in K} p_{m,k}(x)\mu_3^{F_{m,k}} - \sum_{k\in K} b_{m,k}(x)\mu_3^{G_{m,k}} \right|,$$

$$C(x) = \frac{1}{2!} \left| \sum_{k\in K} p_{m,k}(x)\mu_2^{F_{m,k}} - \sum_{k\in K} b_{m,k}(x)\mu_2^{G_{m,k}} \right|,$$

$$\delta_1(x) = \left(\sum_{k\in K} p_{m,k}(x) \left(\phi^{F_{m,k}} - x\right)^2 \right)^{1/2}$$

and

$$\delta_2(x) = \left(\sum_{k\in K} b_{m,k}(x) \left(\phi^{G_{m,k}} - x\right)^2 \right)^{1/2}.$$

Proof. First of all, by using Lemma 1, we get

$$\begin{aligned} |(U_m - V_m)(f, x)| &\leq \left| \sum_{k\in K} p_{m,k}(x)F_{m,k}(f) - \sum_{k\in K} b_{m,k}(x)G_{m,k}(f) \right| \\ &\leq \sum_{k\in K} p_{m,k}(x) \left| F_{m,k}(f) - f(\phi^{F_{m,k}}) - \frac{\mu_2^{F_{m,k}}}{2!} f''(\phi^{F_{m,k}}) - \frac{\mu_3^{F_{m,k}}}{3!} f'''(\phi^{F_{m,k}}) \right| \\ &+ \sum_{k\in K} b_{m,k}(x) \left| G_{m,k}(f) - f(\phi^{G_{m,k}}) - \frac{\mu_2^{G_{m,k}}}{2!} f''(\phi^{G_{m,k}}) - \frac{\mu_3^{G_{m,k}}}{3!} f'''(\phi^{G_{m,k}}) \right| \\ &+ \left| \sum_{k\in K} p_{m,k}(x) \frac{\mu_2^{F_{m,k}}}{2!} - \sum_{k\in K} b_{m,k}(x) \frac{\mu_2^{G_{m,k}}}{2!} \right| \cdot \|f''\| \end{aligned}$$

$$
\begin{aligned}
&+\left|\sum_{k\in K} p_{m,k}(x)\frac{\mu_3^{F_{m,k}}}{3!}-\sum_{k\in K} b_{m,k}(x)\frac{\mu_3^{G_{m,k}}}{3!}\right|\cdot\|f'''\|\\
&+\sum_{k\in K} p_{m,k}(x)|f(\phi^{F_{m,k}})-f(x)|+\sum_{k\in K} b_{m,k}(x)|f(\phi^{G_{m,k}})-f(x)|\\
\le\ &\frac{1}{4!}\left(\sum_{k\in K} p_{m,k}(x)\mu_4^{F_{m,k}}+\sum_{k\in K} b_{m,k}(x)\mu_4^{G_{m,k}}\right)\|f^{(iv)}\|\\
&+\left|\sum_{k\in K} p_{m,k}(x)\frac{\mu_2^{F_{m,k}}}{2!}-\sum_{k\in K} b_{m,k}(x)\frac{\mu_2^{G_{m,k}}}{2!}\right|\cdot\|f''\|\\
&+\left|\sum_{k\in K} p_{m,k}(x)\frac{\mu_3^{F_{m,k}}}{3!}-\sum_{k\in K} b_{m,k}(x)\frac{\mu_3^{G_{m,k}}}{3!}\right|\cdot\|f'''\|\\
&+\sum_{k\in K} p_{m,k}(x)|f(\phi^{F_{m,k}})-f(x)|+\sum_{k\in K} b_{m,k}(x)|f(\phi^{G_{m,k}})-f(x)|\\
=\ &A(x)\|f^{(4)}\|+B(x)\|f^{(3)}\|+C(x)\|f^{(2)}\|\\
&+\left(1+\frac{\sum_{k\in K} p_{m,k}(x)\left(\phi^{F_{m,k}}-x\right)^2}{\delta_1^2(x)}\right)\omega_1(f,\delta_1(x))\\
&+\left(1+\frac{\sum_{k\in K} b_{m,k}(x)\left(\phi^{G_{m,k}}-x\right)^2}{\delta_2^2(x)}\right)\omega_1(f,\delta_2(x))\\
=\ &A(x)\|f^{(4)}\|+B(x)\|f^{(3)}\|+C(x)\|f^{(2)}\|+2\omega_1(f,\delta_1(x))+2\omega_1(f,\delta_2(x)).
\end{aligned}
$$

This completes the proof of Theorem 2. □

Remark 1. *Let*

$$\nu_1(x)=\left(U_m\left((e_1-x)^2;x\right)\right)^{\frac{1}{2}}$$

and

$$\nu_2(x)=\left(V_m\left((e_1-x)^2;x\right)\right)^{\frac{1}{2}}.$$

Then, by using the result of Shisha and Mond [26], we find that

$$
\begin{aligned}
|(U_m-V_m)(f;x)| &\le |U_m(f;x)-f(x)|+|V_m(f;x)-f(x)|\\
&\le 2\omega_1(f,\nu_1(x))+2\omega_1(f,\nu_2(x)).
\end{aligned}
$$

Since

$$F_{m,k}^2(e_1)\le F_{m,k}(e_1^2)$$

and

$$G_{m,k}^2(e_1)\le G_{m,k}(e_1^2),$$

it follows that

$$\delta_i(x)\le\nu_i(x)\qquad(i=1,2).$$

2. Applications of Theorem 2

As applications of the Theorem 2, in this section, we give estimates of the differences between some of the most used positive linear operators in Approximation Theory. The considered examples involve the Baskakov type operators, the Szász–Mirakyan type operators, and the Durrmeyer type operators. We also show for the Durrmeyer type operators that, in some particular cases, our result improves the estimates obtained by using the classical result of Shisha and Mond [26].

2.1. Difference Between the Baskakov and the Szász–Mirakyan Operators

The Szász–Mirakyan operators are defined by

$$S_m(f,x) = \sum_{k=0}^{\infty} p_{m,k}(x) F_{m,k}(f), \tag{1}$$

where

$$p_{m,k}(x) = e^{-mx}\frac{(mx)^k}{k!} \quad \text{and} \quad F_{m,k}(f) = f\left(\frac{k}{m}\right).$$

Lemma 2. *The moments of S_m satisfy the following relation:*

$$S_m(e_{n+1},x) = \frac{x}{m} S'_m(e_n,x) + x S_m(e_n,x).$$

In particular,

$$S_m(e_0,x) = 1, \quad S_m(e_1,x) = x \quad \textit{and} \quad S_m(e_2,x) = x^2 + \frac{x}{m}$$

and

$$S_m(e_3,x) = x^3 + \frac{3x^2}{m} + \frac{x}{m^2} \quad \textit{and} \quad S_m(e_4,x) = x^4 + \frac{6x^3}{m} + \frac{7x^2}{m^2} + \frac{x}{m^3}.$$

Remark 2. *We have*

$$\phi^{F_{m,k}} = F_{m,k}(e_1) = \frac{k}{m}$$

and, for $r \in \mathbb{N}$, we get

$$\mu_r^{F_{m,k}} := F_{m,k}(e_1 - \phi^{F_{m,k}} e_0)^r = 0.$$

The Baskakov operators are defined by

$$V_m(f;x) = \sum_{k=0}^{\infty} v_{m,k}(x) G_{m,k}(f), \tag{2}$$

where

$$v_{m,k}(x) = \binom{m+k-1}{k} \frac{x^k}{(1+x)^{m+k}} \quad \text{and} \quad G_{m,k}(f) = f\left(\frac{k}{m}\right).$$

Lemma 3. *The moments satisfy the following relation:*

$$V_m(e_{n+1},x) = \frac{x(1+x)}{m} V'_m(e_n,x) + x V_m(e_n,x).$$

The moments of the Baskakov operators up to order 4 are listed below:

$$\begin{aligned}
V_m(e_0,x) &= 1\\
V_m(e_1,x) &= x\\
V_m(e_2,x) &= \frac{x^2(m+1)+x}{m}\\
V_m(e_3,x) &= \frac{x^3(m+1)(m+2)+3x^2(m+1)+x}{m^2}\\
V_m(e_4,x) &= \frac{x^4(m+1)(m+2)(m+3)+6x^3(m+1)(m+2)+7x^2(m+1)+x}{m^3}.
\end{aligned}$$

Remark 3. *We have*

$$\phi^{G_{m,k}} = G_{m,k}(e_1) = \frac{k}{m},$$

and, for $r \in N$, we get

$$\mu_r^{G_{m,k}} := G_{m,k}(e_1 - \phi^{G_{m,k}} e_0)^r = 0.$$

Now, as an application of Theorem 2, the difference of V_m and S_m defined, respectively, by Equations (1) and (2), can be given as Proposition 1 below.

Proposition 1. *Let $I = [0, \infty)$, $f \in D(I)$ and $f^{(s)} \in E_B(I)$ $(s = 1, 2, 3, 4)$. Then, for each $x \in [0, \infty)$, it is asserted that*

$$|(V_m - S_m)(f, x)| \leq 2\omega_1\left(f, \sqrt{\frac{x(1+x)}{m}}\right) + 2\omega_1\left(f, \sqrt{\frac{x}{m}}\right).$$

The proof of Proposition 1 follows from Remarks 2 and 3, Lemmas 2 and 3, and Theorem 2. We, therefore, omit the details involved.

2.2. *Difference Between the Baskakov and the Szász–Mirakyan–Baskakov Operators*

In the year 1983, Prasad et al. [27] introduced a class of the Szász–Mirakyan–Baskakov type operators. These operators were subsequently improved by Gupta [28] as follows:

$$M_m(f, x) = \sum_{k=0}^{\infty} p_{m,k}(x) H_{m,k}(f), \tag{3}$$

where

$$H_{m,k}(f) = (m-1)\int_0^{\infty} v_{m,k}(t) f(t) dt.$$

Here $p_{m,k}$ and $v_{m,k}$ are defined in Equations (1) and (2), respectively.

Remark 4. *Since*

$$H_{m,k}(e_r) = (m-1)\int_0^{\infty} \binom{m+k-1}{k} \frac{t^k}{(1+t)^{m+k}} t^r dt = \frac{(k+r)!(m-r-2)!}{k!(m-2)!},$$

we get

$$\phi^{H_{m,k}} = H_{m,k}(e_1) = \frac{k+1}{m-2}$$

and

$$\begin{aligned}
\mu_2^{H_{m,k}} &= H_{m,k}(e_1 - \phi^{H_{m,k}} e_0)^2 \\
&= H_{m,k}(e_2) + \left(\frac{k+1}{m-2}\right)^2 - 2H_{m,k}(e_1)\left(\frac{k+1}{m-2}\right) \\
&= \frac{(k+2)(k+1)}{(m-2)(m-3)} - \left(\frac{k+1}{m-2}\right)^2 \\
&= \frac{k^2 + mk + m - 1}{(m-2)^2(m-3)},
\end{aligned}$$

$$\begin{aligned}
\mu_3^{H_{m,k}} &= H_{m,k}(e_1 - \phi^{H_{m,k}} e_0)^3 \\
&= H_{m,k}(e_3) - 3H_{m,k}(e_2)\left(\frac{k+1}{m-2}\right) \\
&\quad + 3H_{m,k}(e_1)\left(\frac{k+1}{m-2}\right)^2 - H_{m,k}(e_0)\left(\frac{k+1}{m-2}\right)^3 \\
&= \frac{4k^3 + 6mk^2 + (2m^2 + 4m - 4)k + 2m(m-1)}{(m-2)^3(m-3)(m-4)}
\end{aligned}$$

and

$$\begin{aligned}
\mu_4^{H_{m,k}} &= H_{m,k}(e_1 - \phi^{H_{m,k}} e_0)^4 \\
&= H_{m,k}(e_4) - 4H_{m,k}(e_3)\left(\frac{k+1}{m-2}\right) + 6H_{m,k}(e_2)\left(\frac{k+1}{m-2}\right)^2 \\
&\quad - 4H_{m,k}(e_1)\left(\frac{k+1}{m-2}\right)^3 + H_{m,k}(e_0)\left(\frac{k+1}{m-2}\right)^4 \\
&= \frac{(k+1)(k+2)(k+3)(k+4)}{(m-5)(m-4)(m-3)(m-2)} - 4\frac{(k+1)(k+2)(k+3)}{(m-4)(m-3)(m-2)}\left(\frac{k+1}{m-2}\right) \\
&\quad + 6\frac{(k+1)(k+2)}{(m-3)(m-2)}\left(\frac{k+1}{m-2}\right)^2 - 4\frac{(k+1)}{(m-2)}\left(\frac{k+1}{m-2}\right)^3 + \left(\frac{k+1}{m-2}\right)^4.
\end{aligned}$$

In Proposition 2 below, a quantitative result concerning the estimate of the difference between M_m and V_m is proved.

Proposition 2. *If $f \in D([0,\infty))$ with $f^{(i)} \in C_B[0,\infty)$ $(i = 2,3,4)$, then, for each $x \in [0,\infty)$, it is asserted that*

$$|(M_m - V_m)(f,x)| \le A(x)\|f^{(4)}\| + B(x)\|f^{(3)}\| + C(x)\|f^{(2)}\| + 2\omega_1(f,\delta_1(x)) + 2\omega_1(f,\delta_2(x)),$$

where

$$\begin{aligned}
A(x) &= \frac{1}{8(m-5)(m-4)(m-3)(m-2)^4}\Big\{x^2(x+1)^2 m^5 \\
&\quad + x(4x^3 + 14x^2 + 14x + 5)m^4 \\
&\quad + (x+1)(24x^2 + 5x + 3)m^3 + 28x^2 + 7x - 8)m^2\Big\},
\end{aligned}$$

$$B(x) = \frac{x(x+1)(2x+1)m^3 + (2x+1)(3x+1)m^2 - m}{3(m-2)^2(m-3)(m-4)},$$

$$C(x) = \frac{x(1+x)m^2 + (x+1)m - 1}{2(m-2)^2(m-3)},$$

$$\delta_1(x) = \sqrt{\frac{x(1+x)}{m}}$$

and

$$\delta_2(x) = \frac{\sqrt{4x^2 + (4+m)x + 1}}{(m-2)}.$$

Proof. Applying Remarks 3 and 4, together with Lemma 2, we find that

$$
\begin{aligned}
A(x) &= \frac{1}{4!}\sum_{k\in K}\left(p_{m,k}(x)\mu_4^{H_{m,k}} + v_{m,k}(x)\mu_4^{G_{m,k}}\right) \\
&= \frac{1}{4!}\sum_{k=0}^{\infty} p_{m,k}(x)\left[\frac{(k+1)(k+2)(k+3)(k+4)}{(m-5)(m-4)(m-3)(m-2)}\right. \\
&\quad -4\frac{(k+1)(k+2)(k+3)}{(m-4)(m-3)(m-2)}\left(\frac{k+1}{m-2}\right) \\
&\quad \left. +6\frac{(k+1)(k+2)}{(m-3)(m-2)}\left(\frac{k+1}{m-2}\right)^2 - 4\frac{(k+1)}{(m-2)}\left(\frac{k+1}{m-2}\right)^3 + \left(\frac{k+1}{m-2}\right)^4\right] \\
&= \frac{1}{8(m-5)(m-4)(m-3)(m-2)^4}\left\{x^2(x+1)^2m^5\right. \\
&\quad + x(4x^3+14x^2+14x+5)m^4 \\
&\quad \left. + (x+1)(24x^2+5x+3)m^3 + 28x^2+7x-8)m^2\right\}
\end{aligned}
$$

and

$$
B(x) = \frac{1}{3!}\left|\sum_{k=0}^{\infty} p_{m,k}(x)\mu_3^{H_{m,k}}\right.
$$

$$
\left. - \sum_{k=0}^{\infty} v_{m,k}(x)\mu_3^{G_{m,k}}\right| = \frac{x(x+1)(2x+1)m^3 + (2x+1)(3x+1)m^2 - m}{3(m-2)^2(m-3)(m-4)}.
$$

Furthermore, we have

$$
\begin{aligned}
C(x) &= \frac{1}{2}\left|\sum_{k=0}^{\infty} p_{m,k}(x)\mu_2^{H_{m,k}} - \sum_{k=0}^{\infty} v_{m,k}(x)\mu_2^{G_{m,k}}\right| \\
&= \frac{x(1+x)m^2 + (x+1)m - 1}{2(m-2)^2(m-3)},
\end{aligned}
$$

$$
\begin{aligned}
\delta_1(x) &= \left(\sum_{k=0}^{\infty} v_{m,k}(x)(\phi^{G_{m,k}} - x)^2\right)^{1/2} \\
&= \left(\sum_{k=0}^{\infty} v_{m,k}(x)\left(\frac{k}{m} - x\right)^2\right)^{1/2} \\
&= \sqrt{\frac{x(1+x)}{m}}
\end{aligned}
$$

and

$$
\begin{aligned}
\delta_2(x) &= \left(\sum_{k=0}^{\infty} p_{m,k}(x)(\phi^{H_{m,k}} - x)^2\right)^{1/2} \\
&= \left(\sum_{k=0}^{\infty} p_{m,k}(x)\left(\frac{k+1}{m-2} - x\right)^2\right)^{1/2} \\
&= \frac{\sqrt{4x^2 + (4+m)x + 1}}{(m-2)}.
\end{aligned}
$$

Now, by using Theorem 2, Proposition 2 is proved. □

2.3. Difference Between the Baskakov and the Szász–Mirakyan–Kantorovich Operators

Let $p_{m,k}$ be the Szász–Mirakyan basis function defined in Equation (1). In addition, let

$$J_{m,k}(f) = m\int_{k/m}^{(k+1)/m} f(t)dt.$$

The Szász–Mirakyan–Kantorovich operators are defined by

$$K_m(f;x) = \sum_{k=0}^{\infty} p_{m,k}(x)J_{m,k}(f). \tag{4}$$

Remark 5. *The following result can be obtained by simple computation:*

$$\phi^{J_{m,k}} = J_{m,k}(e_1) = \frac{k}{m} + \frac{1}{2m}.$$

Moreover, we have

$$\begin{aligned}
\mu_2^{J_{m,k}} &= J_{m,k}(e_1 - \phi^{J_{m,k}}e_0)^2 \\
&= J_{m,k}(e_2) - 2\left(\frac{k}{m} + \frac{1}{2m}\right)^2\left(\frac{k}{m} + \frac{1}{2m}\right)^2 \\
&= \frac{1}{12m^2},
\end{aligned}$$

$$\begin{aligned}
\mu_3^{J_{m,k}} &:= J_{m,k}(e_1 - \phi^{J_{m,k}}e_0)^3 \\
&= J_{m,k}(e_3) - 3J_{m,k}(e_2)\left(\frac{k}{m} + \frac{1}{2m}\right) \\
&+ 3J_{m,k}(e_1)\left(\frac{k}{m} + \frac{1}{2m}\right)^2 - J_{m,k}(e_0)\left(\frac{k}{m} + \frac{1}{2m}\right)^3 \\
&= 0
\end{aligned}$$

and

$$\begin{aligned}
\mu_4^{J_{m,k}} &:= J_{m,k}(e_1 - \phi^{J_{m,k}}e_0)^4 \\
&= J_{m,k}(e_4) - 4J_{m,k}(e_3)\left(\frac{k}{m} + \frac{1}{2m}\right) + 6J_{m,k}(e_2)\left(\frac{k}{m} + \frac{1}{2m}\right)^2 \\
&- 4J_{m,k}(e_1)\left(\frac{k}{m} + \frac{1}{2m}\right)^3 + J_{m,k}(e_0)\left(\frac{k}{m} + \frac{1}{2m}\right)^4 \\
&= \frac{1}{80m^4}.
\end{aligned}$$

The following quantitative result concerning the difference between K_m and V_m is proved next.

Proposition 3. *Let $I = [0,\infty)$. If $f \in D(I)$ with $f^{(i)} \in E_B(I) \quad (i = 2,3,4)$, then, for each $x \in [0,\infty)$, it is asserted that*

$$|(K_m - V_m)(f,x)| \leq A(x)\|f^{(4)}\| + C(x)\|f^{(2)}\| + 2\omega_1(f,\delta_1) + 2\omega_1(f,\delta_2),$$

where

$$A(x) = \frac{1}{1920m^4} \quad \text{and} \quad C(x) = \frac{1}{24m^2}$$

and

$$\delta_1(x) = \sqrt{\frac{x(1+x)}{m}} \quad \textit{and} \quad \delta_2(x) = \frac{\sqrt{4mx+1}}{2m}.$$

Proof. Applying Remarks 3 to 5 and Lemma 2, we get

$$\begin{aligned} A(x) &:= \frac{1}{4!}\sum_{k=0}^{\infty}(p_{m,k}(x)\mu_4^{J_{m,k}} + v_{m,k}(x)\mu_4^{G_{m,k}}) \\ &= \frac{1}{4!}\sum_{k=0}^{\infty} p_{m,k}(x)\frac{1}{80m^4} \\ &= \frac{1}{1920m^4} \end{aligned}$$

and

$$\begin{aligned} B(x) &= \frac{1}{3!}\left|\sum_{k=0}^{\infty} p_{m,k}(x)\mu_3^{J_{m,k}} - \sum_{k=0}^{\infty} v_{m,k}(x)\mu_3^{G_{m,k}}\right| \\ &= 0. \end{aligned}$$

Furthermore, we have

$$\begin{aligned} C(x) &= \frac{1}{2!}\left|\sum_{k=0}^{\infty} p_{m,k}(x)\mu_2^{J_{m,k}} - \sum_{k=0}^{\infty} v_{m,k}(x)\mu_2^{G_{m,k}}\right| \\ &= \frac{1}{24m^2}, \end{aligned}$$

$$\begin{aligned} \delta_1(x) &= \left(\sum_{k=0}^{\infty} v_{m,k}(x)(\phi^{G_{m,k}} - x)^2\right)^{1/2} \\ &= \left(\sum_{k=0}^{\infty} v_{m,k}(x)\left(\frac{k}{m} - x\right)^2\right)^{1/2} \\ &= \sqrt{\frac{x(1+x)}{m}} \end{aligned}$$

and

$$\begin{aligned} \delta_2(x) &= \left(\sum_{k=0}^{\infty} p_{m,k}(x)(\phi^{J_{m,k}} - x)^2\right)^{1/2} \\ &= \left(\sum_{k=0}^{\infty} p_{m,k}(x)\left(\frac{k}{m} + \frac{1}{2m} - x\right)^2\right)^{1/2} \\ &= \frac{\sqrt{4mx+1}}{2m}. \end{aligned}$$

Upon collecting the above estimates and by using Theorem 2, the proof of Proposition 3 is completed. □

2.4. Difference of Two Genuine-Durrmeyer Type Operators

Let $\rho > 0$ and $f \in C[0,1]$. Suppose also that

$$F^{\rho}_{m,k}(f) = \begin{cases} f(0) & (k=0) \\ \displaystyle\int_0^1 \frac{t^{k\rho-1}(1-t)^{(m-k)\rho-1}}{B(k\rho,(m-k)\rho)} f(t)dt & (k \neq 0,1) \\ f(1) & (k=1). \end{cases}$$

Păltănea and Gonska (see [29–31]) introduced and studied a new class of the Bernstein–Durrmeyer type operators defined by

$$U^{\rho}_m : C[0,1] \to \Pi_m \quad \text{and} \quad U^{\rho}_m(f;x) := \sum_{k=0}^{m} F^{\rho}_{m,k}(f) p_{m,k}(x),$$

where

$$p_{m,k}(x) = \binom{m}{k} x^k (1-x)^{m-k}.$$

Neer and Agrawal [32] introduced a class of the genuine-Durrmeyer type operators as follows:

$$\tilde{U}^{\rho}_m(f;x) = \sum_{k=0}^{m} F^{\rho}_{m,k}(f) p^{<\frac{1}{m}>}_{m,k}(x),$$

where

$$p^{<\frac{1}{m}>}_{m}(x) = \frac{2 \cdot m!}{(2m)!} \binom{m}{k} (mx)_k (m-mx)_{m-k}.$$

Proposition 4 below provides an estimate of the difference between U^{ρ}_m and $\tilde{U}^{\rho}_m$.

Proposition 4. *Let $f \in C^4[0,1]$. Then the following inequality holds true:*

$$\left|\left(U^{\rho}_m - \tilde{U}^{\rho}_m\right)(f;x)\right| \leq A(x)\|f^{(4)}\| + B(x)\|f^{(3)}\| + C(x)\|f^{(2)}\| + 2\omega_1(f,\delta_1(x)) + 2\omega_1(f,\delta_2(x)),$$

where

$$A(x) := \frac{x(1-x)(n-1)}{8m^3(m\rho+1)(m\rho+2)(m\rho+3)(m+1)(m+2)(m+3)} \cdot \Big\{ m\rho(3m^4+5m^3+7m^2-5m-6) + 4m^5 + 4m^4 + 4m^3 - 30m^2 + 30m + 36 + x(1-x)(m-2)(m-3)(m\rho-6)(2m^3+6m^2+11m+6),$$

$$B(x) := \frac{x(1-x)|1-2x|(m-2)(m-1)(3m+2)}{3(m\rho+1)(m\rho+2)m^2(m+1)(m+2)},$$

$$C(x) := \frac{x(1-x)(m-1)}{2(m\rho+1)m(m+1)},$$

$$\delta_1(x) := \sqrt{\frac{x(1-x)}{m}}$$

and

$$\delta_2(x) := \sqrt{\frac{2x(1-x)}{m+1}}.$$

Proof. In Theorem 2, we set

$$F_{m,k}(f) = G_{m,k}(f) = F^{\rho}_{m,k}(f),$$

so that we have

$$\phi^{F_{m,k}} = \phi^{G_{m,k}} = \frac{k}{m};$$

$$\mu_2^{F_{m,k}} = \mu_2^{G_{m,k}} = F_{m,k}\left(e_1 - \phi^{F_{m,k}}\right)^2 = \frac{k(m-k)}{m^2(m\rho+1)};$$

$$\begin{aligned}\mu_3^{F_{m,k}} &= \mu_3^{G_{m,k}} = F_{m,k}\left(e_1 - \phi^{F_{m,k}}\right)^3 \\ &= \frac{2k(2k^2 - 3km + m^2)}{m^3(m\rho+1)(m\rho+2)}\end{aligned}$$

and

$$\begin{aligned}\mu_4^{F_{m,k}} &= \mu_4^{G_{m,k}} = F_{m,k}\left(e_1 - \phi^{F_{m,k}}\right)^4 \\ &= \frac{3k(k^3 m\rho - 2k^2m^2\rho + km^3\rho - 6k^3 + 12k^2m - 8km^2 + 2m^3)}{m^4(m\rho+1)(m\rho+2)(m\rho+3)}.\end{aligned}$$

Now, by considering the following relations:

$$\sum_{k=0}^{n} p_{m,k}(x) = 1,$$

$$\sum_{k=0}^{n} \frac{k}{m} p_{m,k}(x) = x,$$

$$\sum_{k=0}^{m} \left(\frac{k}{m}\right)^2 p_{m,k}(x) = \frac{x(mx - x + 1)}{m},$$

$$\sum_{k=0}^{m} \left(\frac{k}{m}\right)^3 p_{m,k}(x) = \frac{x(m^2x^2 - 3mx^2 + 3mx + 2x^2 - 3x + 1)}{m^2},$$

$$\begin{aligned}\sum_{k=0}^{m} \left(\frac{k}{m}\right)^4 p_{m,k}(x) = \frac{x}{m^3}\Big(&m^3x^3 - 6m^2x^3 + 6m^2x^2 + 11mx^3 \\ &-18mx^2 - 6x^3 + 7mx + 12x^2 - 7x + 1\Big),\end{aligned}$$

$$\sum_{k=0}^{m} p_{m,k}^{<\frac{1}{m}>}(x) = 1,$$

$$\sum_{k=0}^{m} \left(\frac{k}{m}\right) p_{m,k}^{<\frac{1}{m}>}(x) = x,$$

$$\sum_{k=0}^{m}\left(\frac{k}{m}\right)^2 p_{m,k}^{<\frac{1}{m}>}(x) = x^2 + \frac{2x(1-x)}{m+1},$$

$$\sum_{k=0}^{m}\left(\frac{k}{m}\right)^3 p_{m,k}^{<\frac{1}{m}>}(x) = x^3 + \frac{6mx^2(1-x)}{(m+1)(m+2)} + \frac{6x(1-x)}{(m+1)(m+2)}$$

and

$$\sum_{k=0}^{m}\left(\frac{k}{m}\right)^4 p_{m,k}^{<\frac{1}{m}>}(x) = x^4 + \frac{12(m^2+1)x^3(1-x)}{(m+1)(m+2)(m+3)} + \frac{12(3m-1)x^2(1-x)}{(m+1)(m+2)(m+3)} + \frac{2(13m-1)x(1-x)}{m(m+1)(m+2)(m+3)},$$

the proof of Proposition 4 is completed. □

Example 1. *Applying Proposition 2 for* $f(x) = \frac{x}{x^2+1}$, $x \in [0,1]$ *and* $\rho = 2$, *we get the following estimate:*

$$|(U_m^\rho - \tilde{U}_m^\rho)(f;x)| \le E_m(f), \tag{5}$$

where

$$E_m(f) = K_1\|f^{(4)}\| + K_2\|f^{(3)}\| + K_3\|f^{(2)}\| + 2(\delta_1+\delta_2)\|f'\| \quad \text{and} \quad f \in C^4[0,1]$$

and

$$K_1 := \frac{(m-1)}{64m^3(2m+1)(m+1)^2(2m+3)(m+2)(m+3)} \cdot \left(\frac{1}{2}(m-2)(m-3)^2(2m^3+6m^2+11m+6) + 10m^5 + 14m^4 + 18m^3 - 40m^2 + 18m + 36\right),$$

$$K_2 := \frac{(m-2)(m-1)(3m+2)}{24(2m+1)(m+1)^2m^2(m+2)},$$

$$K_3 := \frac{m-1}{8(2m+1)m(m+1)},$$

$$\delta_1 := \frac{1}{2\sqrt{m}}$$

and

$$\delta_2 := \frac{1}{\sqrt{2(m+1)}}.$$

Now, by using the result of Shisha and Mond (see [26]; see also Remark 1), we get the following estimate:

$$|(U_m^\rho - \tilde{U}_m^\rho)(f;x)| \le E_m^{(SM)}(f), \tag{6}$$

where

$$E_m^{(SM)}(f) = \left(\sqrt{\frac{3}{2m+1}} + \sqrt{\frac{5m+1}{(m+1)(2m+1)}}\right)\|f'\|, \; f \in C^1[0,1].$$

Table 1 below contains the values of $E_m(f)$ and $E_m^{(SM)}(f)$ for certain given values of n. We note here that, for this particular case, the estimate in Equation (5) is better than the estimate given by the Shisha–Mond result in Equation (6).

Table 1. Estimates for the difference of $U_m^{\rho} f$ and $\tilde{U}_m^{\rho} f$.

m	$E_m(f)$	$E_m^{(SM)}(f)$
10	0.74402776700	0.84783596730
10^2	0.24073382150	0.27926364330
10^3	0.07632064786	0.08868768037
10^4	0.02414170100	0.02805750320
10^5	0.00763444237	0.00887294116
10^6	0.00241421554	0.00280588236
10^7	0.00076343666	0.00088729829
10^8	0.00024142458	0.00028058836

2.5. *Difference of the Durrmeyer Operators and the Lupaş–Durrmeyer Operators*

Durrmeyer [33] and, independently, Lupaş [34] defined the Durrmeyer operators by

$$M_m(f,x) = (m+1)\sum_{k=0}^{m} p_{m,k}(x) \int_0^1 p_{m,k}(t)\, f(t)\, dt \qquad (x \in [0,1]). \tag{7}$$

Gupta et al. [35] introduced a modification of the operator in Equation (7) as follows:

$$D_m^{<\frac{1}{m}>}(f;x) = (m+1)\sum_{k=0}^{m} p_{m,k}^{<\frac{1}{m}>} \int_0^1 p_{m,k}(t) f(t) dt \qquad (f \in C[0,1]). \tag{8}$$

Finally, the difference between M_m and $D_m^{<\frac{1}{m}>}$ is provided in the estimate asserted by Proposition 5 below.

Proposition 5. *Let $f \in C^4[0,1]$. Then the following inequality holds true:*

$$\left|\left(M_m - D_m^{<\frac{1}{m}>}\right)(f;x)\right| \le A(x)\|f^{(4)}\| + B(x)\|f^{(3)}\| + C(x)\|f^{(2)}\| + 2\omega_1(f,\delta_1(x)) + 2\omega_1(f,\delta_2(x)),$$

where

$$\begin{aligned} A(x) := & \frac{1}{8(m+1)(m+2)^5(m+3)^2(m+4)(m+5)} \\ & \cdot \{x(1-x)m(m-1)\,[x(1-x)(m-2)(m-3)(m-4) \\ & \quad \cdot (2m^3+6m^2+11m+6) + 11m^5 + 41m^4 + 77m^3 + 25m^2 + 26m + 24\big] \\ & + (m+1)^2(m+2)(m+3)(3m^2+5m+4)\Big\}, \end{aligned}$$

$$B(x) := \frac{m|x(1-x)(1-2x)(m-1)(m-2)(3m+2) + m^3 + 4m^2 + 5m + 2|}{3(m+1)(m+2)^3(m+3)(m+4)},$$

$$C(x) := \frac{m(m-1)x(1-x) + (m+1)^2}{2(m+1)(m+2)^2(m+3)},$$

$$\delta_1(x) := \frac{\sqrt{x(1-x)m + (2x-1)^2}}{m+2}$$

and

$$\delta_2(x) := \frac{\sqrt{2x(1-x)m^2 + (1-2x)^2(m+1)}}{(m+2)\sqrt{m+1}}.$$

Proof. In Theorem 2, we let

$$F_{m,k}(f) = G_{m,k}(f) = (m+1)\int_0^1 p_{m,k}(t)f(t)dt,$$

so that we have

$$\phi^{F_{m,k}} = \phi^{G_{m,k}} = \frac{k+1}{m+2},$$

$$\mu_2^{F_{m,k}} = \mu_2^{G_{m,k}} = F_{m,k}\left(e_1 - \phi^{F_{m,k}}\right)^2 = \frac{(k+1)(m-k+1)}{(m+2)^2(m+3)},$$

$$\mu_3^{F_{m,k}} = \mu_3^{G_{m,k}} = F_{m,k}\left(e_1 - \phi^{F_{m,k}}\right)^3 = \frac{2(k+1)(2k^2-3km+m^2-2k+m)}{(m+2)^3(m+3)(m+4)}$$

and

$$\mu_4^{F_{m,k}} = \mu_4^{G_{m,k}} = F_{m,k}\left(e_1 - \phi^{F_{m,k}}\right)^4 = \frac{3(k+1)(k-m-1)(k^2m-km^2-4k^2+4km-3m^2-5m-4)}{(m+2)^4(m+3)(m+4)(m+5)}.$$

Now, by applying the relations from the proof of Proposition 2, the resulting estimate of the difference of the Durrmeyer operator and the Lupaş–Durrmeyer operator is as asserted by Proposition 5. □

Example 2. *By pplying Proposition 5 for $f(x) = \cos(2\pi x)$ for $x \in [0,1]$, we get the following estimate:*

$$\left|\left(M_m - D_m^{<\frac{1}{m}>}\right)(f;x)\right| \leq E_m(f), \tag{9}$$

where

$$E_m(f) = K_1\|f^{(4)}\| + K_2\|f^{(3)}\| + K_3\|f^{(2)}\| + 2(\delta_1+\delta_2)\|f'\|,\ f \in C^4[0,1]$$

and

$$K_1 := \frac{1}{8(m+1)(m+2)^5(m+3)^2(m+4)(m+5)} \cdot \left\{\frac{1}{4}m(m-1)\left[\frac{1}{4}(m-2)(m-3)(m-4)(2m^3+6m^2+11m+6) + 11m^5+41m^4+77m^3+25m^2+26m+24\right] +(m+1)^2(m+2)(m+3)(3m^2+5m+4)\right\},$$

$$K_2 := \frac{m}{3(m+1)(m+2)^3(m+3)(m+4)} \cdot \left\{\frac{1}{4}(m-1)(m-2)(3m+2) + m^3+4m^2+5m+2\right\},$$

$$K_3 := \frac{1}{2(m+2)^2(m+3)(m+1)}\left[\frac{1}{4}m(m-1)+(m+1)^2\right],$$

$$\delta_1 := \frac{\sqrt{m+4}}{2(m+2)}$$

and

$$\delta_2 := \frac{\sqrt{m^2+2(m+1)}}{(m+2)\sqrt{2(m+1)}}.$$

Thus, by using the result of Shisha and Mond (see [26]; see also Remark 1), we get the following estimate:

$$\left|\left(M_m - D_m^{<\frac{1}{m}>}\right)(f;x)\right| \leq E_m^{(SM)}(f), \tag{10}$$

where

$$E_m^{(SM)}(f) = 2\left(\sqrt{\frac{m+1}{2(m+2)(m+3)}}+\sqrt{\frac{3m^2+3m+2}{4(m+1)(m+2)(m+3)}}\right)\|f'\|,\ f \in C^1[0,1].$$

Table 2 below gives the values of $E_m(f)$ and $E_m^{(SM)}(f)$ for certain specific values of m. We also note that, for this particular case, the estimate in Equation (9) is better than the estimate given by the Shisha–Mond result in Equation (10).

Table 2. Estimates for the difference of $M_m f$ and $D_m^{<\frac{1}{m}>} f$.

m	$E_m(f)$	$E_m^{(SM)}(f)$
10^2	1.5210054310	1.9330219770
10^3	0.4794548855	0.6237181803
10^4	0.1516781199	0.1976406539
10^5	0.0479680333	0.0625122598
10^6	0.0151689380	0.0197685170
10^7	0.0047968431	0.0062513668
10^8	0.0015168951	0.0019768561

Remark 6. *The earlier works [36,37] proposed certain general families of positive linear operators which reproduce only constant functions. Recently, as a continuation of these works, in [38] some positive linear operators reproducing linear functions were introduced and studied. Analogous further researches for this class of operators are possible.*

3. Conclusions

The studies of the differences of positive linear operators has become an interesting area of research in Approximation Theory. The present paper deals with the estimates of the differences of various positive linear operators, which are defined on bounded or unbounded intervals, in terms of the modulus of continuity. In several earlier papers, the results of the type which we have presented here were obtained for a class of positive linear operators constructed with the same fundamental functions. The novelty of this paper is that the fundamental functions of the positive linear operators can chosen to be different. Our present study makes use of the Baskakov type operators, the Szász–Mirakyan type operators, and the Durrmeyer type operators. In some illustrative numerical examples, we have shown that the estimates obtained in this study are better than the estimates given by the classical Shisha–Mond result. For a future work, we propose to obtain estimates for these operators involving some suitably weighted modulus of smoothness.

Author Contributions: The authors contributed equally to this work. All authors have read and agreed to the published version of the manuscript.

Funding: Project financed by Lucian Blaga University of Sibiu & Hasso Plattner Foundation Research Grants LBUS-IRG-2019-05.

Conflicts of Interest: The authors declare no conflict of interest.

References

1. Yalcin, F.; Simsek, Y. A new class of symmetric Beta type distributions constructed by means of symmetric Bernstein type basis functions. *Symmetry* **2020**, *12*, 779. [CrossRef]
2. Ansari, K.J.; Ahmad, I.; Mursaleen, M.; Hussain, I. On some statistical approximation by (p,q)-Bleimann, Butzer and Hahn Operators. *Symmetry* **2018**, *10*, 731. [CrossRef]
3. Srivastava, H.M. Operators of basic (or q-) calculus and fractional q-calculus and their applications in geometric function theory of complex analysis. *Iran. J. Sci. Technol. Trans. A Sci.* **2020**, *44*, 327–344. [CrossRef]
4. Alda, F.; Rubinstein, B.I.P. The Bernstein mechanism: Function release under differential privacy. In Proceedings of the Thirty-First AAAI Conference on Artificial Intelligence (AAAI17), San Francisco, CA, USA, 4–9 February 2017; pp. 1705–1711.
5. Zhou, D.-X.; Jetter, K. Approximation with polynomial kernels and SVM classifiers. *Adv. Comput. Math.* **2006**, *25*, 323–344. [CrossRef]
6. Lupaş, A. The approximation by means of some linear positive operators. In *Approximation Theory*; Müller, M.W., Felten, M., Mache, D.H., Eds.; Akademie-Verlag: Berlin, Germany, 1995; pp. 201–227.
7. Gonska, H.; Piţul, P.; Raşa, I. On Peano's form of the Taylor remainder, Voronovskaja's theorem and the commutator of positive linear operators. In Proceedings of the 2006 International Conference on Numerical Analysis and Approximation Theory NAAT, Cluj-Napoca, Romania, 5–8 July 2006; Agratini, O., Blaga, P., de Stinta, C.C., Eds.; pp. 55–80.
8. Gonska, H.; Piţul, P.; Raşa, I. On differences of positive linear operators. *Carpathian J. Math.* **2006**, *22*, 65–78.
9. Gonska, H.; Raşa, I. Differences of positive linear operators and the second order modulus. *Carpathian J. Math.* **2008**, *24*, 332–340.
10. Gonska, H.; Raşa, I.; Rusu, M. Applications of an Ostrowski-type inequality. *J. Comput. Anal. Appl.* **2012**, *14*, 19–31.
11. Acu, A.M.; Rasa, I. New estimates for the differences of positive linear operators. *Numer. Algorithms* **2016**, *73*, 775–789. [CrossRef]
12. Garg, T.; Acu, A.M.; Agrawal, P.N. Weighted approximation and GBS of Chlodowsky-Szász-Kantorovich type operators. *Anal. Math. Phys.* **2019**, *9*, 1429–1448. [CrossRef]
13. Srivastava, H.M.; Finta, Z.; Gupta, V. Direct results for a certain family of summation-integral type operators. *Appl. Math. Comput.* **2007**, *190*, 449–457. [CrossRef]
14. Srivastava, H.M.; Zeng, X.-M. Approximation by means of the Szász-Bézier integral operators. *Int. J. Pure Appl. Math.* **2004**, *14*, 283–294.
15. Srivastava, H.M.; Ícoz, G.; Çekim, B. Approximation properties of an extended family of the Szász-Mirakjan Beta-type operators. *Axioms* **2019**, *8*, 111. [CrossRef]
16. Aral, A.; Inoan, D.; Raşa, I. On differences of linear positive operators. *Anal. Math. Phys.* **2019**, *9*, 1227–1239. [CrossRef]
17. Acu, A.M.; Manav, N.; Sofonea, F. Approximation properties of λ-Kantorovich operators. *J. Inequal. Appl.* **2018**, *2018*, 202. [CrossRef]
18. Srivastava, H.M.; Özger, F.; Mohiuddine, S.A. Construction of Stancu-type Bernstein operators based on Bézier bases with shape parameter λ. *Symmetry* **2019**, *11*, 316. [CrossRef]
19. Acu, A.M.; Rasa, I. Estimates for the differences of positive linear operators and their derivatives. *Numer. Algorithms* **2019**. [CrossRef]
20. Acu, A.M.; Hodiş, S.; Raşa, I. A survey on estimates for the differences of positive linear operators. *Construct. Math. Anal.* **2018**, *1*, 113–127. [CrossRef]
21. Gupta, V. Differences of operators of Lupaş type. *Construct. Math. Anal.* **2018**, *1*, 9–14. [CrossRef]
22. Gupta, V. On difference of operators with applications to Szász type operators. *Rev. Real Acad. Cienc. Exactas Fís. Natur. Ser. A Mat.* **2019**, *113*, 2059–2071. [CrossRef]

23. Gupta, V.; Acu, A.M. On difference of operators with different basis functions. *Filomat* **2019**, *33*, 3023–3034. [CrossRef]
24. Gupta, V.; Rassias, T.M.; Agrawal, P.N.; Acu, A.M. Estimates for the differences of positive linear operators. In *Recent Advances in Constructive Approximation Theory*; Springer Optimization and Its Applications; Springer: Cham, Switzerland, 2018; Volume 138.
25. Gupta, V.; Tachev, G. A note on the differences of two positive linear operators. *Construct. Math. Anal.* **2019**, *2*, 1–7. [CrossRef]
26. Shisha, O.; Mond, B. The degree of convergence of linear positive operators. *Proc. Nat. Acad. Sci. USA* **1968**, *60*, 1196–1200. [CrossRef] [PubMed]
27. Prasad, G.; Agrawal, P.N.; Kasana, H.S. Approximation of functions on $[0,\infty]$ by a new sequence of modified Szász operators. *Math. Forum* **1983**, *6*, 1–11.
28. Gupta, V. A note on modified Szász operators. *Bull. Inst. Math. Acad. Sin.* **1993**, *21*, 275–278.
29. Gonska, H.; Pǎltǎnea, R. Quantitative convergence theorems for a class of Bernstein-Durrmeyer operators preserving linear functions. *Ukrain. Math. J.* **2010**, *62*, 913–922. [CrossRef]
30. Gonska, H.; Pǎltǎnea, R. Simultaneous approximation by a class of Bernstein-Durrmeyer operators preserving linear functions. *Czechoslovak Math. J.* **2010**, *60*, 783–799. [CrossRef]
31. Pǎltǎnea, R. A class of Durrmeyer type operators preserving linear functions. *Ann. Tiberiu Popoviciu Sem. Funct. Equ. Approx. Convex.* **2007**, *5*, 109–117.
32. Neer, T.; Agrawal, P.N. A genuine family of Bernstein-Durrmeyer type operators based on Pólya basis functions. *Filomat* **2017**, *31*, 2611–2623. [CrossRef]
33. Durrmeyer, J.L. *Une Formule d'inversion de la Transforme de Laplace: Applications a la Theorie des Moments*; These de 3e cycle; Faculte des Sciences de l'Universite de Paris: Paris, Francce, 1967.
34. Lupaş, A. Die Folge der Betaoperatoren. Ph.D. Thesis, Universität Stuttgart, Stuttgart, Germany, 1972.
35. Gupta, V.; Rassias, T.M. Lupaş-Durrmeyer operators based on Pólya distribution. *Banach J. Math. Anal.* **2014**, *8*, 146–155. [CrossRef]
36. Gupta, V. A large family of linear positive operators. *Rend. Circ. Mat. Palermo (Ser. II)* **2019**. [CrossRef]
37. Srivastava, H.M.; Gupta, V. A certain family of summation-integral type operators. *Math. Comput. Model.* **2003**, *37*, 1307–1315. [CrossRef]
38. Gupta, V.; Srivastava, H.M. A general family of the Srivastava-Gupta operators preserving linear functions. *Eur. J. Pure Appl. Math.* **2018**, *11*, 575–579. [CrossRef]

MDPI
St. Alban-Anlage 66
4052 Basel
Switzerland
Tel. +41 61 683 77 34
Fax +41 61 302 89 18
www.mdpi.com

Symmetry Editorial Office
E-mail: symmetry@mdpi.com
www.mdpi.com/journal/symmetry

www.ingramcontent.com/pod-product-compliance
Lightning Source LLC
LaVergne TN
LVHW070723100526
838202LV00013B/1153